Cell Culture and Tissue Engineering

Wujie Zhang

Cell Culture and Tissue Engineering

A Textbook with Lab Activities

Springer

Wujie Zhang
Chemical and Biomolecular Engineering Program
Department of Physics and Chemistry
Milwaukee School of Engineering
Milwaukee, WI, USA

ISBN 978-3-032-32095-7 ISBN 978-3-032-32096-4 (eBook)
https://doi.org/10.1007/978-3-032-32096-4

This Springer imprint is published by the registered company Springer Nature Switzerland AG
The registered company address is: Gewerbestrasse 11, 6330 Cham, Switzerland

Preface

Cell culture and tissue engineering are becoming increasingly important across multiple disciplines, including chemical and biomolecular engineering and biomedical engineering. Because these topics are closely interconnected, addressing them together in a single textbook offers significant educational advantages: (1) both cellular and tissue engineering concepts are integrated, and (2) students develop a strong foundation in cell culture that prepares them to understand tissue engineering more easily and deeply.

Despite the growing demand, there remains an unmet need in educational resources. Many existing books are reference-focused rather than designed as instructional textbooks; others cover either cell culture or tissue engineering but not both. Additionally, many available texts are not ideally suited for undergraduate instruction, and most lack hands-on, accessible laboratory activities.

With more than 20 years of research experience in cell and tissue engineering and over 14 years of teaching experience, the author undertook this project to help fill those gaps. This textbook aims to be state-of-the-art yet concise, providing comprehensive coverage of fundamental and applied topics—from the basics of cell culture to tissue engineering design and clinical and industrial translation. To support practical learning, this book also includes multiple hands-on laboratory activities in cell culture, cell engineering, and tissue engineering. These activities are easy to adopt and have been successfully implemented and refined over multiple years at the author's institution.

On a personal note, the writing of this book coincided with the 20-year anniversary of the author's introduction to cell culture and tissue engineering. Although the author completed a master's degree in Food Science at the University of Shanghai for Science and Technology, his thesis research was conducted at Shanghai Ninth People's Hospital, affiliated with Shanghai Jiao Tong University School of Medicine. During that period, the author performed mammalian and stem cell culture, cell microencapsulation, and bone tissue engineering for the first time—experiences that inspired him to pursue a Ph.D. in biomedical engineering with a focus on biomaterials, drug delivery, and tissue engineering.

This book emphasizes the foundational components of tissue engineering, as opposed to an application-driven approach organized by specific tissue types. Additional content on particular tissue engineering applications may be integrated as desired. Each chapter in this book is designed to stand alone, providing instructors with flexibility in selecting material that best supports their courses. Eligible instructors may contact the author to request supplementary resources, including lecture slides.

The author gratefully acknowledges the students and colleagues whose encouragement and feedback supported this project. Special thanks are extended to Dr. Michael Navin for his thoughtful proofreading and suggestions.

Finally, the author expresses heartfelt appreciation to his family, both in the United States and China, for their unwavering love and support.

Milwaukee, WI, USA Wujie Zhang

Competing Interests The author has no competing interests to declare that are relevant to the content of this manuscript.

Contents

About the Author

Wujie Zhang is a Professor of Chemical and Biomolecular Engineering at the Milwaukee School of Engineering (MSOE). He earned his Ph.D. in Biomedical Engineering from the University of South Carolina. His scholarly work encompasses biomaterials, tissue engineering, drug delivery, and cancer treatment, with cell- and tissue-engineering-related research focusing on cell encapsulation and the development of novel bioinks for bioprinting applications.

Dr. Zhang has published more than 50 peer-reviewed journal articles in leading journals, including *Acta Biomaterialia*, *Tissue Engineering Part A*, *International Journal of Artificial Organs*, *Polymers*, *Biomedical Physics & Engineering Express*, and *Journal of Applied Biomaterials & Functional Materials*. He is also the author of the books *Fundamentals and Design of Drug Delivery Systems: A Textbook with Lab Activities*, *Biopolymer Engineering*, and *Nanotechnology for Bioengineers*.

At MSOE, Dr. Zhang developed and teaches the Tissue Engineering and Cell Culture courses and has designed numerous laboratory activities related to cell and tissue engineering. He has been recognized nationally and regionally for his contributions, including being featured in ASEE's *Prism* magazine as 1 of 20 high-achieving researchers and educators under 40 (2018) and receiving multiple awards, such as STEM Forward's Young Engineer of the Year (2022), the CUR Engineering Division Undergraduate Research Mentoring Award (2022), and MSOE's Oscar Werwath Distinguished Teacher Award (2023).

Part I

Cell Culture Basics

Biosafety and Cell Culture Facility Design

1

Learning Objectives
- Understand and discuss the fundamental principles of biosafety
- Examine the fundamentals of BSL-2 facility design and equipment layout

1.1 Introduction to Biosafety, Biohazards, and Biosafety Levels

Biosafety refers to the containment principles, technologies, and practices implemented to prevent unintentional exposure to biological agents or their inadvertent release [1]. The fundamentals of containment include the microbiological practices, safety equipment, and facility safeguards that protect laboratory workers, the environment, and the public from exposure to infectious microorganisms that are handled and stored in the laboratory [2]. The emphasis of biosafety is on the safety of personnel, as opposed to protecting the sample from contamination [3]. In the context of cell cultures, workers who handle or manipulate human or animal cells and tissues are at risk of exposure to potentially infectious latent and adventitious agents that may be present in those cells and tissues. Examples include the reactivation of herpes viruses from latency and persistence of human immunodeficiency virus (HIV), hepatitis B virus (HBV), and hepatitis C virus (HCV) within infected individuals [2].

Biohazards, or biological hazards, are disease-producing agents (pathogens) that can be transmitted to individuals through various routes of exposure (modes of transmission) [4]. Sources of biohazard risk are known pathogens, tissue and body fluids, and cell lines [5]. Exposure to these hazards may result in acute or chronic health conditions. For example, HBV and HIV can be transmitted through blood and body fluids, while *Mycobacterium Tuberculosis* is transmitted *via* small-particle aerosols (airborne). Examples of the unsafe work practices include:

W. Zhang, *Cell Culture and Tissue Engineering*,
https://doi.org/10.1007/978-3-032-32096-4_1

- Improper disinfection of equipment and surfaces;
- Improper use of personal protective equipment (PPE);
- Lack of proper hand hygiene practices;
- Use of cell phones in restricted areas, creating opportunities for injury such as due to device contamination;
- Consumption of food and/or drink and the application of cosmetics in laboratory or clinical areas;
- Recapping of syringes;
- Lack of biohazard labeling;
- Improper transport of instruments or specimens [4].

Laboratory biosecurity refers to the protection, control and accountability of valuable biological materials (VBM) within laboratories to prevent unauthorized access, loss, theft, misuse, diversion or intentional release. VBM are biological materials that require administrative oversight, control, accountability, and specific protective and monitoring measures in laboratories to safeguard their economic and historical (archival) value, and/or the population from their potential to cause harm [6]. Biosecurity focuses on protecting biological agents from theft, loss, or misuse. A comprehensive biosecurity plan encompasses three major components: physical security (preventing unauthorized access to facilities and ensuring that only appropriate people within the facility can access agents), personnel reliability (ensuring that all staff are responsible and are suitable to work with sensitive materials), and information security (ensuring all electronic information is safe) [7]. Biosafety and laboratory biosecurity are complementary. In fact, the implementation of specific biosafety activities often covers some biosecurity aspects. Laboratory biosecurity should be built upon a firm foundation of good laboratory biosafety practices [6].

Biological agents are classified into "risk" or "hazard groups" based on their characteristics and epidemiological profiles. According to the World Health Organization (WHO), the classification consists of: Risk Group 1 (no or low individual and community risk), Risk Group 2 (moderate individual risk, low community risk), Risk Group 3 (high individual risk, low community risk), and Risk Group 4 (high individual and community risk) [1]. The four biohazard risk groups correlate with four biosafety (containment) levels, but they may not agree completely [5]. For example, HIV, a Risk Group 3 agent, can be handled under BSL-2 conditions with good practices and procedures [8]. Table 1.1 summarizes the U.S. biosafety levels (BSL) 1–3. Although the basic principles of BSL-1 through BSL-4 are widely recognized, the specific practices, equipment, and facility requirements may vary depending on the institution, the particular agent involved, and local regulations.

Table 1.1 United States biosafety levels comparison[a]

	BSL-1	BSL-2	BSL-3
Safety equipment	Pipetting devices (mouth pipetting is prohibited)	1. Class I or II Biological Safety Cabinets (BSCs) or other physical containment devices 2. Lab coats, gloves, face protection as needed	1. Class I or II Biological Safety Cabinets (BSCs) or other physical containment devices 2. Protective lab clothing, gloves, respiratory protection as needed
Agents	Not generally associated with disease in healthy people	Associated with human disease	1. Associated with human disease and cause illness through the air (aerosol) 2. Cause diseases that may have serious or lethal consequences
Facilities	1. Open lab bench or table 2. Sink for hand washing	1. Open bench-top 2. Sink for hand washing 3. Autoclave available 4. Self-closing doors	BSL-2 plus 1. Physical separation from access corridors 2. Two sets of self-closing doors 3. Exhaust air is not recirculated 4. Negative airflow into laboratory 5. Design includes back up/redundant systems
Laboratory practices	Standard microbiological practices	Access to the laboratory is restricted when work is being conducted	1. Access to the laboratory is restricted and controlled at all times 2. Laboratorians are under medical surveillance and might receive immunizations for microbes they work with

[a]Information from https://www.niaid.nih.gov/research/biosafety-labs-needed and https://www.cdc.gov/training/quicklearns/biosafety/

1.2 Cell Culture Facility Design

For mammalian cell cultures, a BSL-2 lab is common. It is prudent to treat all mammalian cell lines as potentially hazardous and to handle them under BSL-2 practices, even if they are classified as BSL-1 (such as NIH/3T3 and MDA-MB-231) [9]. A layout of a medium-sized cell culture lab is shown in Fig. 1.1. When designing cell culture facilities (BSL-2 or higher), two fundamental considerations are contamination and safety. Location, ventilation, gases access (such as CO_2), and cost should be prioritized [10]. Ideally, a dedicated cell culture laboratory should be used, either newly constructed or modified from existing laboratory rooms, along with an adjacent preparation area/room [11]. BSCs and incubators are key pieces of equipment in a cell culture facility. An autoclave doesn't need to be located within

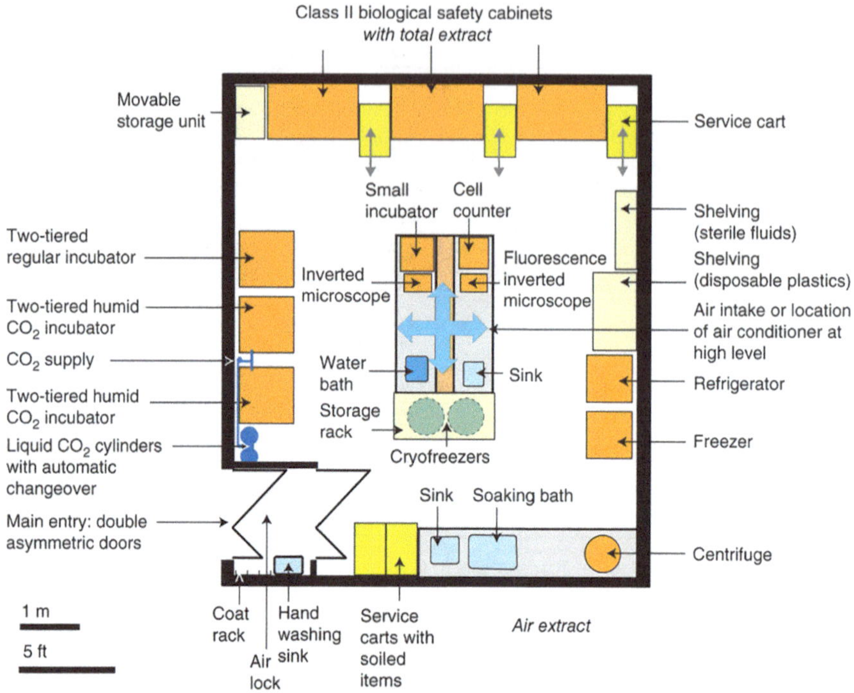

Fig. 1.1 A typical layout of a medium-sized cell culture laboratory. Suitable for five or six persons, with a washing and preparation facility located elsewhere. (Adapted from [11] with permission from John Wiley and Sons)

the laboratory itself but should be easily accessible. Basic cell culture also requires the following:

Centrifuge;
Refrigerator and freezer;
Microscope and hemocytometer;
Water bath.

The layout of the laboratory should be carefully planned to ensure that equipment generating drafts or dust is placed as far as possible from the sterile handling areas. Likewise, equipment that may harbor fungal contamination should be arranged away from the sterile handling areas. Specific considerations include: placing BSCs on the opposite end from the door to minimize air currents; avoiding the placement of a centrifuge on the same bench as a microscope or similar equipment which are prone to potential vibration; and positioning microscopes near BSCs for convenience [11].

From an operational standpoint, the facility should be regularly maintained, and certain equipment (such as BSCs) should undergo regular certification. In addition, routine training, the use of standard operating procedures, and centralized management of all aspects of cell culture are essential to ensure safety and efficiency [10].

Review and Practice Questions

1. What is biosafety, and what are biohazards?
2. What is laboratory biosecurity?
3. True/False:
   (a) Risk Group 2 biological agents should only be used in the biosafety level (BSL)-2 laboratory.
   (b) All *E-coil* strains may be used in BSL-1 laboratory.
   (c) All mammalian cells are classified as BSL-2.
4. Compare biosafety levels 1, 2, and 3.
5. What are the two fundamental considerations in the design of cell culture facilities?
6. Provide examples of layout considerations for equipment placement within a cell culture facility.

References

1. World Health Organization. *Laboratory Biosafety Manual Forth Edition*. Available from: https://iris.who.int/bitstream/handle/10665/337956/9789240011311-eng.pdf.
2. Centers for Disease Control and Prevention. *Biosafety in Microbiological and Biomedical Laboratories*. Available from: https://www.cdc.gov/labs/pdf/SF__19_308133-A_BMBL6_00-BOOK-WEB-final-3.pdf.
3. Micou, M.K. and D. Kilkenny, *Lab Safety*, in *A Laboratory Course in Tissue Engineering*. 2013, CRC Press.
4. The national Institue for Occupational Safety andHealth. *Module 2: Work-Related Risks and Hazards*. Available from: https://www.cdc.gov/niosh/learning/safetyculturehc/module-2/1.html.
5. Capes-Davis, A., et al., *6. Safety and Bioethics*. Eighth edition. ed. Freshney's culture of animal cells: a manual of basic technique and specialized applications. 2021, Wiley Blackwell.
6. World Health Organization. *Laboratory biosecurity guidance*. 2006; Available from: https://iris.who.int/bitstream/handle/10665/69390/WHO_CDS_EPR_2006.6_eng.pdf.
7. Administration for Stragtegic Preparedness & Response. *Biosecurity*. Available from: https://aspr.hhs.gov/S3/Pages/Biosecurity.aspx.
8. Administration for Stragtegic Preparedness & Response. Available from: https://aspr.hhs.gov/S3/Pages/Risk-Groups.aspx.
9. ATCC. *Biocontainment and Safety*. Available from: https://www.atcc.org/the-science/culturing-cells/biocontainment-and-safety.
10. O'Connor, R. and L. O'Driscoll, *Chapter 1—Setting up a Cell Culture Laboratory*, in *Cell Biology (Third Edition)*, J.E. Celis, Editor. 2006, Academic.
11. Capes-Davis, A., et al., *4. Laboratory Design and Layout*. Eighth edition. ed. Freshney's culture of animal cells: a manual of basic technique and specialized applications. 2021, Wiley Blackwell.

Basics of Cell Culture

2

Learning Objectives

- Understand and explain the fundamental principles of cell culture
- Discuss and analyze the key events in the history of cell culture

2.1 Basic Concepts of Cell Culture

Cell culture is generally regarded as a technique in which cells are cultivated outside a living organism under controlled conditions (e.g., temperature, pH, and nutrient levels) [1]. According to the *Merriam-Webster dictionary*, culture (verb) means "to grow in a prepared medium" or "to start a culture from." As a noun, it refers to the act or process of cultivating living materials [2]. Tissue culture is also used in the field of cell culture, referring to the culture of cells initially derived from a tissue sample [3] or the culture of intact tissue [4]. Regardless of the definition, three core components characterize cell culture: (1) it is performed *in vitro* (in a culture vessel outside a living organism); (2) cells are maintained in a viable state; and (3) cells are maintained in a metabolically active state [5]. It is worth noting that during cell culture, not all cells are able to proliferate, or proliferate readily.

Primary cell culture refers to the culture of cells directly obtained from tissues/organs by mechanical or chemical disintegration, or by enzymatic digestion. These cells usually exhibit a low growth rate and are heterogenous. Secondary cell culture refers to primary cultures that have been passaged or subcultured and grown for an extended period in fresh medium [6]. Subculture or passaging is the transfer of cells from one culture vessel to another [3]. There are two basic systems for cell culture—adherent and suspension. Adherent cells attached to the state (culture vessel) and generally grow to a monolayer. Because the surface area of the vessel is limited, passaging is required. In contrast, cells are not attached to the substrate in suspension culture (Fig. 2.1 and Table 2.1).

W. Zhang, *Cell Culture and Tissue Engineering*,
https://doi.org/10.1007/978-3-032-32096-4_2

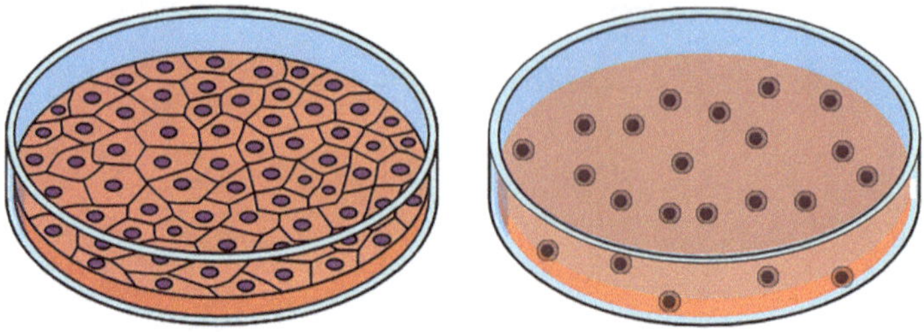

Fig. 2.1 Adherent (left) versus suspension (right) cell culture

Table 2.1 Comparison between adherent and suspension cell culture

Adherent cell culture	Suspension cell culture
Cells are dissociated either enzymatically or mechanically	Cells are not dissociated
Growth is limited by growth surface area	Growth is limited by cell concentration in the culture medium
Require tissue culture-treated vessels to promote cell adhesion	Do not require tissue culture-treated vessels but agitation may be required

The primary culture, when subcultured, becomes a cell line or cell strain. Depending on the lifespan of the culture, a cell line may be finite or continuous. Finite cell lines have a limited number of generations, whereas continuous cell lines, also known as immortalized or indefinite cell lines, can divide indefinitely [6, 7]. Human cells in culture do not divide indefinitely but reach a limit (Hayflick limit) of replication and stop all further division. As cells approach this limit, they slow their divisions and enter cellular senescence, a dormant period [8]. The molecular basis of this phenomenon is the gradual shortening of telomeric DNA [9]. Genetically mutated cells derived from cancerous tissues are a common source for establishing immortal cell lines. Noncancer-derived cultured cells may spontaneously acquire genetic mutations that confer unlimited growth, or they may be deliberately immortalized through the introduction of viral oncogenes or by inducing telomerase activity (e.g., *via* transduction of human telomerase reverse transcriptase, hTERT) [10]. For example, two spontaneously immortalized canine embryonic fibroblasts (CaEF) cell lines (CGFR-Ca-1 and -2; derived from two different canine embryos) and an immortal CaEF cell line generated by exogenous introduction of a catalytic telomerase subunit (CGFR-Ca-3) were established from normal CaEF cells. Primary CaEF cells typically reached senescent stage at passage 8. As shown in Fig. 2.2, primary CaEF cells at senescence showed enlarged, flat, β-galactosidase-positive, and highly vacuolated morphologies, phenotypical characteristics of senescent fibroblast. Immortalized CaEF cells, arising from the cellular colonies within

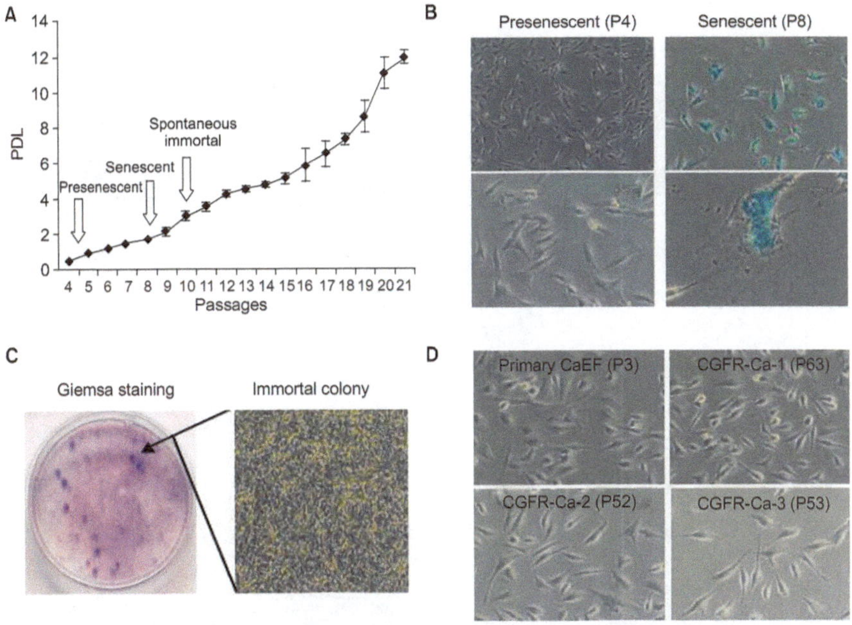

Fig. 2.2 Growth kinetics and spontaneous immortalization of primary canine embryonic fibroblasts (CaEF). (**a**) Population doubling levels (PDL) of primary CaEF. (**b**) Representative images (upper photos, 40×; lower photos, 100× magnification) of presenescent passage 4 and senescent passage 8 CaEF stained with X-gal solution to detect senescence-associated β-galactosidase activity. (**c**) Passage 10 senescent CaEF stained with Giemsa solution (left) and representative immortalized cells arising from a colony (right). (**d**) Morphology of primary CaEF compared with three immortal cell lines. (Adapted from [11] with permission from Springer Nature)

the senescent cell population, appeared relatively smaller, brighter and more uniform compared with presenescent passage 3 CaEF cells [11]. A widely used system for immortalization is lentivirus-mediated simian virus 40 Large T (SV40LT) system, which binds and inactivates p53 and Rb proteins (cell-cycle regulating) [9]. Cells types immortalized using this method include canine corneal epithelial cell line (CCEC, Fig. 2.3) [12], Egyptian Rousettus bat fibroblasts [13], human adenomyosis ectopic lesion cells [14], and human aortic valve interstitial cell lines [15]. However, immortalized cells often display altered cellular properties compared to their primary counterparts, such as increased tolerance to oxidative damage [16].

The purposes of cell culture can be broadly divided into two categories: basic and applied. Basic applications include studying intracellular activities and cell-cell interactions. Applied uses involve bioproduction, drug testing, cellular therapies, and tissue engineering (Fig. 2.4) [3, 6].

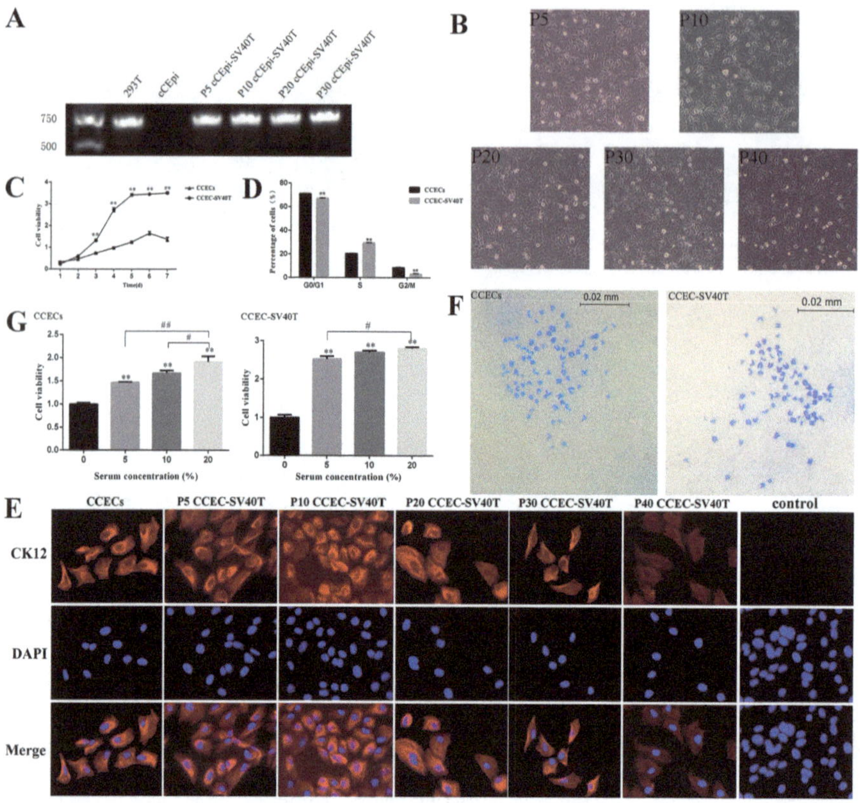

Fig. 2.3 Characteristics of the CCEC-SV40T line. (**a**) PCR assays of SV40T mRNA. (**b**) Morphology of CCEC-SV40T cells across generations. (**c**) Comparison of proliferation ability between CCEC-SV40T cells and CCECs. (**d**) Comparison of cell cycle profiles between CCECs and CCEC-SV40T cells. (**e**) Immunofluorescence detection of cytokeratin (CK) 12 between CCEC-SV40T cells and CCECs. (**f**) Karyotype analysis of CCEC-SV40T cells and CCECs. (**g**) Serum dependence comparison between CCEC-SV40T cells and CCECs. CCEC: canine corneal epithelial cells, SV40T: Simian vacuolating virus 40 large T. (Adapted from [12])

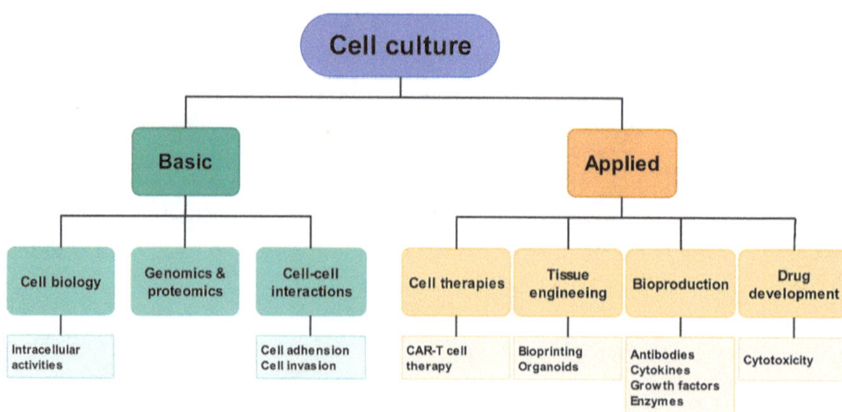

Fig. 2.4 Basic and applied purposes of cell culture

2.2 Historical Aspects of Cell Culture

Ross Harrison's article, *"The outgrowth of the nerve fiber as a mode of protoplasmic movement,"* is one of the most important articles in cell culture history. He invented the method of tissue culture and used it to prove the neuron doctrine [17, 18]. In 1923, Alexis Carrel developed a range of culture flasks designed to "maintain tissues in a condition of uninterrupted growth in a medium which does not deteriorate spontaneously" (Fig. 2.5) [19]. In 1926, differentiation of embryonic tissues *in vivo* and *in vitro* was reported [20]. The earliest mammalian cell lines, established by Earle et al. in 1943, were derived from the subcutaneous tissue of a C3H mouse strain [21, 22]. In 1952, the first human (also first immortal human) cell line, HeLa, was established from a cervix cancer biopsy obtained from **He**nrietta **La**cks (Fig. 2.6). Since then, HeLa cells have been used extensively to test the effects of drugs, study the human genome, and contribute to vaccine development, including COVID-19 vaccines [23–25].

Both the modified Carrel flask (T-flask) and spinner culture systems were developed in the 1950s [3]. In 1967, the suspension culture of both cell strains and primary cells using microcarriers was first reported [26]. Mouse embryonic stem cells were first isolated and propagated in culture in 1981 [27], whereas the culture of human embryonic stem (ES) cells was reported in 1998 by researchers at the University of Wisconsin-Madison, led by Dr. James Thomson [28]. In 2006, induced pluripotent stem cells (iPSCs) were generated from mouse embryonic and adult fibroblasts using defined factors, Oct3/4, Sox2, c-Myc, and Klf4, under ES cell culture conditions [29]. In 2014, a cellular reprogramming phenomenon termed stimulus-triggered acquisition of pluripotency (STAP) was reported. In STAP, strong external stimuli, such as transient low-pH stressor, were claimed to convert somatic cells into pluripotent cells [30]. Unfortunately, the results could not be produced and the article was subsequently retracted [31, 32].

Fig. 2.5 A technician manipulating the specially designed "Carrel flask" and the aspiration device designed by Alexis Carrel in the 1920s, widely used into the 1950s. (Courtesy of The Rockefeller Archive Center)

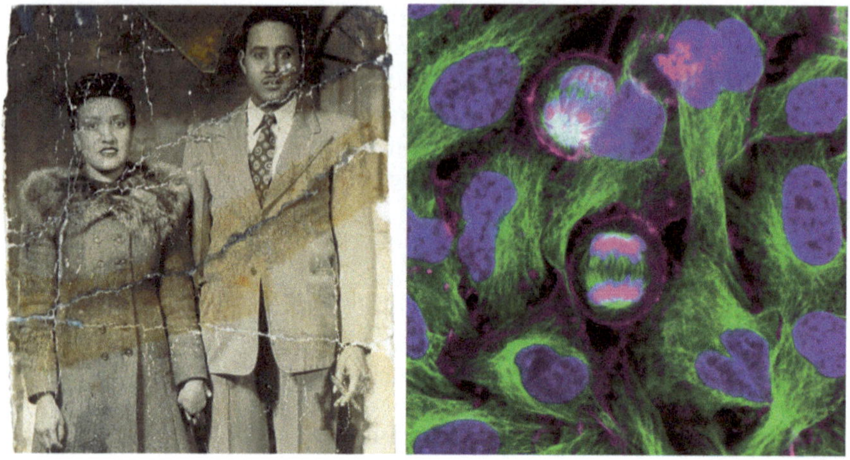

Fig. 2.6 Left: photo of Henrietta Lacks with her husband. (Adapted from [25] with permission from Springer Nature). Right: dividing HeLa cells. Credit: Kevin Mackenzie, University of Aberdeen. (Source: Wellcome Collection)

Review and Practice Questions

1. What is cell culture? What is primary cell culture?
2. Compare adherent and suspension cell cultures.
3. Provide examples of applications of cell culture.
4. Discuss key events in the history of cell culture.

References

1. Lin, Y.H., T.M. Pan, and M.H. Wu, *5.12—Microfluidic Technology and Its Biological Applications*, in *Comprehensive Biotechnology (Second Edition)*, M. Moo-Young, Editor. 2011, Academic.
2. Merriam-Webster Dictionary. *Culture*. Available from: https://www.merriam-webster.com/dictionary/culture.
3. Capes-Davis, A., et al., *1. Introduction*. Eighth edition. ed. Freshney's culture of animal cells: a manual of basic technique and specialized applications. 2021, Wiley Blackwell.
4. Merriam-Webster Dictionary. *Tissue culture*. Available from: https://www.merriam-webster.com/dictionary/tissue%20culture.
5. Yoon, J.-Y., *Cell Culture*, in *Tissue Engineering: A Primer with Laboratory Demonstrations*. 2022, Springer.
6. Verma, A., M. Verma, and A. Singh, *Chapter 14—Animal tissue culture principles and applications*, in *Animal Biotechnology (Second Edition)*, A.S. Verma and A. Singh, Editors. 2020, Academic.
7. Weiskirchen, S., et al., *A Beginner's Guide to Cell Culture: Practical Advice for Preventing Needless Problems*. Cells, 2023. **12**(5):682.
8. Hoopes, L.M., *Aging and Cell Division*. Nature Education, 2010. **3**(9):55.
9. Chalak, M., et al., *Cell Immortality: In Vitro Effective Techniques to Achieve and Investigate Its Applications and Challenges*. Life, 2024. **14**(3):417.

10. Hayden, P.J., *Chapter 2—Cell sources and methods for producing organotypic in vitro human tissue models*, in *Organ-on-a-chip*, J. Hoeng, D. Bovard, and M.C. Peitsch, Editors. 2020, Academic Press.
11. You, S., et al., *Cellular characteristics of primary and immortal canine embryonic fibroblast cells.* Experimental & Molecular Medicine, 2004. **36**(4):325–335.
12. Guo, L., et al., *Immortalization effect of SV40T lentiviral vectors on canine corneal epithelial cells.* BMC Veterinary Research, 2022. **18**(1):181.
13. Bai, L., et al., *Establishment of immortalized Egyptian Rousettus bat cell lines.* FEBS Open Bio, 2024. **14**(4):598–612.
14. Wang, Z., et al., *Establishment of an immortalized cell line derived from human adenomyosis ectopic lesions.* Tissue and Cell, 2024. **86**:102284.
15. Wang, Z., et al., *Establishment and characterization of a novel immortalized human aortic valve interstitial cell line.* Scientific Reports, 2025. **15**(1):10917.
16. Alcock, D., et al., *Publisher Correction: Generating bat primary and immortalised cell-lines from wing biopsies.* Scientific Reports, 2024. **14**(1):30530.
17. Harrison, R.G., *The outgrowth of the nerve fiber as a mode of protoplasmic movement.* Journal of Experimental Zoology, 1959. **142**(1):5–73.
18. Keshishian, H., *Ross Harrison's "The outgrowth of the nerve fiber as a mode of protoplasmic movement".* Journal of Experimental Zoology Part A: Comparative Experimental Biology, 2004. **301A**(3):201–203.
19. Millet, L.J. and M.U. Gillette, *Over a century of neuron culture: from the hanging drop to microfluidic devices.* Yale Journal of Biology and Medicine, 2012. **85**(4):501–21.
20. Strangeways, T.S.P. and H.B. Fell, *Experimental studies on the differentiation of embryonic tissues growing in vivo and in vitro.—I. The development of the undifferentiated limb-bud (a) when subcutaneously grafted into the post-embryonic chick and (b) when cultivated in vitro.* Proceedings of the Royal Society of London. Series B, Containing Papers of a Biological Character, 1926. **99**(698):340–366.
21. Almeida, J.L. and C.T. Korch, *Authentication of Human and Mouse Cell Lines by Short Tandem Repeat (STR) DNA Genotype Analysis*, in *Assay Guidance Manual*, S. Markossian, et al., Editors. 2004, Eli Lilly & Company and the National Center for Advancing Translational Sciences.
22. Earle, W.R., et al., *Production of Malignancy in Vitro. IV. The Mouse Fibroblast Cultures and Changes Seen in the Living Cells.* JNCI: Journal of the National Cancer Institute, 1943. **4**(2):165–212.
23. *The Legacy of Henrietta Lacks.* Available from: https://www.hopkinsmedicine.org/henrietta-lacks.
24. O'Brien, S.J., *Cell Culture Forensics.* Proceedings of the National Academy of Sciences of the United States of America, 2001. **98**(14):7656–7658.
25. Callaway, E., *Deal done over HeLa cell line.* Nature, 2013. **500**(7461):132–133.
26. Van Wezel, A.L., *Growth of Cell-strains and Primary Cells on Micro-carriers in Homogeneous Culture.* Nature, 1967. **216**(5110):64–65.
27. Evans, M.J. and M.H. Kaufman, *Establishment in culture of pluripotential cells from mouse embryos.* Nature, 1981. **292**(5819):154–156.
28. Thomson, J.A., et al., *Embryonic Stem Cell Lines Derived from Human Blastocysts.* Science, 1998. **282**(5391):1145–1147.
29. Takahashi, K. and S. Yamanaka, *Induction of Pluripotent Stem Cells from Mouse Embryonic and Adult Fibroblast Cultures by Defined Factors.* Cell, 2006. **126**(4):663–676.
30. Obokata, H., et al., *RETRACTED ARTICLE: Stimulus-triggered fate conversion of somatic cells into pluripotency.* Nature, 2014. **505**(7485):641–647.
31. Cyranoski, D., *Papers on 'stress-induced' stem cells are retracted.* Nature, 2014. Available from: https://doi.org/10.1038/nature.2014.15501.
32. Cyranoski, D., *Failed replications put STAP stem-cell claims to rest.* Nature, 2015. Available from: https://doi.org/10.1038/nature.2015.18412.

Biology of Cultured Cells 3

Learning Objectives

- Discuss cell metabolism, with particular emphasis on aerobic respiration
- Evaluate the functions of key cell culture media components and supplements
- Compare major adhesion molecules and different types of cell culture substrates, including the use of feeder cells and microcarriers
- Interpret the cell cycle and perform calculations related to first-order growth kinetics
- Discuss the classifications and mechanisms of cell death

3.1 Cell Metabolism

Cell metabolism is the series of biochemical reactions that enable cells to produce the energy required for their survival and maintenance [1]. Physiological cellular metabolism consists of specific metabolic reactions that convert a carbon source into the building blocks needed for macromolecule biosynthesis, energy production, and cellular homeostasis [2]. Cell metabolism can be divided into two parts: (1) catabolic reactions, which breakdown complex molecules to generate energy while generating small molecules as waste or by-product, and (2) anabolic reactions, which are the biosynthesis of large molecules that consumes energy [3]. Adenosine triphosphate (ATP) is the main energy currency of the cell (Fig. 3.1). During ATP hydrolysis, high energy phosphate bond(s) are cleaved to release energy. To produce ATP, one glucose molecule can yield about 30 ATPs in eukaryotes through aerobic respiration (glycolysis, pyruvate oxidation, citric acid cycle (also known as Krebs cycle or tricarboxylic acid (TCA) cycle), electron transport chain & chemiosmosis) (Table 3.1 and Fig. 3.2) [4]. Glutamine, a key nutrient, can also be used for energy production through glutaminolysis and citric acid cycle (Fig. 3.2) [3, 5].

W. Zhang, *Cell Culture and Tissue Engineering*,
https://doi.org/10.1007/978-3-032-32096-4_3

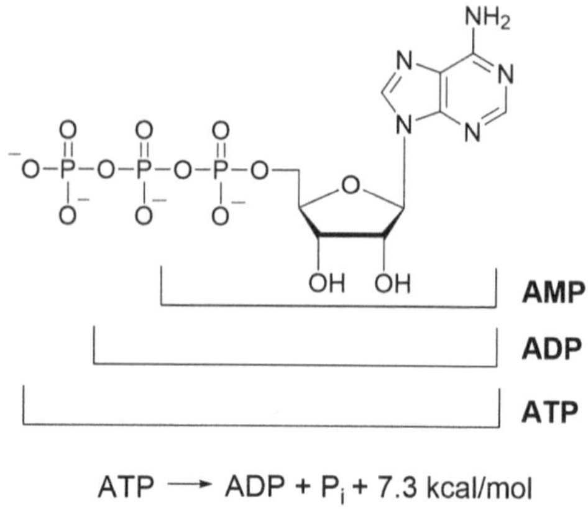

Fig. 3.1 Structure of ATP, ADP and AMP. ATP: adenosine triphosphate, ADP: adenosine diphosphate, AMP: adenosine monophosphate, P_i: inorganic phosphate

Table 3.1 An overview of aerobic respiration

Stage	Location	Input	Output	# of ATP produced (net)
Glycolysis	Cytoplasm	1 glucose 2 ATP	2 pyruvate 4 ATP 2 NADH	2
Pyruvate oxidation	Mitochondria matrix	2 pyruvate	2 acetyl CoA 2 NADH CO_2	0
Citric acid cycle	Mitochondria matrix	2 acetyl CoA	2 ATP 6 NADH 2 $FADH_2$ CO_2	2
Electron transport chain + chemiosmosis	Mitochondria inner membrane	10 NADH 2 $FADH_2$ ADP + P_i O_2	28 ATP H_2O NAD^+ FAD	28

Note: Total net ATP yield is 32 per glucose molecule (30 in eukaryotes since it takes 2 ATPs to transport 2 NADHs generated by glycolysis in the cytoplasm into the mitochondria). ATP: adenosine triphosphate, CoA: coenzyme A, NAD^+/NADH: nicotinamide adenine dinucleotide, FAD/$FADH_2$: flavin adenine dinucleotide

To obtain optimal metabolic activities, secondary requirements such as growth factors and cofactors should also be provided. Additionally, proper temperature, pH, and oxygen level should also be maintained [3]. Oxygen levels, which significantly affect culture phenotype, can be influenced by multiple factors, including the culture vessel, seeding density, medium volume, partial pressure of oxygen, and tissue culture media composition. Unfortunately, this is not routinely taken into account [6]. It is also common to induce hypoxia (a reduction or lack of oxygen) in cell culture, since hypoxia can be either physiologic or pathologic, as in the case of solid tumors [7]. Hypoxia can lead to acidification and increased oxidative stress, both of which significantly impact cell physiology [8]. With respect to pH, buffering

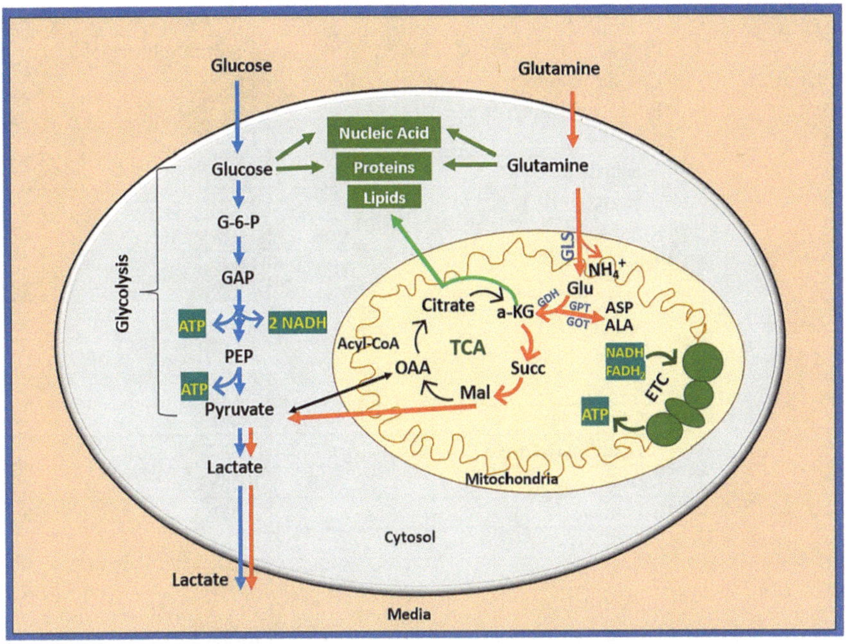

Fig. 3.2 Glycolysis and Glutaminolysis metabolic pathway overview. G-6-P: glucose-6-phosphate, GAP: glyceraldehyde 3-phosphate, PEP: phosphoenolpyruvate, GLS: glutaminase, α-KG: α-ketoglutarate, GLU: glutamine, GDH: glutamate dehydrogenase (GDH), GOT: glutamate oxaloacetate transaminase, GPT: glutamate pyruvate transaminase, ASP: aspartate, ALA: alanine, Succ: succinate, Mal: malate, Acetyl-COA, acetyl coenzyme A, OAA: Oxaloacetate, ETC: electron transport chain, TCA: tricarboxylic acid. (Adapted from [5])

systems such as $CO_2/NaHCO_3$ system ($CO_2 \leftrightarrow H_2CO_3 \leftrightarrow HCO_3^-$) or chemical buffers such as HEPES (4-(2-hydroxyethyl)-1-piperazineethanesulfonic acid) are commonly used to maintain the pH, generally within the range of 7.2–7.4. During cell culture, the pH decreases over time, and medium renewal is necessary to help maintain proper pH as well as oxygenation [3, 6]. For basic cell culture, an incubator and cell culture medium are generally enough for meeting these requirements. The incubator can help maintain the temperature, specialized media provide nutrients and such for cell metabolism, and together they help maintain the pH (*via* the $CO_2/NaHCO_3$ buffering system).

A complete cell culture medium typically consists (by volume) of 89% of base medium, 10% serum, and 1% antibiotics. Depending on the cell type, an appropriate cell culture medium should be selected. Common basal cell culture media include Dulbecco's Modified Eagle Medium (DMEM), Eagle's Minimum Essential Medium (EMEM), and Roswell Park Memorial Institute (RPMI) 1640. For example, DMEM (high glucose) contains glucose, amino acids, vitamins, inorganic salts, and Phenol Red (Table 3.2). D-Glucose is the primary energy source. Glutamine is an important amino acid that supplies carbon and nitrogen to fuel biosynthesis [9].

Table 3.2 Formulation of Dulbecco's Modified Eagle's Medium (DMEM)

	Component	Concentration (mg/L)
Vitamins	Choline chloride	4.0
	D-Calcium pantothenate	4.0
	Folic acid	4.0
	Niacinamide	4.0
	Pyridoxine•HCl	4.0
	Riboflavin	0.4
	Thiamine•HCl	4.0
	i-Inositol	7.2
Amino acids	Glycine	30.0
	L-Arginine•HCl	84.0
	L-Cystine•2 HCl	63.0
	L-Glutamine	584.0
	L-Histidine•HCl•H_2O	42.0
	L-Isoleucine	105.0
	L-Leucine	105.0
	L-Lysine•HCl	146.0
	L-Methionine	30.0
	L-Phenylalanine	66.0
	L-Serine	42.0
	L-Threonine	95.0
	L-Tryptophan	16.0
	L-Tyrosine disodium salt•2 H_2O	104.0
	L-Valine	94.0
Inorganic salts	$CaCl_2$ (anhydrous)	200.0
	$Fe(NO_3)_3$•9 H_2O	0.1
	$MgSO_4$ (anhydrous)	97.7
	KCl	400.0
	$NaHCO_3$	3700.0
	NaCl	6400.0
	NaH_2PO_4•H_2O	125.0
Others	D-Glucose	4500.0
	Phenol Red	15.0

Note: data is from https://www.thermofisher.com/us/en/home/technical-resources/media-formulation.8.html

Inorganic salts, particularly NaCl, help maintain the osmotic balance. $NaHCO_3$ serves as a buffering component. Phenol Red functions as a pH indicator; however, for imaging and biological assays, Phenol Red-free medium is preferred. Additional concerns with using Phenol Red in cell culture include its potential estrogenic activity and redox activity [10–12].

Serum, most commonly fetal bovine serum (FBS), contains hormones, lipids, trace elements, and growth factors that are critical for cell growth and proliferation. However, because of its large variability in composition and quality, as well as the ethical concerns over the harvest and collection of fetal bovine serum, the use of serum-free medium and chemically defined medium is becoming increasingly common in cell culture [13]. Serum-free media are typically cell type-specific and usually require supplementation with specific factors. Chemically defined media contain no proteins, growth factors, hydrolysates, or components of unknown composition; instead, they consist solely of substances whose chemical compositions and structures are fully known [14].

Antibiotics are often added to prevent contamination—for example, a 100X penicillin/streptomycin solution is commonly used. Penicillin, a β-lactam antibiotic, is effective primarily against Gram-positive bacteria, while streptomycin-sulfate, an aminoglycoside antibiotic, is effective against Gram-negative bacteria and *Mycobacteria* spp. [15]. However, for more experienced personnel, the use of antibiotics is generally discouraged. The rationales include the following: routine usage of antibiotics may develop slow growing persistent/resistant bacterial contaminants that can cause subtle alterations of cell differentiation and behavior; certain antibiotics can significantly alter gene expression and regulation, potentially affecting experimental results; and antibiotics may mask low-level contamination [12].

3.2 Cell Adhesion

Most cells grow as monolayers that adhere to the substrate. Adherent cells are classified as epithelia-like (rounded shape and tendency to grow in patches) or fibroblast-like (elongated shape, either bipolar or stellate in appearance) [16]. Figure 3.3 shows microscopic images of both epithelial cells and fibroblasts. Cells adhere to each other and to substrates, including the extracellular matrix (ECM; composed of collagens, fibronectin, laminin, etc., which will be further discussed in Chap. 9), through adhesion molecules (CAMs) such as integrins, selectins, cadherins, members of the immunoglobulin superfamily, and others such as mucins. CAMs are mostly transmembrane receptor proteins composed of three domains: intracellular, transmembrane, and extracellular. The intracellular domain interacts with the cytoskeleton for signaling, while the extracellular domain mediates interactions with other CAMs or the ECM (Figs. 3.4 and 3.5). Integrins typically bind to the ECM (focal adhesion), while selectins and cadherins are primarily responsible for cell-cell adhesion. An integrin molecule consists of two noncovalently associated transmembrane glycoprotein subunits—α and β (18 integrin α subunits and 8 β subunits are so far known, which can heterodimerize into 24 different integrins). Cadherins are Ca-dependent transmembrane proteins that play a vital role in junctional adhesions, maintaining their intracellular cohesion, holding cells within tissues together, and preserving tissue architecture [17–19]. The cytoskeleton—including actin filaments (microfilaments), intermediate filaments, and microtubules—carries out three broad functions: spatially organizes the contents of the cell, connects the cell physically and biochemically to the external environment, and generates coordinated forces that enable cell movement and shape changes (Fig. 3.6) [20]. The actin

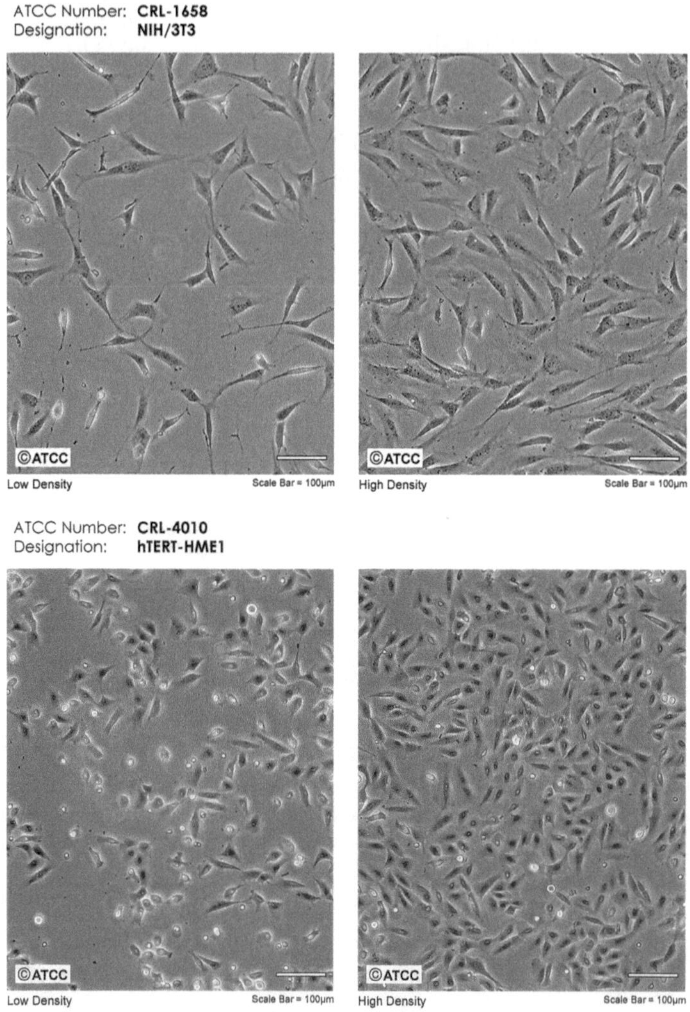

Fig. 3.3 Morphology of NIH/3T3 fibroblasts (fibroblast morphology; top) and hTRET-HME1 cells (epithelial morphology; bottom). DCAPC: dairy cow anterior pituitary cell. (Adapted with permission from the American Type Culture Collection (ATCC))

cytoskeleton (a meshwork of filamentous actin (F-actin) and various accessory proteins), is particularly crucial for both cell-cell and cell-ECM adhesions [21]. For example, studies have demonstrated that directed actin polymerization drives epithelial cell-cell adhesion [22].

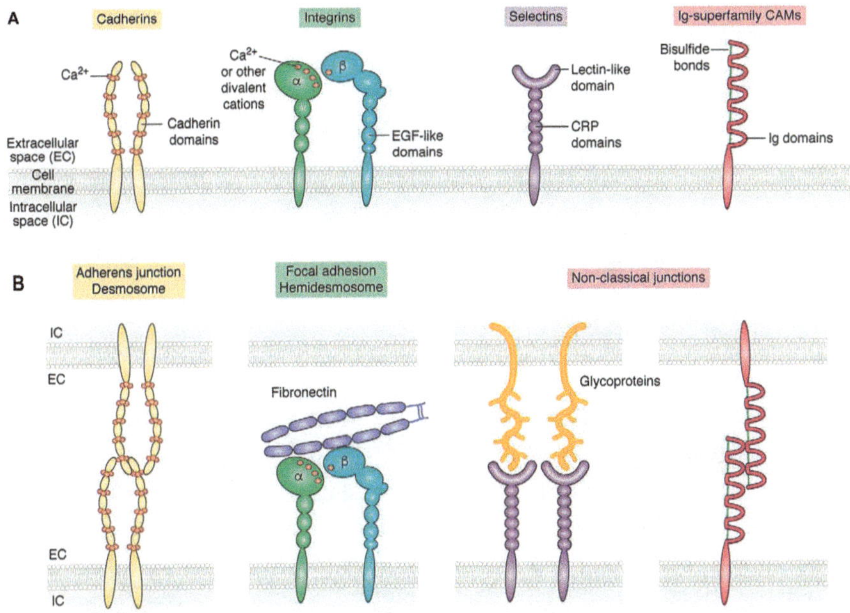

Fig. 3.4 (**a**) Schematic representation of the structure of four major classes of cell adhesion molecules. (**b**) Interactions between cell adhesion molecules on the cell surface create different types of cellular adhesions, including adherens junctions, gap junctions, desmosomes, and hemidesmosomes. EGF: epidermal growth factor, CRP: complement regulatory protein domains. (Adapted from [19])

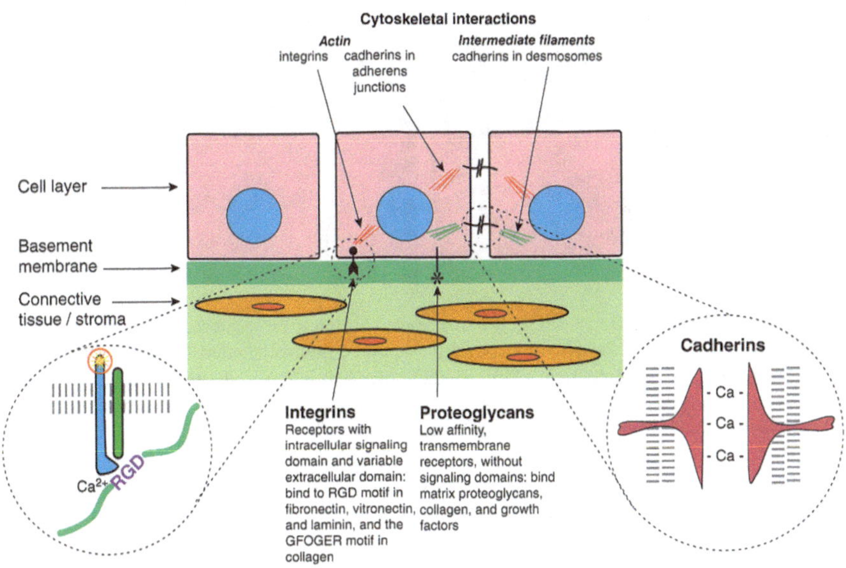

Fig. 3.5 Diagrammatic representation of a layer of epithelial cells above connective tissue (stroma) containing fibrocytes, separated from it by the basement membrane. Cadherins are depicted between like cells, while integrins and proteoglycans are shown between the epithelial cells and the ECM. (Adapted from [16] with permission from John Wiley and Sons)

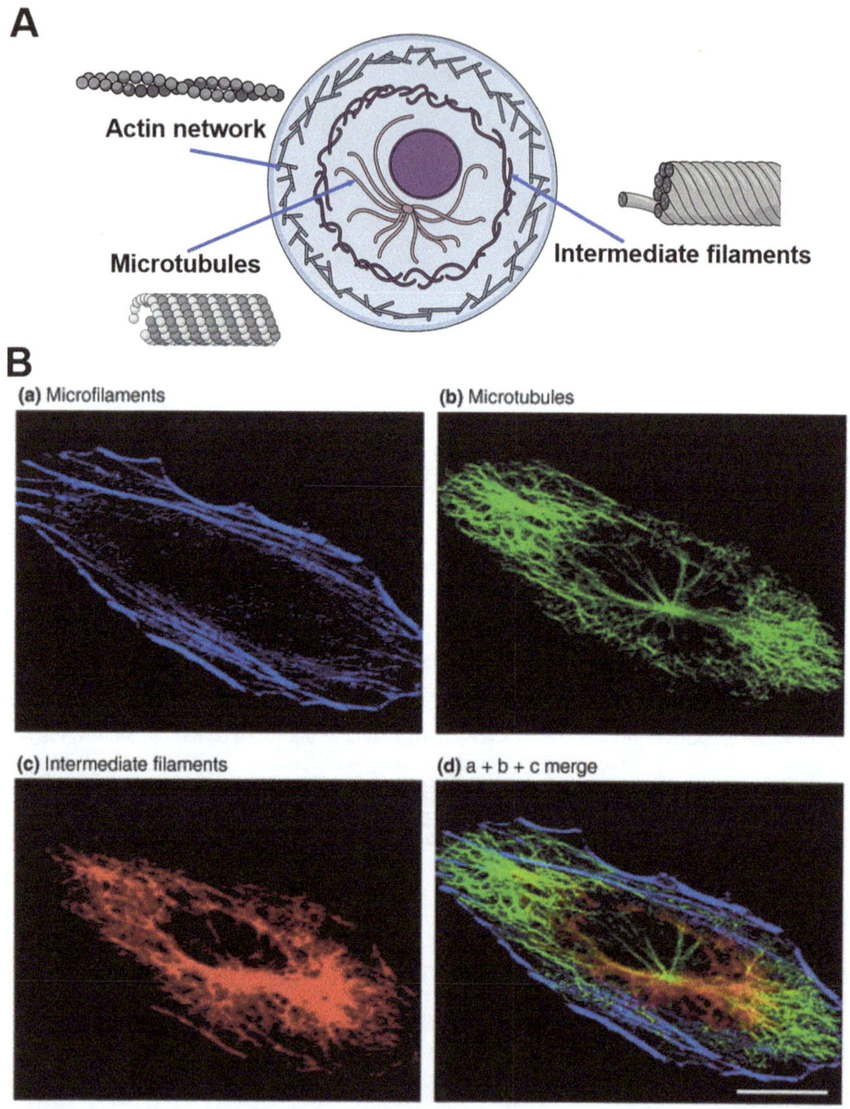

Fig. 3.6 A: schematic showing different cytoskeleton structures. B: fluorescence imaging of the three major components of the cytoskeleton. Huh7 cultured human hepatocytes were triple-stained with (**a**) phalloidin to visualize microfilaments (actin) (blue), (**b**) anti-tubulin antibody to visualize microtubules (green), and (**c**) anti-K8/K18 antibody to visualize IFs (red). A superimposed image is shown in (**d**). Scale bar = 10 μm. (Adapted from M. Bishr Omary, Nam-On Ku, Guo-Zhong Tao, Diana M. Toivola, Jian Liao, 'Heads and tails' of intermediate filament phosphorylation: multiple sites and functional insights, Trends in Biochemical Sciences, 31(7) 2006, 383–394, https://doi.org/10.1016/j.tibs.2006.05.008, with permission from Elsevier)

3.3 Cell Culture Substrates

As shown in Fig. 3.7, common cell culture vessels include Petri dishes, T-flasks, and multi-well plates. These vessels are mostly made of plastic, typically polystyrene. For cells to attach and spread, the surface has to be treated—for example, with plasma gas, which imparts negative charges on the surface and makes the surface more hydrophilic, facilitating the attachment of anchorage-dependent cells [23]. As discussed in Chap. 1, the first cell culture vessels were made of glass; however, they are no longer commonly used for cell culture. For imaging purposes, glass substrates are preferred over plastic due to their superior optical clarity.

For suspension cell culture, spinner flasks are commonly used. Even adherent cells, especially for scale-up, can be cultured in spinner flasks combined with microcarriers (Fig. 3.8) [24–27]. For example, Feng Yang et al. optimized parameters for cell proliferation and differentiation using spinner flasks and microcarriers (Fig. 3.9) [28]. Botao Zhang et al. demonstrated the significance of impeller design optimization for enhancing mesenchymal stem cells cultivation on microcarriers in spinner flasks (Fig. 3.10) [29].

To promote cell adhesion, substrate coating has been widely used. Common coating materials include gelatin, collagen, laminin, fibronectin, poly-*L*-lysine, poly-*D*-lysine, and poly-*L*-ornithine. The choice of coating material significantly impacts cell adherence, morphology, and behavior (Fig. 3.11) [30]. Interestingly, cells themselves have also been used as substrates, serving as feeder cells. These cells, treated with mitomycin, γ-irradiation, or electric pulses, are unable to divide but provide extracellular secretions that support the proliferation of other cells [31]. For example, stromal cells as feeder cells are used to recapitulate biomimetic niches of hematopoietic stem cells in adult bone marrow [32]. It has been demonstrated that human feeder cells, such as inactivated human adult uterine endometrial cells (hUECs), human adult breast parenchymal cells (hBPCs) and embryonic fibroblasts (hEFs), can support the proliferation of human embryonic stem cells (hESCs) without inducing differentiation, offering the advantage of supporting more passages compared to mouse embryonic fibroblasts [33]. Feeder cells have also been shown to facilitate telomere maintenance and chromosomal stability of ESCs (Fig. 3.12) [34].

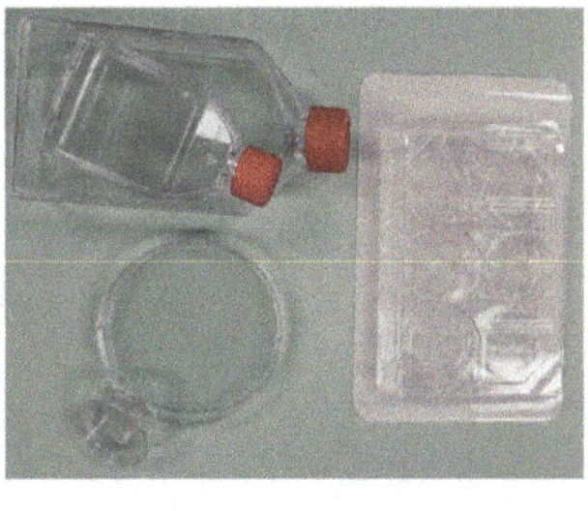

Fig. 3.7 Basic plastic-based cell culture vessels

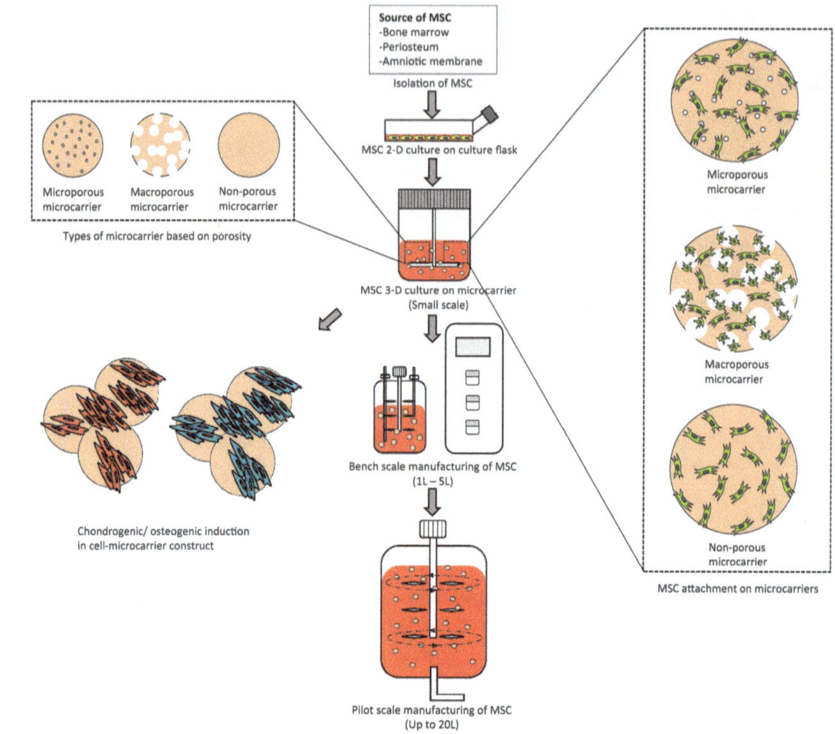

Fig. 3.8 Schematic illustrating the basic flow of up-scaling MSC culture from laboratory scale into manufacturing scale. MSC: mesenchymal stem cells. (Adapted from [27])

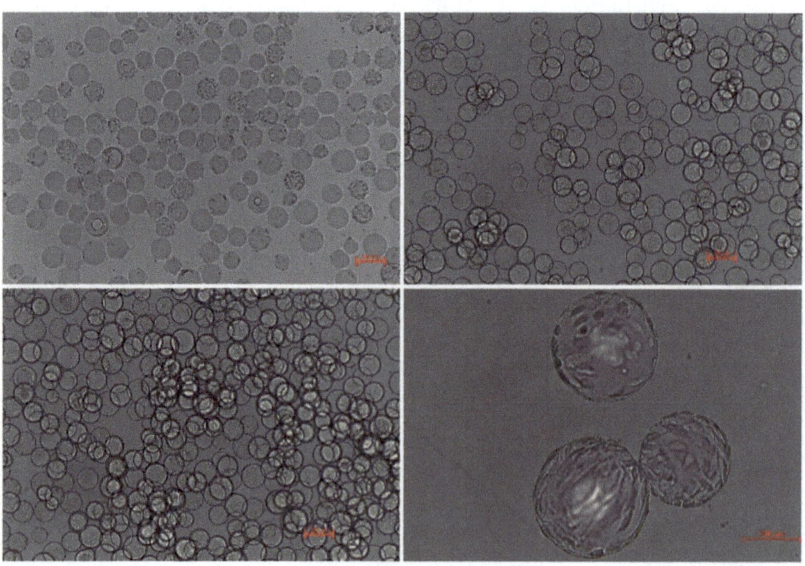

Fig. 3.9 Comprehensive evaluation of optimized conditions for promoting cell proliferation in the culture system. (Adapted from [28])

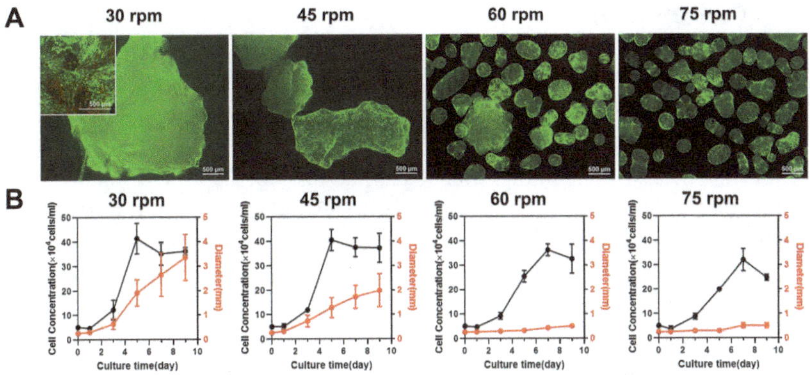

Fig. 3.10 (**a**) Cell viability staining with Calcein–AM/PI on day 9 under different rotational speeds. Scale bar: 500 μm. (**b**) Dynamic curves of cell density and aggregate size. PI: propidium iodide. (Adapted from [29])

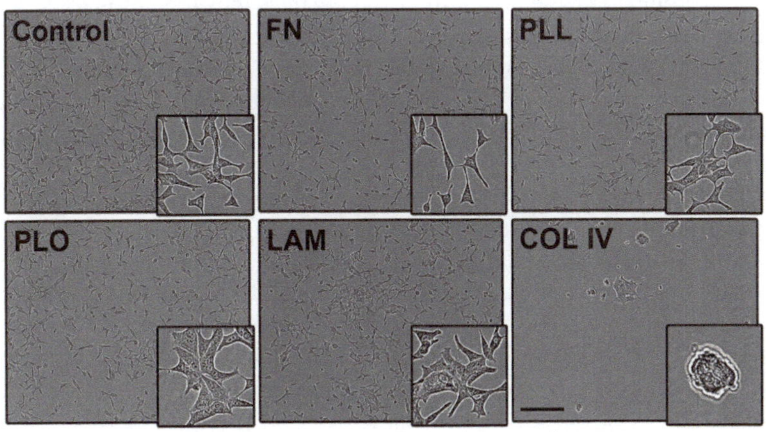

Fig. 3.11 LNCaP cells imaged after 96 h. Scale bar = 300 μm. FN: fibronectin, PLL: poly-L-lysine, PLO: poly-L-ornithine, LAM: laminin, COL IV: collagen type IV. *Adapted from* [30]

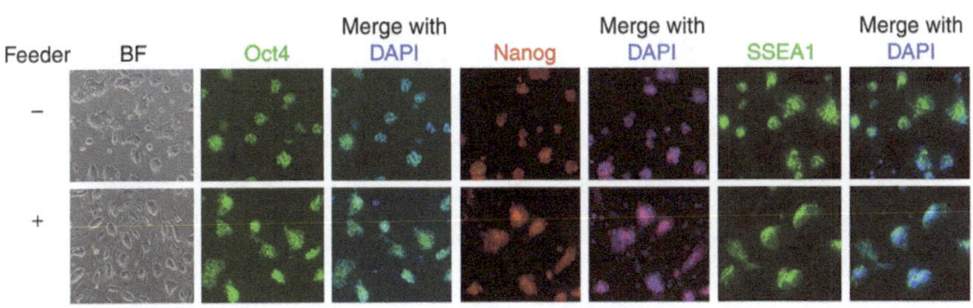

Fig. 3.12 Morphology of F1 ESCs cultured in feeder or feeder-free conditions under bright field (BF) with phase-contrast optics and immunofluorescence of pluripotency-associated markers Oct4, Nanog, and SSEA1. DAPI: diamidino-2-phenylindole, Oct4: Octamer binding transcription factor4, Nanog: after 'Tir nan Og', SSEA1: stage-specific embryonic antigen1. Scale bar, 100 μm. (Adapted from [34])

3.4 Cell Division

For somatic cells, the most basic function of the cell cycle is to accurately duplicate the vast amount of DNA in chromosomes and then segregate the copies precisely into two genetically identical daughter cells. As shown in Fig. 3.13, the eukaryotic cell cycle is typically divided into four sequential phases: G1, S, G2, and M. During the S (synthesis) phase, DNA replication occurs, while during the M phase, mitosis (nuclear division) and cytokinesis (cytoplasmic division) are completed. The two gap phases, G1 and G2, allow cells to monitor the internal and external environments to ensure the conditions are suitable and that necessary preparations are complete before S phase and mitosis. Cells do not always complete the cell cycle; they can delay progress through G1 and enter a resting phase (G0) for varying lengths of times [35]. Cell division is highly regulated, and the cell cycle can be arrested at specific checkpoints.

Mitosis is classically divided into multiple stages: prophase, prometaphase, metaphase, and anaphase. As shown in Fig. 3.14, from the traditional view, spindles made of microtubules capture, transport and organize the chromosomes during the early stages of mitosis in a somatic cell [36]. Interestingly, a recent study showed that actin filaments (F-actin) anchor the chromosomes at the nuclear periphery, transport the chromosomes to the center of the cell, and then form a sheath around the metaphase spindle that limits spindle size in early mouse embryos (Fig. 3.14) [36, 37].

Typically, mammalian cells in continuous culture divide every 1–3 days [16]. Finite cell lines proliferate for a limited number of generations before reaching senescence, generally between 20 and 80 population doublings (PDL; the total number of population doublings that the cells have undergone in culture). After passaging, cells generally grow exponentially (log phase) after a lag period and then reach plateau (stationary phase) (Figs. 3.15 and 3.16). The exponential growth of mammalian cells can be mathematically represented by the first-order kinetics model:

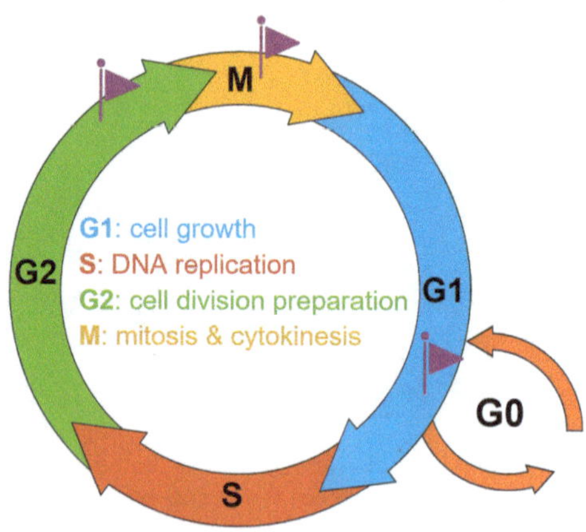

Fig. 3.13 Phases of the cell cycle. Purple flags: checkpoints

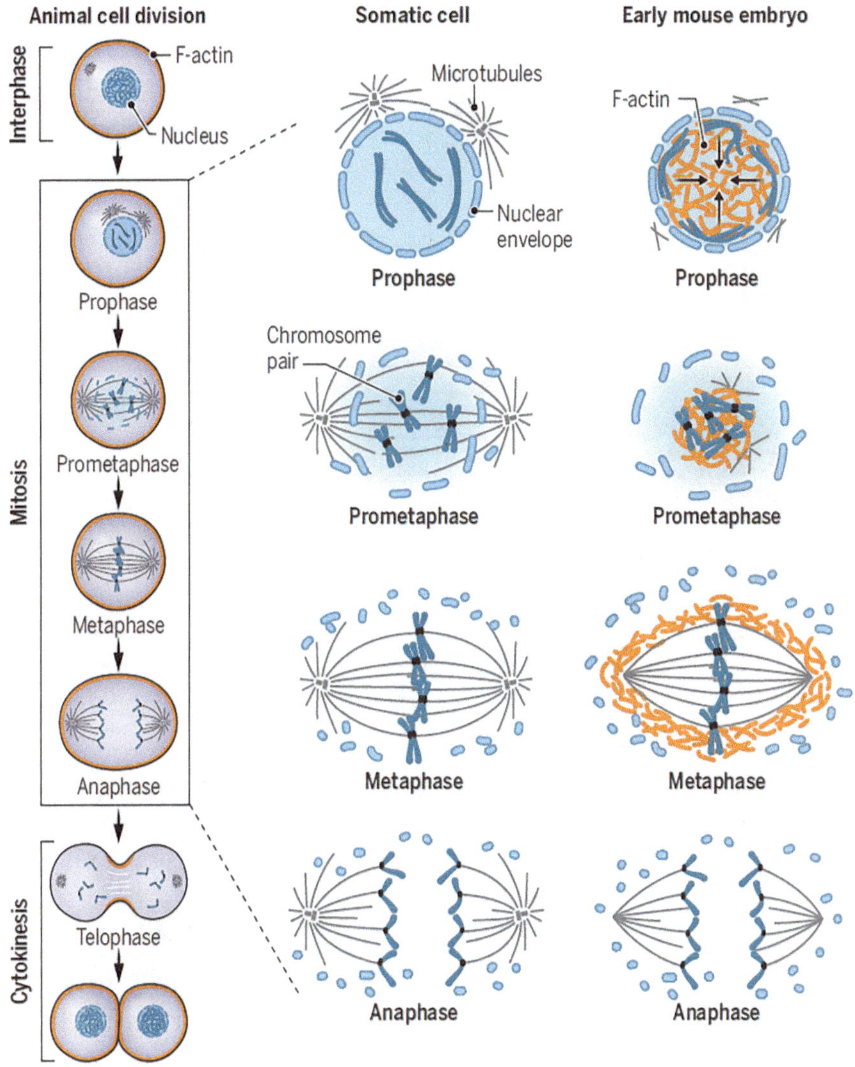

Fig. 3.14 An extra job for actin in cell division. *Adapted from* [36] *with permission from The American Association for the Advancement of Science*

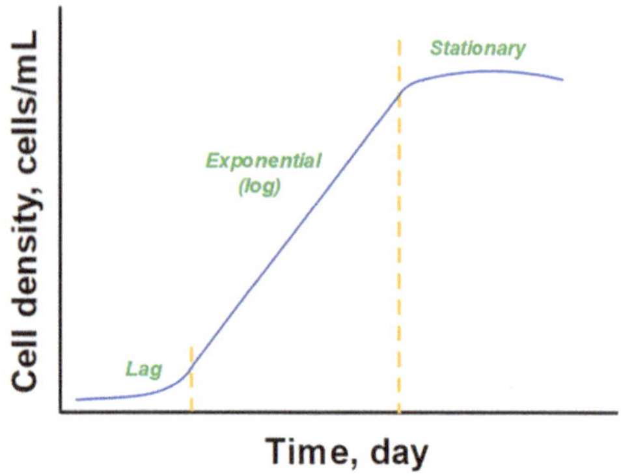

Fig. 3.15 A type growth cure (semilog plot of cell density versus time)

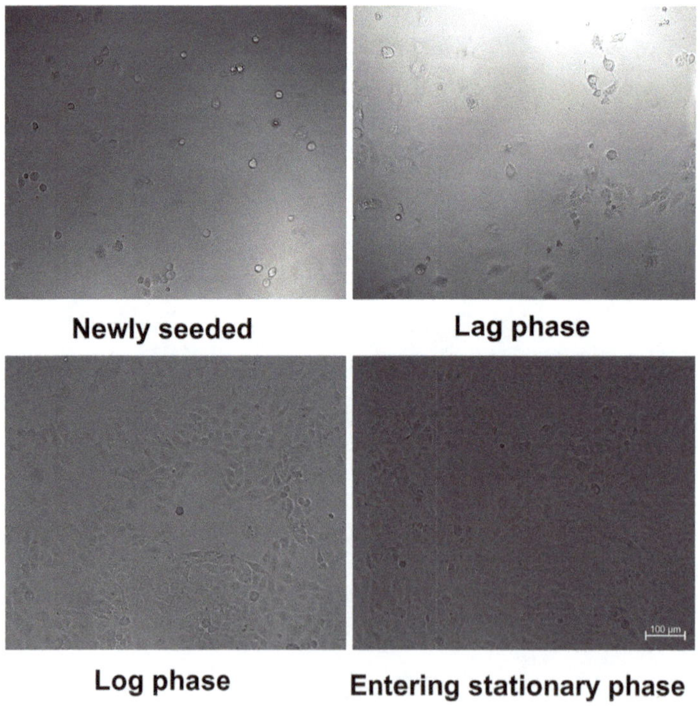

Fig. 3.16 Phases of the growth cycle. Cells: MDA-MB-231. Scale bar: 100 μm. (Images courtesy of Drs. Weizheng Wang and Qingsu Cheng)

$$\frac{dC}{dt} = kt$$

Integrating

$$\int_0^t dC / C = k \int_0^t dt$$

$$lnC_t - lnC_0 = kt$$

$$\ln\left(\frac{C_t}{C_0}\right) = kt$$

$$e^{\ln\left(\frac{C_t}{C_0}\right)} = e^{kt}$$

$$\frac{C_t}{C_0} = e^{kt}$$

$$C_t = C_0 e^{kt}$$

where C is the cell density (cell/mL),
t is time (h),
and k is the growth rate (h^{-1}).
The time for a cell population to double their number, i.e., doubling time (t_d), can also be calculated. At t_d,

$$\ln\left(\frac{C_t}{C_0}\right) = 2$$

$$t_d = \frac{\ln_2}{k} \approx \frac{0.693}{k}.$$

Problem 3.1

C3H10T1/2 cells were maintained in DMEM supplemented with 10% FBS and 1% penicillin/streptomycin. Images were taken daily to quantify cell numbers (cells/field), as shown in the following table. Generate a graph of cell number versus time and a semilog (ln) plot of cell number versus time. Then confirm the cells grew exponentially and calculate t_d.

Time (h)	Cell number (cells/field)
24	38.4
48	101.3
72	288.6
96	627.2

Answer:

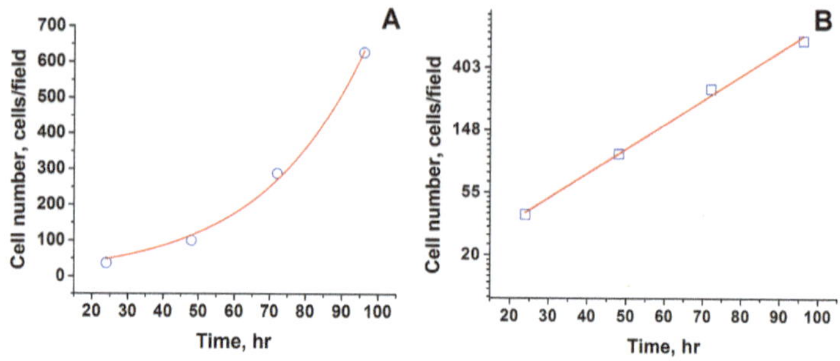

Graph A is the cell number versus time plot, while graph B is the semilog (ln) plot of cell number versus time.

Perform exponential regression of the cell number versus time plot using the equation: $y = ae^{bx}$, and the results are:
$a = 21.17275 \pm 4.22451$,
$b = 0.03537 \pm 0.00218$,
$R^2 = 0.99674$.
Since R^2 is very close to 1, the cells grew exponentially.

Similarly, perform the linear regression of the semilog plot and a high R^2 value is obtained (0.99654).

To calculate t_d:
$C_t = 21.2 \times e^{0.035t}$, $k = 0.035$ hr.$^{-1}$.
$t_d = 0.693/(0.035\ \text{hr.}^{-1}) = 19.6$ hr.

During cell passaging, it is critical to use a proper split ratio. The split ratio is the divisor of the dilution ratio of a cell culture at subculture (e.g., one flask divided into six, or 10 ml into 60 ml, corresponds to a split ratio of 1:6). If the split ratio is too high (i.e., low seeding density), it leads to poor yield and cells may eventually die out. Conversely, low split ratios lead to more frequent cell passaging and potential cell deterioration [38].

3.5 Cell Death

Cell death is the irreversible degeneration of vital cellular functions culminating in the loss of cellular integrity (permanent plasma membrane permeabilization or cellular fragmentation). Cell death can be either accidental (ACD) or regulated (RCD) [39, 40]. ACD is the virtually instantaneous and uncontrollable form of cell death corresponding to the physical disassembly of the plasma membrane caused by extreme physical, chemical, or mechanical cues. RCD is the form of cell death that results from the activation of one or more signal transduction modules and hence can be pharmacologically or genetically modulated (at least kinetically and to some extent). Traditionally, cell death has been classified based on macroscopic morphological alterations [40]. Apoptosis (type I cell death) is the first identified form of RCD [41]. During apoptosis, cells exhibit cytoplasmic shrinkage, chromatin

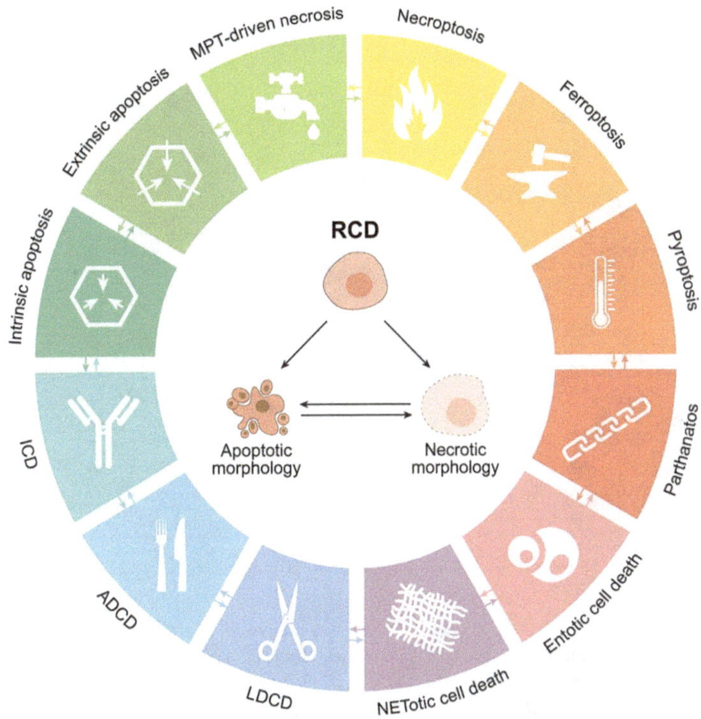

Fig. 3.17 Major cell death subroutines. Regulated cell death (RCD) modes is initiated and propagated by molecular mechanisms that exhibit a considerable degree of interconnectivity. Moreover, each type of RCD can manifest with an entire spectrum of morphological features ranging from fully necrotic to fully apoptotic, and an immunomodulatory profile ranging from anti-inflammatory and tolerogenic to pro-inflammatory and immunogenic. ADCD: autophagy-dependent cell death, ICD: immunogenic cell death, LDCD: lysosome-dependent cell death, MPT: mitochondrial permeability transition. (Adapted from [40] with permission from Springer Nature)

condensation, nuclear fragmentation, and plasma membrane blebbing, culminating in the formation of apoptotic bodies (Fig. 3.17). These vesicles are taken up by cells with phagocytic activity and degraded within lysosomes. Autophagy (type II cell death) is characterized by extensive cytoplasmic vacuolization and culminates in phagocytic uptake followed by lysosomal degradation. Necrosis (type III cell death) involves the disposal of cell corpses without obvious phagocytic and lysosomal involvement (Fig. 3.17) [40]. Early loss of membrane integrity and the consequential expansion of the cytoplasmic compartment are hallmarks of necrosis [42]. An updated classification of cell death focusing on molecular and essential aspects is shown in Fig. 3.17 [40]. Programed cell death (PCD) is regulated cell death mediated by dedicated molecular machines and occurs in strictly physiological scenarios [40, 42]. Representative PCD modalities include apoptosis, necroptosis (a form of caspase-independent programmed necrosis), and pyroptosis (involves activation of inflammasome and caspase activity, and formation of pores in the cell membrane). Apoptosis, the fundamental form of PCD, is characterized by a caspase cascade that occurs without inflammatory responses. It can be intrinsic or extrinsic depending on how the triggers are sensed and integrated by cells (Fig. 3.18) [42, 43].

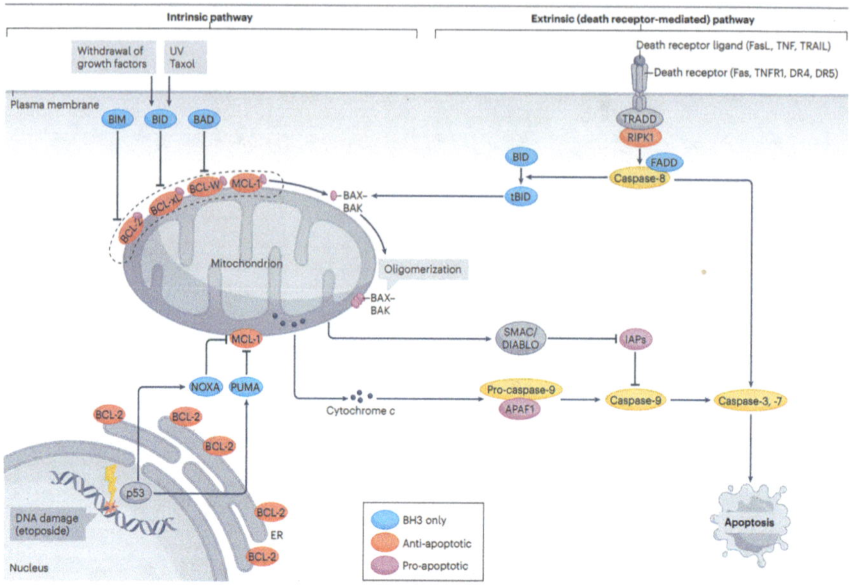

Fig. 3.18 Intrinsic and extrinsic apoptosis. Extrinsic apoptosis (right) is mediated by the activation of plasma membrane-localized death receptors (such as TNFR1, Fas and TRAIL receptors DR4 and DR5) by their cognate ligands (such as TNF, FasL and TRAIL, respectively). Intrinsic apoptosis (left) can be activated by growth factor withdrawal, mitochondrial damage, DNA damage and chemotherapeutic drugs such as taxol. Activation of BH3-only members of the BCL-2 family, such as transcriptional induction of NOXA and PUMA by p53, post-translational modification of BAD and BIM, and cleavage of BID by caspase-8, induces mitochondrial damage by inactivating pro-survival BCL-2 family members, such as BCL-2, BCL-xL and MCL-1, and activating oligomerization of pro-death BCL-2 family members BAX and BAK. Mitochondrial damage leads to the release of cytochrome c and second mitochondria-derived activator of caspase (SMAC; also known as DIABLO) to promote the activation of caspase-9 mediated by APAF1. Activated caspase-9 in turn cleaves downstream caspases, caspase-3 and caspase-7, to mediate the execution of intrinsic apoptosis. Activated caspase-3 and caspase-7 can also exert feedback activation of upstream caspases to allow amplification of caspase cascades. ER: endoplasmic reticulum, UV: ultraviolet. (Adapted from [42] with permission from Springer Nature)

Review and Practice Questions

1. What is cell metabolism? What are the four stages of aerobic respiration, and how many ATPs per glucose are produced through aerobic respiration for eukaryotes?
2. What are the functions of the following components of base cell culture media?
   (a) NaCl
   (b) $NaHCO_3$
   (c) glucose
   (d) glutamine
   (e) Phenol Red

3. You were preparing 100 mL of complete medium but mistakenly added 100 mL of the base medium to the container. How much serum (FBS) should you use to ensure the final medium is "complete"? Show your calculation.
4. What are the major components of serum used in cell culture? Why are the disadvantages of using serum?
5. What are the major disadvantages of using antibiotics in cell culture?
6. What are the common adhesion molecules?
7. Which of the following images shows the typical morphology of a fibroblast? *(Images courtesy of Drs. Weizheng Wang and Qingsu Cheng)*

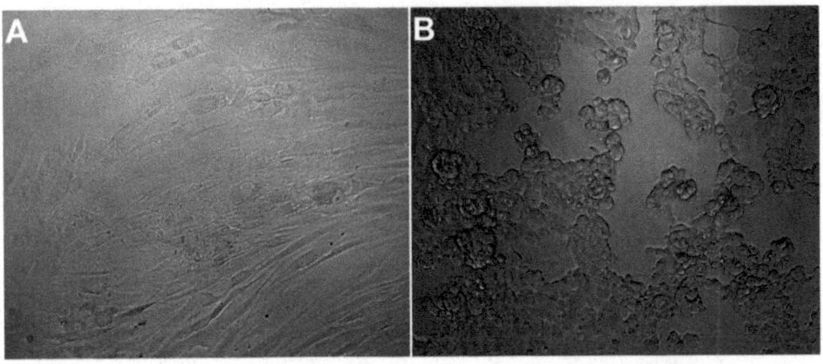

8. How can microcarriers be used as cell culture substrates, particularly in combination with a spinner flask? Provide an example.
9. How can feeder cells be used as cell culture substrates? Provide an example.
10. Describe different phases of cell cycle.
11. Draw of typical semilog plot of cell numbers versus time, indicating the lag, log, and stationary phases.
12. You are constructing a tissue-engineered device in vitro. Here are the design parameters:

Volume = 100 mL.

Only one type of mammalian cells.

Cell density = 10^8 cells/mL.

Doubling time of the given mammalian cell = 12 h.

Assuming the first-order growth model, calculate the mammalian cells' initial number to complete the device (transplant) in one week. *(adapted from* [3] *with permission from Springer Nature).*

13. What is cell death? Compare apoptosis and necrosis.

References

1. Gonçalves, I.G. and J.M. García-Aznar, *Hybrid computational models of multicellular tumour growth considering glucose metabolism.* Computational and Structural Biotechnology Journal, 2023. **21**:1262–1271.
2. Hewitson, T.D. and E.R. Smith, *A Metabolic Reprogramming of Glycolysis and Glutamine Metabolism Is a Requisite for Renal Fibrogenesis—Why and How?* Frontiers in Physiology, 2021. **12:** 645857.

3. Yoon, J.-Y., *Cell Metabolism*, in *Tissue Engineering: A Primer with Laboratory Demonstrations*. 2022, Springer.
4. Mason, K.A., J.B. Losos, and T. Duncan, *How Cells Harvest Energy*, in *Biology*. 2023, McGraw Hill LLC.
5. Mirveis, Z., N. Patil, and H.J. Byrne, *Experimental and computational investigation of the kinetic evolution of the glutaminolysis pathway and its interplay with the glycolysis pathway.* FEBS Open Bio, 2024. **14**(8):1247–1263.
6. Al-Ani, A., et al., *Oxygenation in cell culture: Critical parameters for reproducibility are routinely not reported.* PLoS One, 2018. **13**(10):e0204269.
7. Wu, D. and P. Yotnda, *Induction and testing of hypoxia in cell culture.* J Vis Exp, 2011(54), 54.
8. D'Aiuto, N., et al., *Hypoxia, acidification and oxidative stress in cells cultured at large distances from an oxygen source.* Scientific Reports, 2022. **12**(1):21699.
9. Yoo, H.C., et al., *Glutamine reliance in cell metabolism.* Experimental & Molecular Medicine, 2020. **52**(9):1496–1516.
10. Morgan, A., et al., *Caution for the routine use of phenol red—It is more than just a pH indicator.* Chemico-Biological Interactions, 2019. **310**:108739.
11. Berthois, Y., J.A. Katzenellenbogen, and B.S. Katzenellenbogen, *Phenol red in tissue culture media is a weak estrogen: implications concerning the study of estrogen-responsive cells in culture.* Proceedings of the National Academy of Sciences of the United States of America, 1986. **83**(8):2496–500.
12. Weiskirchen, S., et al., *A Beginner's Guide to Cell Culture: Practical Advice for Preventing Needless Problems.* Cells, 2023. **12**(5):682.
13. Gstraunthaler, G., *Alternatives to the use of fetal bovine serum: serum-free cell culture.* ALTEX—Alternatives to Animal Experimentation, 2003. **20**(4):275–281.
14. Karnieli, O., et al., *A consensus introduction to serum replacements and serum-free media for cellular therapies.* Cytotherapy, 2017. **19**(2):155–169.
15. Tufa, T.B., et al., *Efficacy of Penicillin–Streptomycin Brands against Staphylococcus aureus: Concordance between Veterinary Clinicians' Perception and the Realities.* Antibiotics, 2023. **12**(3):570.
16. Capes-Davis, A., et al., *2. Biology of Cultured Cells.* Eighth edition. ed. Freshney's culture of animal cells: a manual of basic technique and specialized applications. 2021, Wiley Blackwell.
17. Harjunpää, H., et al., *Cell Adhesion Molecules and Their Roles and Regulation in the Immune and Tumor Microenvironment.* Frontiers in Immunology, 2019. **10**:1078.
18. Morwood, A.J., et al., *The Role of Extracellular Matrix (ECM) Adhesion Motifs in Functionalised Hydrogels.* Molecules, 2023. **28**(12):4616.
19. Janiszewska, M., M.C. Primi, and T. Izard, *Cell adhesion in cancer: Beyond the migration of single cells.* Journal of Biological Chemistry, 2020. **295**(8):2495–2505.
20. Fletcher, D.A. and R.D. Mullins, *Cell mechanics and the cytoskeleton.* Nature, 2010. **463**(7280):485–492.
21. Clarke, D.N. and A.C. Martin, *Actin-based force generation and cell adhesion in tissue morphogenesis.* Current Biology, 2021. **31**(10):R667–R680.
22. Vasioukhin, V., et al., *Directed Actin Polymerization Is the Driving Force for Epithelial Cell-Cell Adhesion.* Cell, 2000. **100**(2):209–219.
23. *Cell Culture Vessels and Plastics*. Available from: https://www.thermofisher.com/us/en/home/references/gibco-cell-culture-basics/cell-culture-environment/culture-plastics.html.
24. Leber, J., et al., *Microcarrier choice and bead-to-bead transfer for human mesenchymal stem cells in serum-containing and chemically defined media.* Process Biochemistry, 2017. **59**:255–265.
25. Major, G.S., et al., *Mapping the microcarrier design pathway to modernise clinical mesenchymal stromal cell expansion.* Trends in Biotechnology, 2024. **42**(7):859–876.
26. Bellani, C.F., et al., *Scale-Up Technologies for the Manufacture of Adherent Cells.* Frontiers in Nutrition, 2020. **7**:575146.
27. Koh, B., et al., *Three dimensional microcarrier system in mesenchymal stem cell culture: a systematic review.* Cell & Bioscience, 2020. **10**(1):75.

28. Yang, F., et al., *Physical optimization of cell proliferation and differentiation using spinner flask and microcarriers.* AMB Express, 2022. **12**(1):63.
29. Zhang, B., et al., *Enhancing mesenchymal stem cells cultivated on microcarriers in spinner flasks via impeller design optimization for aggregated suspensions.* Bioresources and Bioprocessing, 2023. **10**(1):89.
30. Liberio, M.S., et al., *Differential Effects of Tissue Culture Coating Substrates on Prostate Cancer Cell Adherence, Morphology and Behavior.* PLOS ONE, 2014. **9**(11): p. e112122.
31. Llames, S., et al., *Feeder Layer Cell Actions and Applications.* Tissue Engineering Part B: Reviews, 2015. **21**(4):345–353.
32. Xiao, Y., et al., *Current insights into the bone marrow niche: From biology in vivo to bioengineering ex vivo.* Biomaterials, 2022. **286**:121568.
33. Lee, J.B., et al., *Available human feeder cells for the maintenance of human embryonic stem cells.* Reproduction, 2004. **128**(6):727–735.
34. Guo, R., et al., *Feeders facilitate telomere maintenance and chromosomal stability of embryonic stem cells.* Nature Communications, 2018. **9**(1):2620.
35. Alberts, B., et al., *An Overview of the Cell Cycle*, in *Molecular Biology of the Cell.* 2002, Garland Science.
36. Bement, W., *Rewriting the rules of cell division.* Science, 2025. **388**(6749):817–818.
37. Hernandez, B., et al., *Actin organizes chromosomes and microtubules to ensure mitotic fidelity in the preimplantation embryo.* Science, 2025. **388**(6749):eads1234.
38. Capes-Davis, A., et al., *14. Subculture and Cell Lines.* Eighth edition. ed. Freshney's culture of animal cells: a manual of basic technique and specialized applications. 2021, Wiley Blackwell.
39. Melo, R.C.N., A.M. Dvorak, and P.F. Weller, *Chapter 6—Eosinophil cell death,* in *Eosinophil Ultrastructure*, R.C.N. Melo, A.M. Dvorak, and P.F. Weller, Editors. 2022, Academic.
40. Galluzzi, L., et al., *Molecular mechanisms of cell death: recommendations of the Nomenclature Committee on Cell Death 2018.* Cell Death & Differentiation, 2018. **25**(3):486–541.
41. Ai, Y., et al., *The biochemical pathways of apoptotic, necroptotic, pyroptotic, and ferroptotic cell death.* Molecular Cell, 2024. **84**(1):170–179.
42. Yuan, J. and D. Ofengeim, *A guide to cell death pathways.* Nature Reviews Molecular Cell Biology, 2024. **25**(5):379–395.
43. Qian, S.e., et al., *Programmed cell death: molecular mechanisms, biological functions, diseases, and therapeutic targets.* MedComm, 2024. **5**(12):e70024.

Stem and Cancer Cells

4

Learning Objectives

- Compare different types of stem cells based on their potency
- Examine and evaluate culture approaches for pluripotent stem cells, including the selection of media and substrates
- Assess and design strategies for embryoid body formation to direct stem cell differentiation
- Discuss the significance of cancer stem(-like) cells and methods for their enrichment

4.1 Stem Cells

There are various ways to define stem cells. Regardless, the two aspects generally incorporated are self-renewal and differentiation (Fig. 4.1). For example, the National Human Genome Research Institute defines a stem cell as a cell with the potential to form many of the different cell types found in the body. When stem cells divide, they can form more stem cells or other cells that perform specialized functions [1]. Self-renewal is the process by which stem cells divide to produce more stem cells, perpetuating the stem cell pool throughout life. It is division with maintenance of the undifferentiated state [2]. Self-renewal of stem cells frequently also entails resistance to the cellular senescence observed in mature differentiated cells [3]. Differentiation is the process by which stem cells become more specialized cell types and acquire new functions through the expression of new genes, mRNA, and proteins. It involves the deactivation of some genes and the activation of a new set of genes [4]. Thus, during cell division, a stem cell may not always produce two cells of the same fate. One mechanism described to control daughter cell fates is asymmetric cell division. This can be mediated by the asymmetric distribution of "cell fate determinants", such as membrane components, organelles, and cytosolic components. The difference must exceed a critical threshold to trigger a cell-fate

W. Zhang, *Cell Culture and Tissue Engineering*,
https://doi.org/10.1007/978-3-032-32096-4_4

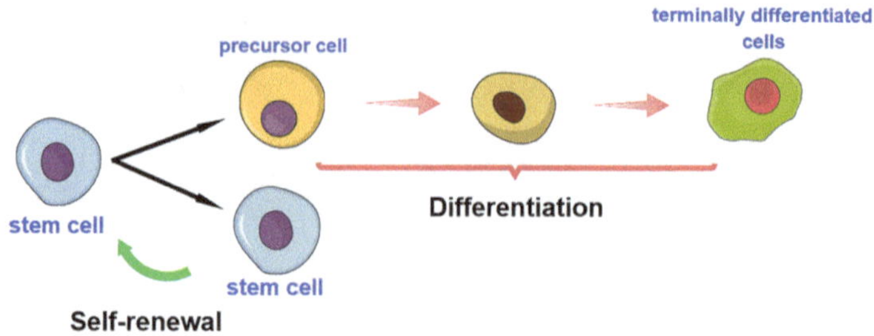

Fig. 4.1 Self-renewal and differentiation of stem cells

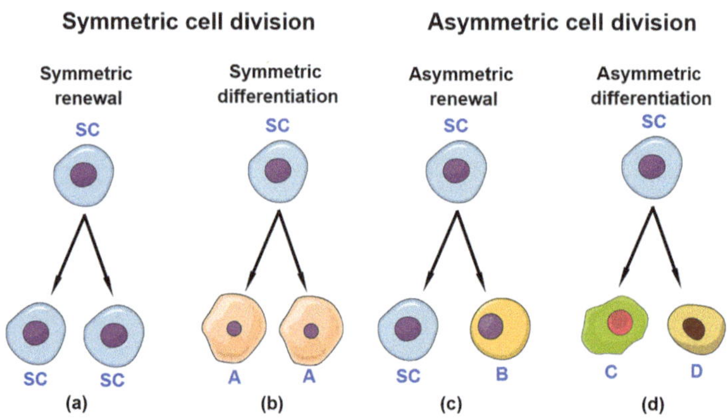

Fig. 4.2 Alternative modes of stem cell divisions. (**a**) symmetric renewal: both daughter cells retain the stem cell (SC) potential of the mother cell; (**b**) symmetric differentiation: both daughter cells are committed to differentiate toward the same lineage (A); (**c**) asymmetric renewal: one daughter cell becomes committed to differentiation (B), while the other daughter cell retains the SC potential; (**d**) asymmetric differentiation: both daughter cells become committed to different fates (C & D). (Image was generated based on [5])

decision. Following asymmetric cell division of a stem cell, the transcriptional and/or epigenetic profile of the daughter cells can change. Asymmetric cell division can be divided into asymmetric renewal (one daughter cell retains stem cell potential) and asymmetric differentiation (two daughter cells with distinct potentials). Similarly, symmetric division can be differentiated into symmetric renewal and symmetric differentiation (Fig. 4.2) [5].

Differentiation is a progressive process. Progression from a stem cell to a particular lineage traditionally implied an increase in commitment, marking the point where a cell or its progeny could no longer transfer to a separate lineage. As a stem cell became more committed, its descendants express a more limited and specialized repertoire of properties [6]. For example, stem cells in adult bone marrow and other depots are capable of replication and can differentiate along distinctive lineage pathways to produce highly specialized phenotypes (Fig. 4.3) [7]. The ability of undifferentiated stem cells to differentiate into a variety of lineages and tissues is referred to as pluripotency [3]. It is important to introduce two relevant terms:

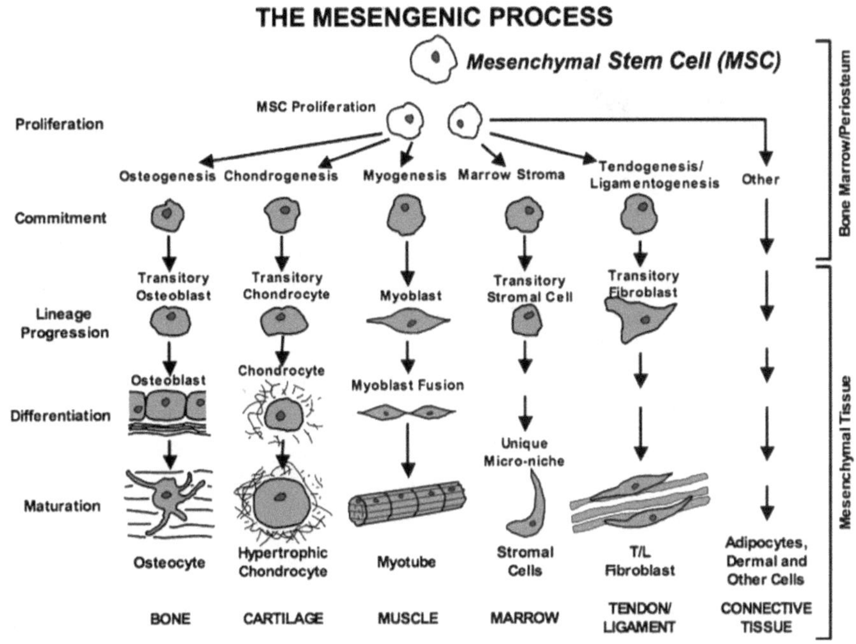

Fig. 4.3 Mesengenesis: a multipotent stem cell found in adult bone marrow and other depots is capable of replicating and producing progeny that differentiate along distinctive lineage pathways, giving rise to highly specialized phenotypes that form bone, cartilage, muscle, marrow stroma, tendon/ligament, and other connective tissues. (Adapted from [7] with permission from John Wiley and Sons)

precursor and progenitor cells. These are not well defined and can be confusing [6, 8]. Generally, they are used to describe transit-amplifying cells as generic terms. It is recommended to use more specific terminology designate a cell's state, its parent, and its daughter in a lineage progression [8]. It is worth noting that stem cells, their surrounding niches, and their signals are heterogeneous, comprising distinct subsets genetically preprogrammed for different functions, behaviors, and fates [9].

Based on potency, the stem cells can be classified as totipotent, pluripotent, multipotent, oligopotent, and unipotent (Fig. 4.4) [10]. During development, cells gradually lose potency, becoming more specialized. Totipotent cells can differentiate into embryonic and extraembryonic cell types. These cells are produced by the fusion of an egg and sperm cell and can give rise to a complete, viable organism. Cells produced by the first few divisions of the fertilized egg are also totipotent [10]. Pluripotent and multipotent stem cells are most commonly applied in tissue engineering. Embryonic stem cells (ESCs) are pluripotent. To understand ESCs, gametes production and embryonic development must first be examined. In humans, gametes, oocytes and sperm, are produced through oogenesis and spermatogenesis, respectively (Fig. 4.5) [11]. Both oogenesis and spermatogenesis are forms of meiosis—a kind of reductive cell division. During oogenesis, a polar body is produced in each cell division. Polar bodies contain one set of chromosomes and minimal cytoplasm, and they degenerate. Hence, one oogonium produces only one oocyte. Moreover,

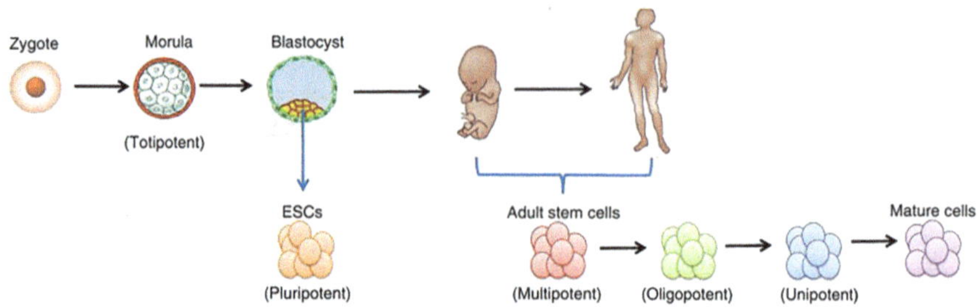

Fig. 4.4 Potency of stem cells during development. ESCs: embryonic stem cells. (Adapted from [10] with permission from Elsevier)

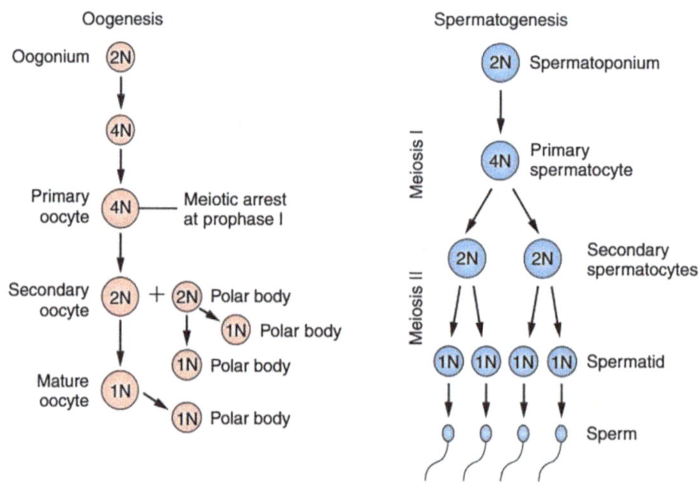

Fig. 4.5 Comparison between oogenesis and spermatogenesis and their meiotic outcomes. (Adapted from [11] with permission from Elsevier)

meiosis II occurs only if the secondary oocyte is fertilized. ESCs can be harvested from the inner cell mass (ICM) of the pre-implantation blastocyst or from the morula stage (Figs. 4.6 and 4.7) [10, 12]. To obtain ESCs from blastocysts, several approaches have been developed: isolating the ICM immunosurgically or mechanically, or culturing the whole blastocysts after removal of the zona pellucida [13].

ESCs cell culture is relatively challenging, both for maintaining pluripotency and for directing differentiation towards specific lineages. Live feeder cells, which are proliferation-arrested but metabolically-active, are commonly used to provide direct cell-cell interactions, produce a bioactive matrix for cell-ECM interactions, and release nutrients/growth factors into the culture medium. These factors help maintain ESC pluripotency and prevent spontaneous differentiation [14]. However, the use of feeder cells raises major concerns for culturing human ESCs (hESCs), particularly for large-scale and/or clinical-grade manufacturing [15]. Feeder-free ESC culture systems have therefore become increasingly popular, especially chemically defined systems. To replace feeder cells, synthetic coatings, synthetic biomolecules, and

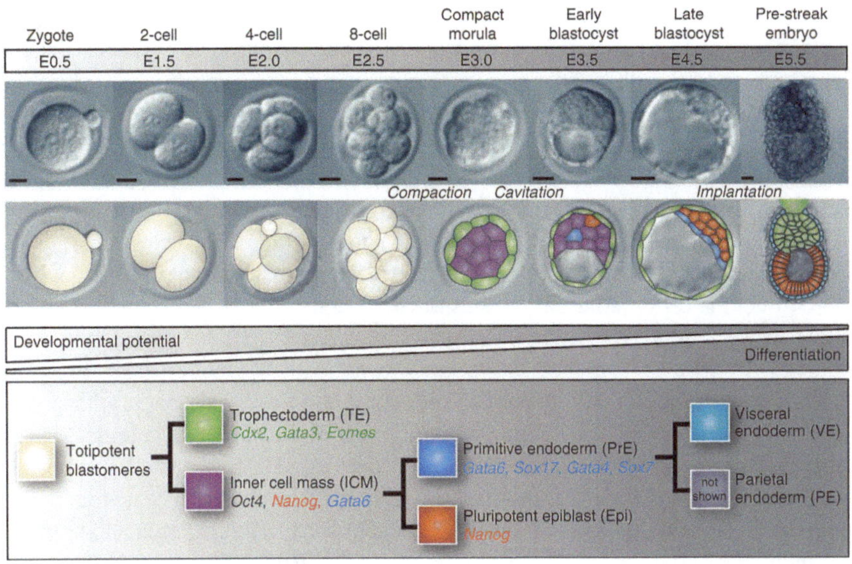

Fig. 4.6 Overview of early embryonic development. Proper lineage segregation before implantation is ensured by two cell-fate decisions: the first giving rise to trophectoderm and inner cell mass, and the second separating primitive endoderm and epiblast. Lineage-associated gene expression is noted below each cell type. After implantation, the PrE differentiates into visceral and parietal endoderm. E: embryonic day. Scale bars: 50 µm. (Adapted from Niakan, K., Schrode, N., Cho, L. et al. Derivation of extraembryonic endoderm stem (XEN) cells from mouse embryos and embryonic stem cells. Nature Protocol 8, 1028–1041 (2013). https://doi.org/10.1038/nprot.2013.049, with permission from Springer Nature)

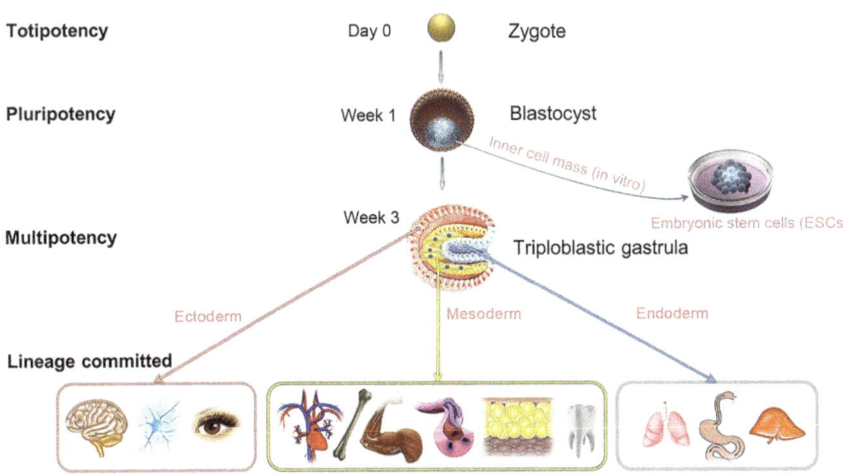

Fig. 4.7 Development progression from zygote to three germ layers. Through mitosis, the zygote undergoes multiple stages to form an individual. ESCs are derived from the inner cell mass of blastocysts, which give rise to embryos. Trophoblasts are derived from the outer cell mass, which develops into placenta. Three germ layers, ectoderm, mesoderm, and endoderm, are formed during the gastrula period, laying the foundation of organs and system development. (Adapted from [10] with permission from Elsevier)

recombinant proteins have been explored [16]. Over time, various culture media/systems have been developed for cultivating pluripotent stem cells. Examples include Knockout medium with Knockout serum replacement and leukemia inhibitory factor (LIF; for retaining self-renewal and pluripotency) [17, 18], as well as the combinations of different inhibitors with LIF (e.g., 2i medium containing PD0325901and CHIR99021; and 5i/L/A medium containing five kinase inhibitors and two growth factors—PD0325901, IM-12, SB590885, WH4–023, Y-27632, Activin A, and recombinant human LIF) [19, 20]. Figure 4.8 shows different culture systems (various combinations of media and substrate) for maintain distinct types of ESCs.

To direct the differentiation of ESCs *in vitro*, aggregates called embryoid bodies (EBs) are commonly formed. EBs structurally resemble the pregastrulation-stage embryos [21]. Several methods have been developed to generate EBs (Fig. 4.9) [22–25]. For example, EBs can be derived from monolayer cultures of pluripotent stem cells using enzymatic treatment to fragment and detach cell colonies. In the hanging-drop method, one of the easiest approaches, cells aggregate by gravity in suspended drops of medium. Alternatively, single cells can be centrifuged into uniquely shaped microwells to promote homogeneous cell aggregation. Other methods include bioprinting and encapsulation of cells in hydrogels [22]. Microscale hydrogel systems, such as microcapsules, have been successfully employed for EB formation [26–28]. For instance, murine R1 ES cells encapsulated in alginate-chitosan-alginate microcapsules maintained >90% viability and formed single EBs per capsule within 7 days [27]. Core-shell structure microcapsules produced *via*

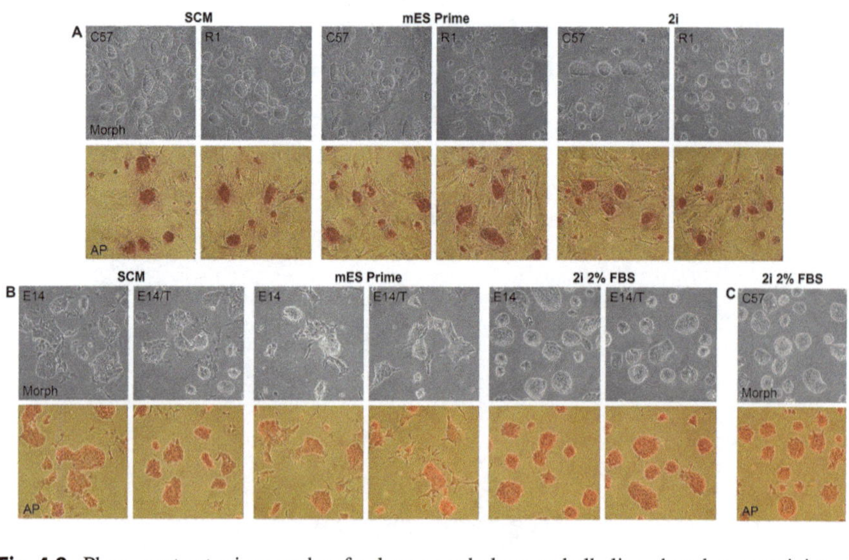

Fig. 4.8 Phase-contrast micrographs of colony morphology and alkaline phosphatase staining of mES cells grown on feeders (**a**) and gelatin (**b**) for 15 passages in SCM, mES Prime Kit, or 2i media. Feeder-dependent mES cell line C57 was maintained in 2i medium on gelatin for at least 15 passages without showing morphological signs of spontaneous differentiation or loss of alkaline phosphatase activity (a key indicator of pluripotent stem cells) (**c**). (Adapted from [17])

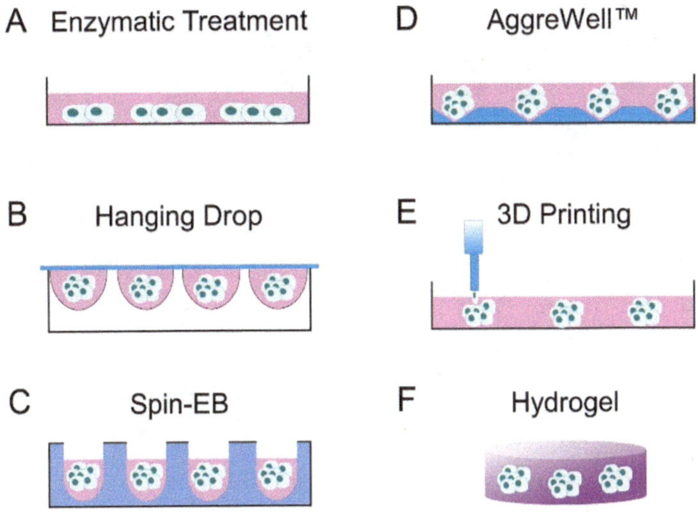

Fig. 4.9 Schematic representation of commonly used methods for embryoid bodies formation. (Adapted from [22])

electrospray and microfluidic devices have also supported EB formation and differentiation into beating cardiomyocytes (Fig. 4.10) [26, 28]. Interestingly, EB size has been shown to significantly affect the differentiation outcomes [29]. Differentiation of ESCs-derived EBs into lineages have been investigated, including neurectoderm and neural progenitors [30], adipocytes [31], hemogenic endothelium and blood progenitors [32], and hepatocyte-like cells [33]. A recent study demonstrated that seeding ESCs-derived EBs onto lung-derived ECM could promote differentiation into lung cells (Fig. 4.11) [34].

Due to ethical concerns and non-patient-specific nature of human embryonic stem cells (hESCs), induced pluripotent stem cells (iPSCs) offer a promising alternative. In 2026, iPSCs was first reported by Kazutoshi Takahashi and Shinya Yamanaka [35], and, in 2012, John Gurdon and Shinya Yamanaka received the Nobel Prize in Physiology and Medicine for "the discovery that mature cells can be reprogrammed to become pluripotent" [36]. In their landmark 2006 article, Takahashi and Yamanaka demonstrated that pluripotent stem cells could be induced from mouse embryonic or adult fibroblasts by introducing four factors, Oct3/4, Sox2, c-Myc, and Klf4, under ESC culture conditions. These induced cells exhibited ESC-like morphology, growth properties, and gene expression. Moreover, they can form EBs and differentiate into multiple lineages (Fig. 4.12) [35]. Since then, numerous reprogramming strategies, both transgene-based and transgene-free, have been developed (Fig. 4.13) [37].

Like ESCs, iPSCs can be cultured with and without feeder cells [38]. However, challenges remain for clinical applications of iPSCs, including genetic instability and oncogenesis risk [39, 40]. Pluripotent stem cells (PSCs) also vary considerably across species [41]. For example, mouse PSCs are thought to exist in a naive state,

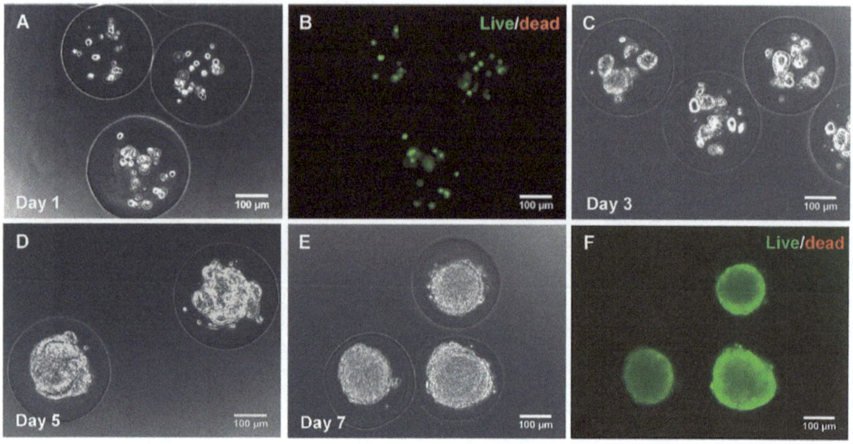

Fig. 4.10 (**a**): phase contrast image showing ~20 embryonic stem cells encapsulated in each microcapsule immediately after microencapsulation; (**b**): the corresponding fluorescence image showing high viability of the encapsulated cells immediately after microencapsulation; (**c–e**): phase contrast images of the encapsulated cells on day 3, 5, and 7, respectively, showing proliferation of the encapsulated ES cells to form a single aggregate in each microcapsule on day 5–7; (**f**): fluorescence image of the ES cell aggregates in the microcapsule on day 7 confirming high viability of the aggregated cells. (Adapted from [28])

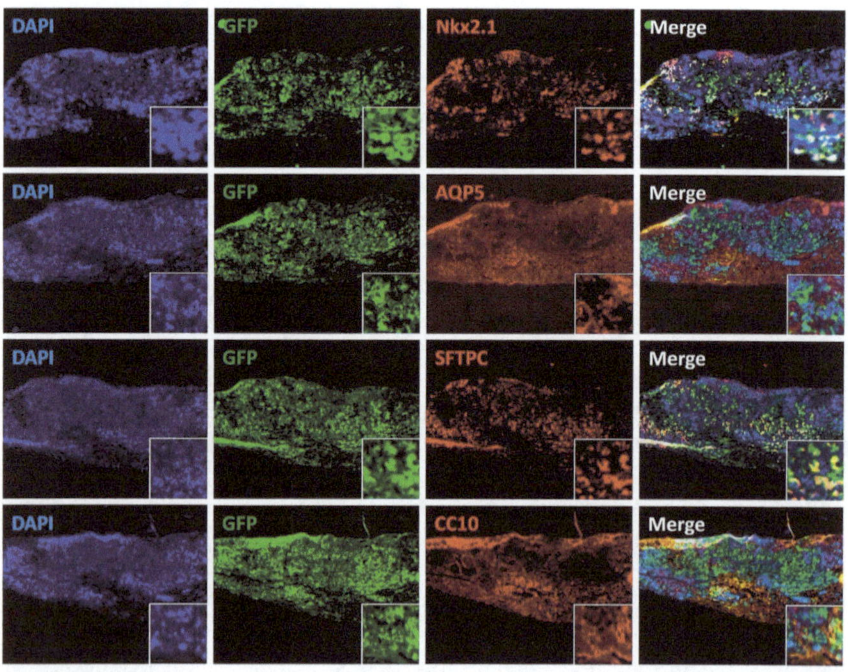

Fig. 4.11 Immunohistochemical analysis of EB outgrowths cultured on lung-derived matrix sheets for determining lung-related cell markers. DAPI: 4′,6-diamidino-2-phenylindole. G4-2 embryonic stem cells expressed enhanced green fluorescent protein (eGFP) under the CAG promotor. Nkx-2.1, AQP5, SFTPC, and CC10 are lung-related cell markers. (Adapted from [34] with permission from Elsevier)

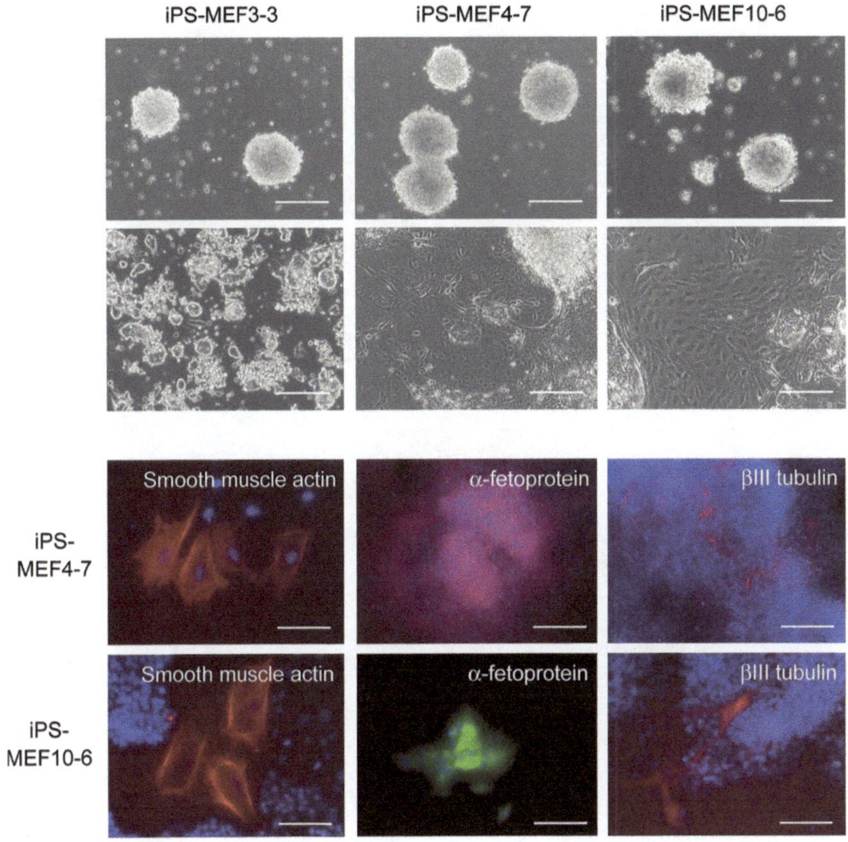

Fig. 4.12 Pluripotency of induced pluripotent stem cells derived from mouse embryonic fibroblasts. Top: *in vitro* embryoid body formation (upper row) and differentiation (lower row). Scale bars = 200 µm. Bottom: immunostaining confirming *in vitro* differentiation into all three germ layers. Scale bars = 100 µm. (Adapted from [35] with permission from Elsevier)

whereas human pluripotent stem cells (hPSCs), hESCs and hiPSCs, exist in a primed state, which is a more committed pluripotent state than a naive state [42]. Recent studies have focused on the establishment and maintenance of naive cell stage of primed hPSCs using naive-conversion media (Fig. 4.14) [42, 43].

Multipotent stem cells can be isolated from many tissues, including bone marrow and umbilical cords. Among these, mesenchymal stem cells (MSCs) are most widely studied. The broad introduction of the term 'mesenchymal stem cell' was in 1999 [44]. MSC is also used to indicate mesenchymal stromal cells. In 2005, the International Society for Cell & Gene Therapy clarified that the two are not equivalent or interchangeable. Mesenchymal stem cells refer to a population with demonstrable progenitor cell functionality, whereas the mesenchymal stromal cells imply a bulk population with notable secretory, immunomodulatory, and homing properties. Minimal criteria for defining multipotent MSCs include plastic adherence, expression of CD73, CD90 and CD105, absence of hematopoietic and endothelial markers (CD11b, CD14, CD19, CD34, CD45, CD79a and HLA-DR), and *in vitro*

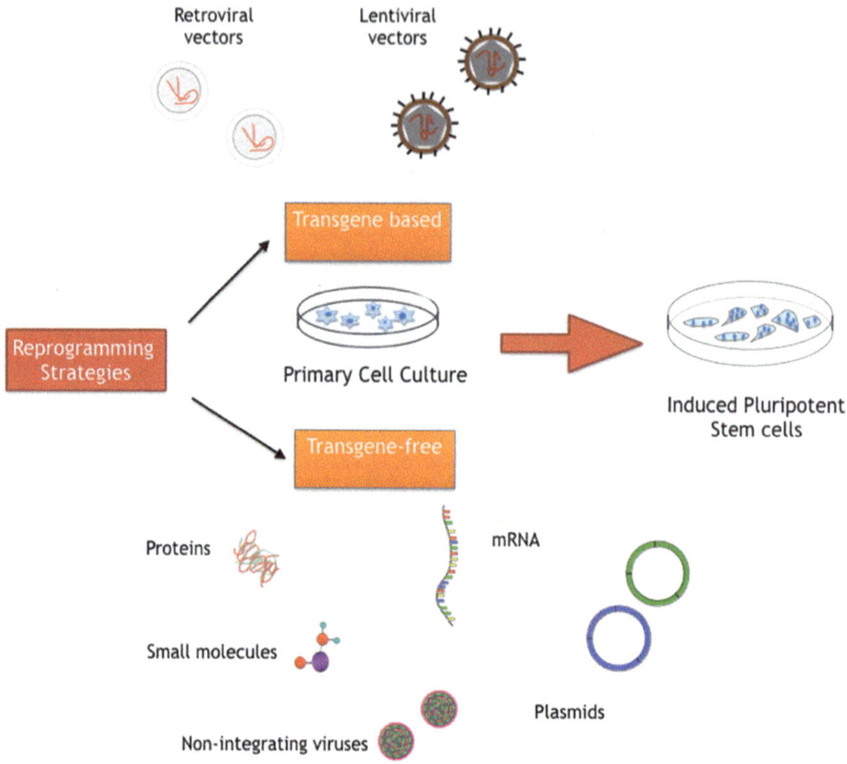

Fig. 4.13 Schematic of various reprogramming strategies for generating induced pluripotent stem cells (iPSCs). (Adapted from [37])

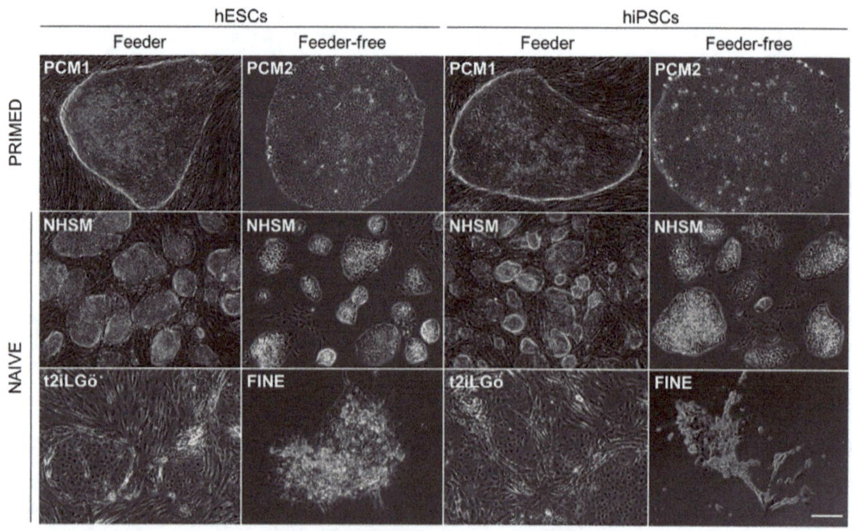

Fig. 4.14 Morphological differences between primed and naive hESCs (human embryonic stem cells) and hiPSCs (human induced pluripotent stem cells) colonies cultured under media conditions. Scale bar: 200 µm. (Adapted from [42])

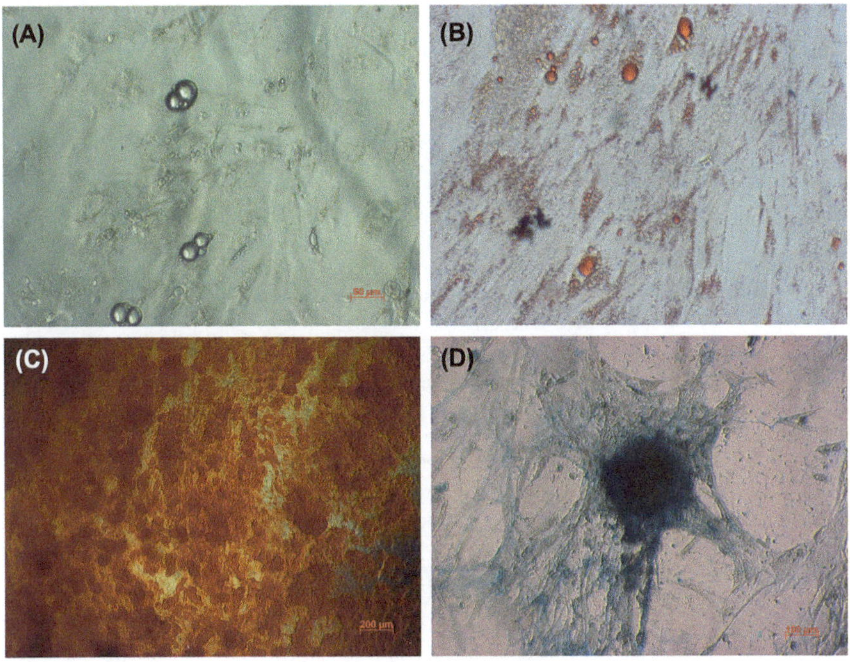

Fig. 4.15 UC-MSCs differentiated into adipocytes (**a**) with Oil Red staining (**b**), osteoblasts with Alizarin red staining (**c**), and chondroblasts that positive with Alcian blue staining (**d**). UC-MSCs: umbilical cord-derived (UC)-MSCs. (Adapted from Le Thi Bich, P., Nguyen Thi, H., Dang Ngo Chau, H. et al. Allogeneic umbilical cord-derived mesenchymal stem cell transplantation for treating chronic obstructive pulmonary disease: a pilot clinical study. Stem Cell Research & Therapy 11, 60 (2020). https://doi.org/10.1186/s13287-020-1583-4)

differentiation into adipocyte, chondrocyte and osteoblast lineages (Fig. 4.15) [45–47]. There have been suggestions to discourage the use of the term 'mesenchymal stem cell' [48]. Recently, minimum criteria have also been proposed for defining induced mesenchymal stem cells (iMSCs; iPSCs derived MSCs) [49]. The trilineage differentiation potential of iMSCs into osteocytes, adipocytes and chondrocytes has also been demonstrated [50].

Compared with PSCs, MSCs are easier to obtain but have limited differentiational potential. Overall, stem cells from different sources each offer unique advantages and limitations (Table 4.1).

Table 4.1 Comparison of different types of stem cells

Source	Potency	Advantages	Disadvantages
Zygote	Totipotent	1. Can differentiate into any type of human cell 2. Can generate a complete organism	1. Major ethical concerns 2. Totipotent state lasts only a very short period
Inner cell mass of blastocyst (ESCs)	Pluripotent	1. Long-term self-renewal 2. Can differentiate into multiple cell types	1. Major ethical concerns 2. Risk of teratoma formation if undifferentiated cells are implanted 3. Generally not patient-specific
Reprogrammed somatic cells (e.g., iPSCs)	Pluripotent	1. No major ethical concerns 2. Can be patient-specific	1. Genetic instability 2. Risk of oncogenesis
Mesenchymal or hematopoietic stem cells	Multipotent	1. Easily accessible 2. No major ethical concerns 3. Can be patient-specific	1. Limited differentiation potential 2. Reduced proliferation capacity

4.2 Cancer Cells

Cancer is a group of diseases characterized by the uncontrolled growth of abnormal cells that can spread to other parts of the body. Common features of cancer cells include genetic changes (mutations), uncontrolled proliferation, tumor formation (although not all cancer cells form tumors, and not all tumors are cancerous (malignant)), and cancer spread (metastasis). There are many types of cancer, but they are generally categorized into two main groups: blood (hematologic) cancers and solid tumor cancers [51, 52]. Regarding the cancer risk, a study by Cristian Tomasetti and Bert Vogelstein demonstrated a strong correlation between tissue-specific cancer risk and the lifetime number of tissue-specific stem cell divisions, leading to the "bad luck" hypothesis [53]. However, a subsequent article argued that cancer risk is heavily influenced by extrinsic factors [54]. In a 2017 follow-up article, Cristian Tomasetti, Lu Li, and Bert Vogelstein highlighted the recognition of a third contributor to cancer—replicative (R) mutations (unavoidable errors associated with DNA replication), in addition to inherited and environmental factors, which doesn't diminish the importance of primary prevention but emphasizes that not all cancer can be prevented by avoiding environmental risk factors [55]. Chromosomal abnormalities are widely observed in tumors, including aneuploidy (gain or loss of whole or partial chromosomes) and polyploidy (genome doubling) (Fig. 4.16) [56]. Cancer cells are typically aneuploid [57], and some researchers have even proposed that cancer is fundamentally a chromosomal, rather than a genetic, disease [58].

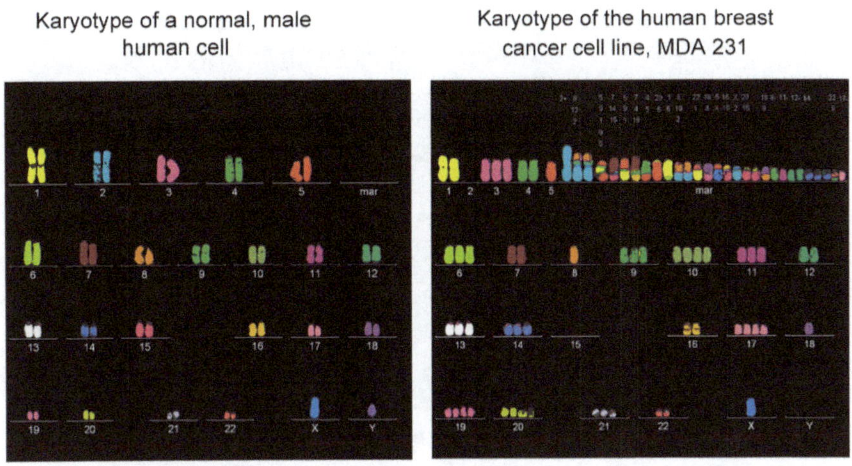

Fig. 4.16 Metaphase chromosomes of the human breast cancer cell line, MDA 231, and of a normal human male cell. The group labeled "mar" (for marker chromosome) shows structurally abnormal chromosomes, which are either intra- or inter-chromosomally rearranged to form hybrid chromosomes. (Adapted from [58])

In vitro cancer cell culture presents challenges, as continuous cancer cell lines are not easy to develop. Furthermore, cancer cell lines have significant limitations: differences in gene expression compared with *in vivo* tumor tissues, lack of interactions with other cell types, and absence of cytokines and signaling molecule influences [59]. Even cancer cell lines established from the same tumor type can differ markedly in appearance. For example, breast cell lines in 3D culture display distinct morphologies that can be divided into four groups: round, mass, grape-like, and stellate (Fig. 4.17) [60].

Cancer stem(-like) cells (CSCs) are defined by their capacity for self-renewal and for generating the heterogeneous lineages of cancer cells within a tumor. The CSC concept remains debated and somewhat controversial [61]. CSCs are often implicated in multidrug resistance. As illustrated in Fig. 4.18, CSCs resist chemotherapy through mechanisms such as quiescence, reduce metabolic activity, and other adaptations [62]. Targeting CSCs with therapeutic agents holds great promise for future antitumor strategies [63]. Identifying and isolating CSCs can aid in assessing the tumor aggressiveness and in evaluating both established and experimental therapies [64]. Several cell surface markers, including CD44, CD24, and CD133, are commonly used for CSC enrichment [65]. CSCs can be propagated in non-adherent (e.g., using ultra-low attachment plates), serum-free, growth factor-supplemented media conditions (e.g., EGF and FGF), forming spherical colonies known as tumorspheres [64]. Microencapsulation has also been shown to enhance CSC enrichment [66]. Regarding the distribution of CSCs within spheroids, current research suggests that they are heterogeneously distributed [67]. Tumorspheres have been used as *in vitro* models to identify predictive biomarkers of chemoresistance and tumor aggressiveness in various cancers [68] (Fig. 4.19).

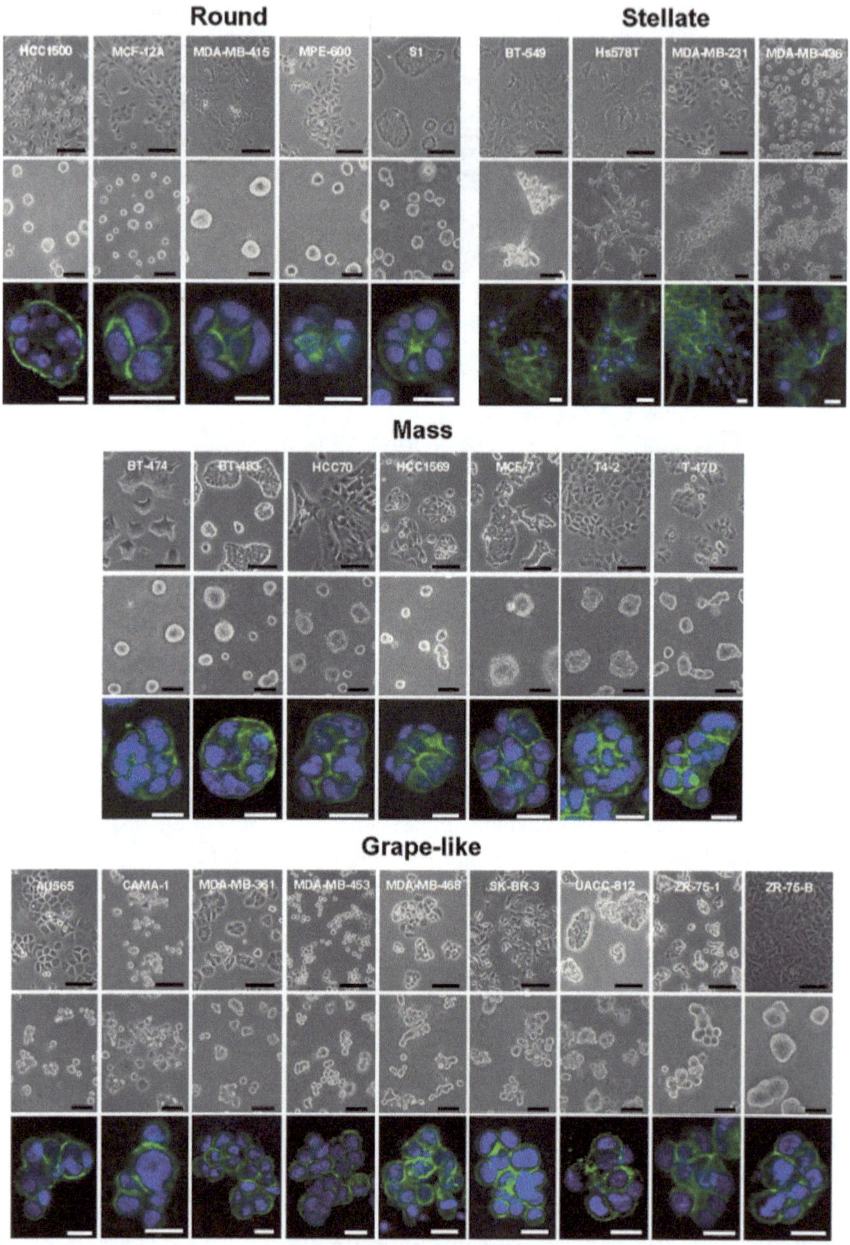

Fig. 4.17 Morphologies of breast cell lines cultured in two- and three-dimensions. Phase contrast images of 25 breast cell lines cultured on tissue culture plastic (top panel) and in the 3D lrECM assay (middle panel) are shown, grouped by 3D morphological classification. 3D cultures were stained for F-actin, and nuclei were counterstained with DAPI. Optical sections of representative colonies are shown for all cell lines (bottom panel), with except MDA-MB-468. Scale bars: top panel, 100 μm; middle panel, 50 μm; bottom panel, 20 μm. lrECM: laminin-rich extracellular matrix, DAPI: 4′,6-Diamidino-2-phenylindole. (Adapted from [60] with permission from John Wiley and Sons)

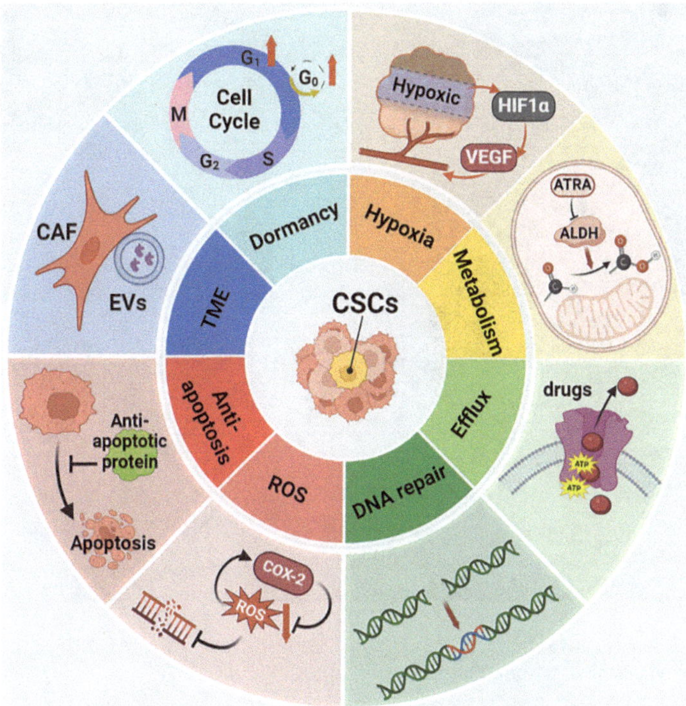

Fig. 4.18 Mechanism of CSC resistance to chemotherapy. HIF1α: hypoxia inducible factor 1α, VEFG: vascular endothelial growth factor, ATRA: all-trans retinoic acid, ALDH: aldehyde dehydrogenease, CSCs: cancer stem(-like) cells, CAF: cancer-associated fibroblasts, EVs: extracellular vesicles, ROS: reactive oxygen species. (Adapted from [62])

Review and Practice Questions

1. What are stem cells? What are the two general aspects that define them?
2. What are the advantages and disadvantages of using feeder cells for embryonic stem cell culture?
3. What is differentiation? And explain its general process.
4. What are some of the approaches for generating embryoid bodies to differentiate embryonic stem cells?
5. Read the article entitled "Induction of Pluripotent Stem Cells from Mouse Embryonic and Adult Fibroblast Cultures by Defined Factors" and answer the following questions:
   (a) What was the main hypothesis of the study?
   (b) How did the authors narrow down the candidate genes from 24 to 4?
6. Compare different types of stem cells—theirs sources, advantages and disadvantages for clinical applications.
7. What are cancer cells?
8. What are cancer stem(-like) cells? Explain their potential mechanisms of drug resistance.

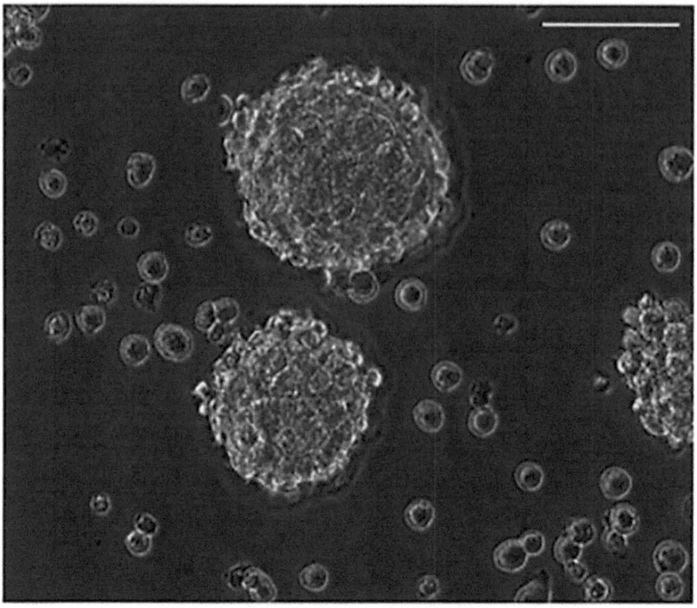

Fig. 4.19 Representative image of tumorspheres derived from MDA-MB-231 cells. (Adapted from [64] with permission from Elsevier)

References

1. National Human Genome Research Institute. *Stem Cell*. Available from: https://www.genome.gov/genetics-glossary/Stem-Cell.
2. He, S., D. Nakada, and S.J. Morrison, *Mechanisms of stem cell self-renewal*. Annual Review of Cell and Development Biology, 2009. **25**:377-406.
3. Rehman, J., *Empowering self-renewal and differentiation: the role of mitochondria in stem cells*. Journal of Molecular Medicine, 2010. **88**(10):981-986.
4. Pittenger, M.F. and C.L. Kerr, *Chapter 2—Stem cells*, in *Tissue Engineering (Third Edition)*, J. De Boer, et al., Editors. 2023, Academic.
5. Murke, F., et al., *Concise Review: Asymmetric Cell Divisions in Stem Cell Biology*. Symmetry, 2015. **7**(4):2025-2037.
6. Capes-Davis, A., et al., *2. Biology of Cultured Cells*. Eighth edition. ed. Freshney's culture of animal cells: a manual of basic technique and specialized applications. 2021, Wiley Blackwell.
7. Caplan, A.I. and J.E. Dennis, *Mesenchymal stem cells as trophic mediators*. Journal of Cellular Biochemistry, 2006. **98**(5):1076-1084.
8. Tajbakhsh, S., *Stem cell: what's in a name?* Nature Reports Stem Cells, 2009. Available from: https://doi.org/10.1038/stemcells.2009.90.
9. Cui, Z., et al., *Stem cell heterogeneity, plasticity, and regulation*. Life Sciences, 2023. **334**: 122240.
10. *Chapter 13—Stem Cells and Regenerative Medicine*, in *Goodman's Medical Cell Biology (Fourth Edition)*, S.R. Goodman, Editor. 2021, Academic.
11. *Chapter 10—Cell Cycle and Cancer*, in *Goodman's Medical Cell Biology (Fourth Edition)*, S.R. Goodman, Editor. 2021, Academic.

12. Yabut, O. and H.S. Bernstein, *The promise of human embryonic stem cells in aging-associated diseases.* Aging, 2011. **3**(5):494-508.
13. Lei, T., et al., *Xeno-free derivation and culture of human embryonic stem cells: current status, problems and challenges.* Cell Research, 2007. **17**(8):682-688.
14. Ueki, M. and Y. Ito, *In Vitro Niches for the Culture of Pluripotent Stem Cells*, in *Encyclopedia of Tissue Engineering and Regenerative Medicine*, R.L. Reis, Editor. 2019, Academic Press.
15. Bigdeli, N., et al., *Upregulation of Adhesion Molecules Sustains Matrix-Free Growth of Human Embryonic Stem Cells.* The Open Stem Cell Journal, 2018. **5**:14-30.
16. Soto, D.A., et al., *Simplification of culture conditions and feeder-free expansion of bovine embryonic stem cells.* Scientific Reports, 2021. **11**(1):11045.
17. Tamm, C., S. Pijuan Galitó, and C. Annerén, *A comparative study of protocols for mouse embryonic stem cell culturing.* PLoS One, 2013. **8**(12):e81156.
18. Cheng, J., et al., *Improved generation of C57BL/6J mouse embryonic stem cells in a defined serum-free media.* genesis, 2004. **39**(2):100-104.
19. Sim, Y.-J., et al., *2i Maintains a Naive Ground State in ESCs through Two Distinct Epigenetic Mechanisms.* Stem Cell Reports, 2017. **8**(5):1312-1328.
20. Fischer, L.A., S.A. Khan, and T.W. Theunissen, *Induction of Human Naïve Pluripotency Using 5i/L/A Medium,* in Rugg-Gunn, P. (eds) *Human Naïve Pluripotent Stem Cells.* Methods in Molecular Biology, vol 2416. 2022, Humana Press.
21. Goh, S.K., P. Olsen, and I. Banerjee, *Extracellular matrix aggregates from differentiating embryoid bodies as a scaffold to support ESC proliferation and differentiation.* PLoS One, 2013. **8**(4):e61856.
22. Zeevaert, K., et al., *Cell Mechanics in Embryoid Bodies.* Cells, 2020. **9**(10):2270.
23. Pettinato, G., X. Wen, and N. Zhang, *Engineering Strategies for the Formation of Embryoid Bodies from Human Pluripotent Stem Cells.* Stem Cells and Development, 2015. **24**(14):1595-609.
24. Lei, X., Z. Deng, and E. Duan, *Uniform Embryoid Body Production and Enhanced Mesendoderm Differentiation with Murine Embryonic Stem Cells in a Rotary Suspension Bioreactor,* in Turksen, K. (eds) *Bioreactors in Stem Cell Biology.* Methods in Molecular Biology, vol 1502. Humana Press.
25. Ouyang, L., et al., *Three-dimensional bioprinting of embryonic stem cells directs highly uniform embryoid body formation.* Biofabrication, 2015. **7**(4):044101.
26. Zhao, S., et al., *Coaxial electrospray of liquid core–hydrogel shell microcapsules for encapsulation and miniaturized 3D culture of pluripotent stem cells.* Integrative Biology, 2014. **6**(9):874-884.
27. Zhang, W., et al., *A novel core–shell microcapsule for encapsulation and 3D culture of embryonic stem cells.* Journal of Materials Chemistry B, 2013. **1**(7):1002-1009.
28. Agarwal, P., et al., *One-step microfluidic generation of pre-hatching embryo-like core–shell microcapsules for miniaturized 3D culture of pluripotent stem cells.* Lab on a Chip, 2013. **13**(23):4525-4533.
29. Moon, S.-H., et al., *Optimizing human embryonic stem cells differentiation efficiency by screening size-tunable homogenous embryoid bodies.* Biomaterials, 2014. **35**(23):5987-5997.
30. Shparberg, R.A., H.J. Glover, and M.B. Morris, *Embryoid Body Differentiation of Mouse Embryonic Stem Cells into Neurectoderm and Neural Progenitors,* in Joglekar, M., Hardikar, A. (eds) *Progenitor Cells.* Methods in Molecular Biology, vol 2029. Humana Press.
31. Olmedo-Suárez, M.A., E. Sánchez-Ramírez, and L. Aguilar-Arnal, *Adipogenic Differentiation from Human Embryonic Stem Cells Through Formation of Embryoid Bodies,* in Pepper, M.S., Ambele, M.A. (eds) *Adipogenesis.* Methods in Molecular Biology, vol 2938. Humana Press.
32. Garcia-Alegria, E., et al., *In vitro differentiation of human embryonic stem cells to hemogenic endothelium and blood progenitors via embryoid body formation.* STAR Protocols, 2021. **2**(1):100367.
33. Greenhough, S., et al., *Development of an embryoid body-based screening strategy for assessing the hepatocyte differentiation potential of human embryonic stem cells following single-cell dissociation.* Cellular Reprogramming, 2013. **15**(1):9-14.

34. Kitamura, T., et al., *Differentiation of embryonic stem cells into lung-like cells using lung-derived matrix sheets.* Biochemical and Biophysical Research Communications, 2023. **686**:149197.
35. Takahashi, K. and S. Yamanaka, *Induction of Pluripotent Stem Cells from Mouse Embryonic and Adult Fibroblast Cultures by Defined Factors.* Cell, 2006. **126**(4):663-676.
36. NobelPrize.org. *The Nobel Prize in Physiology or Medicine 2012.* Available from: https://www.nobelprize.org/prizes/medicine/2012/summary/.
37. Seah, Y.F.S., et al., *Induced Pluripotency and Gene Editing in Disease Modelling: Perspectives and Challenges.* International Journal of Molecular Sciences, 2015. **16**(12):28614-28634.
38. Castro-Viñuelas, R., et al., *Tips and tricks for successfully culturing and adapting human induced pluripotent stem cells.* Molecular Therapy—Methods & Clinical Development, 2021. **23**:569-581.
39. Yoshihara, M., Y. Hayashizaki, and Y. Murakawa, *Genomic Instability of iPSCs: Challenges Towards Their Clinical Applications.* Stem Cell Reviews and Reports, 2017. **13**(1):7-16.
40. Wuputra, K., et al., *Prevention of tumor risk associated with the reprogramming of human pluripotent stem cells.* Journal of Experimental & Clinical Cancer Research, 2020. **39**(1):100.
41. Barry, C., et al., *Species-specific developmental timing is maintained by pluripotent stem cells ex utero.* Developmental Biology, 2017. **423**(2):101-110.
42. Romayor, I., et al., *A Comparative Study of Cell Culture Conditions during Conversion from Primed to Naive Human Pluripotent Stem Cells.* Biomedicines, 2022. **10**(6):1358.
43. Bi, Y., et al., *Optimized derivation and culture system of human naïve pluripotent stem cells with enhanced DNA methylation status and genomic stability.* Protein & Cell, 2025. **16**(10):858-872.
44. Pittenger, M.F., et al., *Multilineage Potential of Adult Human Mesenchymal Stem Cells.* Science, 1999. **284**(5411):143-147.
45. Dominici, M., et al., *Minimal criteria for defining multipotent mesenchymal stromal cells. The International Society for Cellular Therapy position statement.* Cytotherapy, 2006. **8**(4):315-317.
46. Horwitz, E.M., et al., *Clarification of the nomenclature for MSC: The International Society for Cellular Therapy position statement.* Cytotherapy, 2005. **7**(5):393-395.
47. Viswanathan, S., et al., *Mesenchymal stem versus stromal cells: International Society for Cell & Therapy (ISCT®) Mesenchymal Stromal Cell committee position statement on nomenclature.* Cytotherapy, 2019. **21**(10):1019-1024.
48. Robey, P., *"Mesenchymal stem cells": fact or fiction, and implications in their therapeutic use.* F1000Research, 2017. **6**:524.
49. Choudhery, M.S., et al., *Minimum criteria for defining induced mesenchymal stem cells.* Cell Biology International, 2022. **46**(6):986-989.
50. Rajasingh, S., et al., *Comparative analysis of human induced pluripotent stem cell-derived mesenchymal stem cells and umbilical cord mesenchymal stem cells.* Journal of Cellular and Molecular Medicine, 2021. **25**(18):8904-8919.
51. American Cancer Society. *What Is Cancer?* Available from: https://www.cancer.org/cancer/understanding-cancer/what-is-cancer.html.
52. National Cancer Institute. *What Is Cancer?* Available from: https://www.cancer.gov/about-cancer/understanding/what-is-cancer.
53. Tomasetti, C. and B. Vogelstein, *Variation in cancer risk among tissues can be explained by the number of stem cell divisions.* Science, 2015. **347**(6217):78-81.
54. Wu, S., et al., *Substantial contribution of extrinsic risk factors to cancer development.* Nature, 2016. **529**(7584):43-47.
55. Tomasetti, C., L. Li, and B. Vogelstein, *Stem cell divisions, somatic mutations, cancer etiology, and cancer prevention.* Science, 2017. **355**(6331):1330-1334.
56. Baudoin, N.C. and M. Bloomfield, *Karyotype Aberrations in Action: The Evolution of Cancer Genomes and the Tumor Microenvironment.* Genes, 2021. **12**(4):558.

57. Nicholson, J.M. and D. Cimini, *Cancer Karyotypes: Survival of the Fittest.* Frontiers in Oncology, 2013. **3**:148.
58. Duesberg, P., et al., *The Chromosomal Basis of Cancer.* Analytical Cellular Pathology, 2005. **27**(5-6):951598.
59. Mirabelli, P., L. Coppola, and M. Salvatore, *Cancer Cell Lines Are Useful Model Systems for Medical Research.* Cancers, 2019. **11**(8):1098.
60. Kenny, P.A., et al., *The morphologies of breast cancer cell lines in three-dimensional assays correlate with their profiles of gene expression.* Molecular Oncology, 2007. **1**(1):84–96.
61. Finocchiaro, G., *Cancer Stem-Like Cells*, in *Encyclopedia of Cancer*, M. Schwab, Editor. 2009, Springer.
62. Chu, X., et al., *Cancer stem cells: advances in knowledge and implications for cancer therapy.* Signal Transduction and Targeted Therapy, 2024. **9**(1):170.
63. Mai, Y., et al., *The strategies to cure cancer patients by eradicating cancer stem-like cells.* Molecular Cancer, 2023. **22**(1):171.
64. Thakur, R., N. Kumari, and D.P. Mishra, *Chapter 6—Isolation and characterization of cancer stem cells*, in *Protocol Handbook for Cancer Biology*, G. Misra and J. Rajawat, Editors. 2021, Academic.
65. Yu, Z., et al., *Cancer stem cells.* The International Journal of Biochemistry & Cell Biology, 2012. **44**(12):2144-51.
66. Rao, W., et al., *Enhanced enrichment of prostate cancer stem-like cells with miniaturized 3D culture in liquid core-hydrogel shell microcapsules.* Biomaterials, 2014. **35**(27):7762-7773.
67. Fotinós, J., et al., *Assessing the distribution of cancer stem cells in tumorspheres.* Scientific Reports, 2024. **14**(1):11013.
68. Martinez-Bernabe, T., et al., *Tumorspheres as In Vitro Model for Identifying Predictive Chemoresistance and Tumor Aggressiveness Biomarkers in Breast and Colorectal Cancer.* Biology, 2024. **13**(9):724.

Cell Cryopreservation 5

Learning Objectives

- Analyze the processes of cell freezing and vitrification and potential cell injuries during cryopreservation
- Assess the mechanisms of different types of cryoprotectants and evaluate their suitability for specific cryopreservation applications
- Interpret the challenges associated with freeze-drying of mammalian cells

5.1 Introduction to Cell Cryopreservation

Cryopreservation is the process of preserving living cells and tissues at sub-freezing temperatures for extended periods of time. Its major mechanism involves reducing biological and chemical reactions in living cells at low temperatures [1, 2]. It is critical for routine cell culture, cell therapy, tissue engineering, reproductive technologies, and biobanking [3]. For example, cryopreservation is indispensable for the storage and transport of induced pluripotent stem cell (iPSC)-based therapies [4]. One of the major focus areas of the Society for Cryobiology (https://www.societyforcryobiology.org), founded in 1964, is the preservation of cell and tissues for long-term storage. Key journals directly related to cell cryopreservation include *Cryobiology* (https://www.sciencedirect.com/journal/cryobiology) and *Biopreservation and Biobanking* (https://home.liebertpub.com/publications/biopreservation-and-biobanking/110/overview).

Cryopreservation can be broadly divided into freezing and vitrification. Vitrification (or ice-free freezing) is the solidification of a liquid into a non-crystalline or amorphous solid known as a glass [5]. Research into preservation of living cells/tissues dates back to the late 1800s, when scientists used this technology to preserve both spermatozoa and red blood cells [6]. In 1949, C. Polge, A.U. Smith and A.S. Parkes reported on the "chance" discovery of glycerol's cryoprotective function during efforts to preserve avian spermatozoa. In 1950, A. U. Smith extended these

W. Zhang, *Cell Culture and Tissue Engineering*,
https://doi.org/10.1007/978-3-032-32096-4_5

findings by successfully cryopreserving human red blood cells (RBCs) in glycerol. These studies established key principles that shaped the field of biopreservation, including the need for cryoprotective agents (CPAs), the process by which cells could successfully be exposed to penetrating CPAs, and the manner of freezing and thawing [2]. A breakthrough occurred in the 1950s when James Lovelock discovered that the cryopreservation process caused osmotic stress in cells by rapidly freezing the liquid, which directly contributed to the formation of ice crystals in RBCs [6]. In 1972, Peter Mazur, Stanely Leibo, and Ernest Chu published the landmark paper "*A two-factor hypothesis of freezing injury: Evidence from Chinese hamster tissue-culture cells*," which identified two main contributors to cryoinjury: exposure to high salt concentrations at low cooling rates and intracellular ice formation (IIF) at high cooling rates [7, 8]. The two-factor hypothesis describes the relationships between cooling rate and cell survival, with damage arising from either toxic "solution effects" or lethal intracellular ice, depending on conditions [2]. This hypothesis provided a foundation for modern cryopreservation practices and remains one of most significant advances in the field [8]. In the mid-1980s, novel vitrification strategies were developed in which high concentrations of CPA mixtures were used to generate solutions that solidify into a glassy state without ice formation, even at liquid nitrogen temperatures [2, 9, 10].

A cryoprotectant (CPA) is any solute that, when added to cells in suspension, improves post-thaw survival compared with controls lacking the agent. CPAs are generally classified as either permeating or non-permeating. Permeating CPAs can enter cells and directly interact with cellular processes, sometimes causing cytotoxicity. Common examples include glycerol, dimethyl sulfoxide (DMSO), ethylene glycol (EG), and propylene glycol (PG) (Fig. 5.1). They protect cells from mechanical and osmotic stresses by forming hydrogen bonds with water, lowering the freezing point, and in some cases increasing membrane permeability to facilitate water replacement with CPA molecules, which supports vitrification. Non-permeating CPAs, typically sugars and polymers (both natural and synthetic), act extracellularly. Examples include polyethylene glycol (PEG), polyvinylpyrrolidone (PVP), albumin, sucrose, and trehalose [6, 11, 12]. They promote glass formation and dehydration of cells, protect the cell membrane, and suppress ice recrystallization [11, 13, 14]. Synthetic zwitterions have also been shown as efficient non-permeating CPAs [14]. Additional ice-inhibition agents include antifreeze proteins (AFPs), nanomaterials (e.g., graphene oxide (GO) and metal-organic frameworks (MOFs)), and hydrogels. Broadly, CPAs are thought to preserve cells through three key mechanisms: ice-recrystallization inhibition, freezing point depression, and ice shape and growth control [15]. Combining different CPAs can be beneficial. For example, vitrification can be achieved by using non-penetrating CPAs with low toxicities to partially replace permeating CPAs while maintaining the total solute concentrations [16].

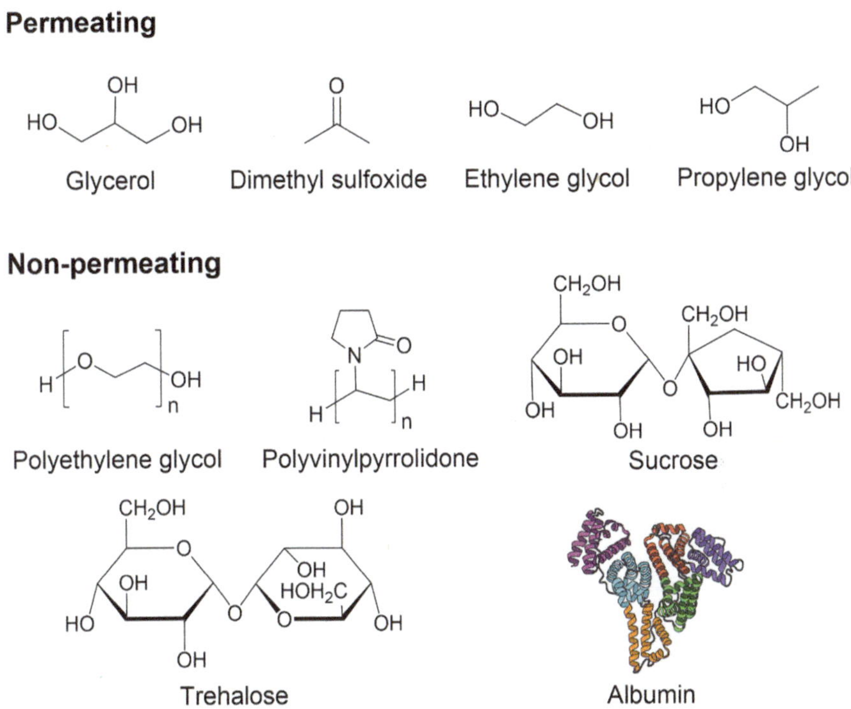

Fig. 5.1 Structures of common cryoprotectants

5.2 Freezing and Vitrification of Cells

Two major aspects in cryopreservation are cooling rate control and the use of CPAs. Starting with cooling rate, the combined effects of IIF and cell dehydration (i.e., solute effect) result in the classical inverted U-shaped relationship between cell survival and cooling rate during (relatively) slow freezing. In the case of vitrification, when the cooling rate exceeds a critical value (CR_v), cells undergo vitrification, leading to a high cell survival rate (Fig. 5.2) [17]. Conversely, when cooling rates fall below CR_{SF}, extracellular ice formation induces cellular dehydration, resulting in biochemical and biophysical alterations that become the dominant cause of cell injury. On the other hand, when cooling rates exceed CR_{SF}, both intracellular and extracellular ice form, with IIF being the primary cause of injury (Fig. 5.3). Therefore, it is crucial to carefully regulate the cooling rate to balance cellular dehydration and ice crystallization, thereby minimizing the formation of large ice crystals [17, 18]. Successful vitrification requires high viscosity, rapid cooling and warming rates, and small sample volumes [19].

Seeding extracellular ice (i.e., ice seeding) can be utilized to reduce IIF. Ice seeding involves the induction of ice crystal formation by introducing an ice crystal, adding an ice nucleator, or simply touching the sample with a very cold small object (e.g., syringe needle). This causes ice to form at a controlled temperature rather than at a random one,

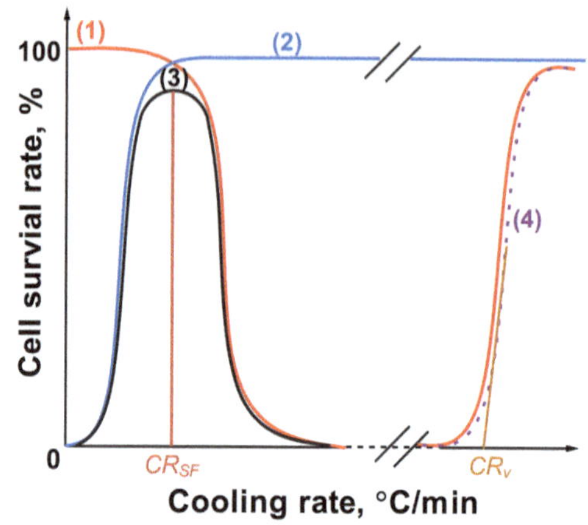

Fig. 5.2 Effects of cooling rate on cel survival rate during cryopreservation. (1): effect of intracellular ice formation (IIF) only; (2): effect of cell dehydration (solute effect) only; (3): combined effect of IIF and cell dehydration at low cooling rates; (4): combined effect of IIF and cell dehydration at high cooling rates. CR_{SF}: optimal cooling rate for slow freezing, CR_V: critical cooling rate for vitrification. (Generated based on [17])

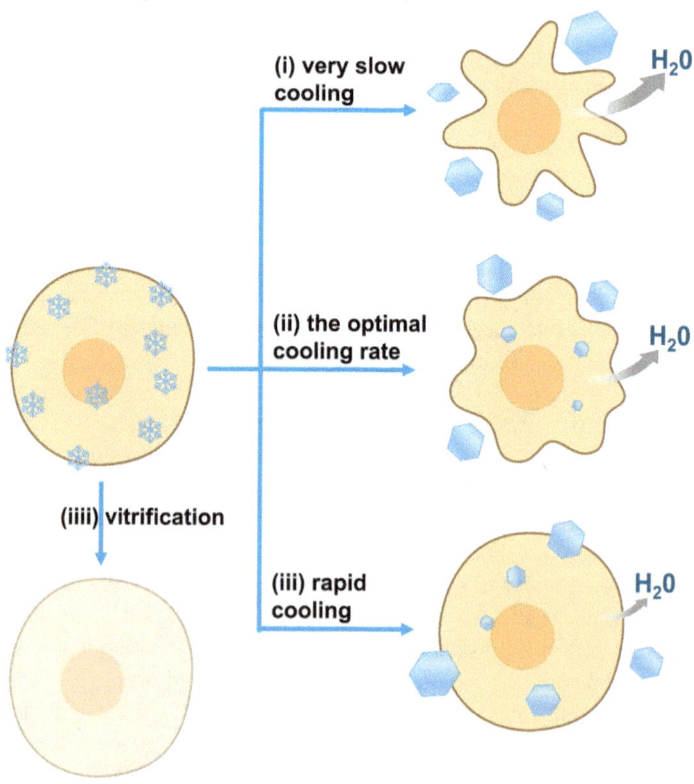

Fig. 5.3 Cell injuries in response to cooling rate. (Adapted from [18] with permission from Elsevier)

allowing extracellular ice to induce cellular dehydration and thereby preventing intracellular ice formation. Optimal freezing processes cool the cells quickly enough to retain intracellular water while still allowing controlled dehydration [20]. A recent study demonstrated that inducing ice nucleation at temperatures close to the equilibrium freezing point of cryoformulations led to greater intracellular dehydration, less IIF, and improved cell membrane integrity and viability. Moreover, controlled ice nucleation ensures more uniform vial-to-vial temperature profiles, thereby improving quality consistency [21]. It is useful to examine the typical freezing profile of water (Fig. 5.4). Normally, rapid cooling can cause water to fall below its melting point (0 °C) without freezing (i.e., supercooling). At a certain degree of supercooling, nucleation occurs suddenly, followed by crystal growth. This phase transition releases latent heat, briefly raising the sample temperature. The final stage is the complete crystallization or solidification [22]. Even under identical protocols, spontaneous nucleation temperatures can vary substantially and occur far below the equilibrium freezing point [16, 21]. Importantly, the greater the extent of supercooling, the greater the immediate stresses imposed on suspended cells at the onset of ice formation [23].

To achieve vitrification, samples must be cooled rapidly enough to pass both the melting temperature (T_m) and the glass transition temperature (T_g). Aqueous solutions, such as cell culture medium-based cryopreservation formulations containing salts, amino acids, proteins, and CPAs, can exhibit different T_m and T_g values depending on their composition and respective concentrations. The presence of salts, proteins, CPAs and other molecules increases viscosity, thereby lowering T_m but raising T_g (Fig. 5.5) [24]. In principle, a sufficiently concentrated solution can vitrify at a given cooling rate. However, high concentrations are often associated with significant toxicity. Thus, it is critical to achieve vitrification without relying on excessively

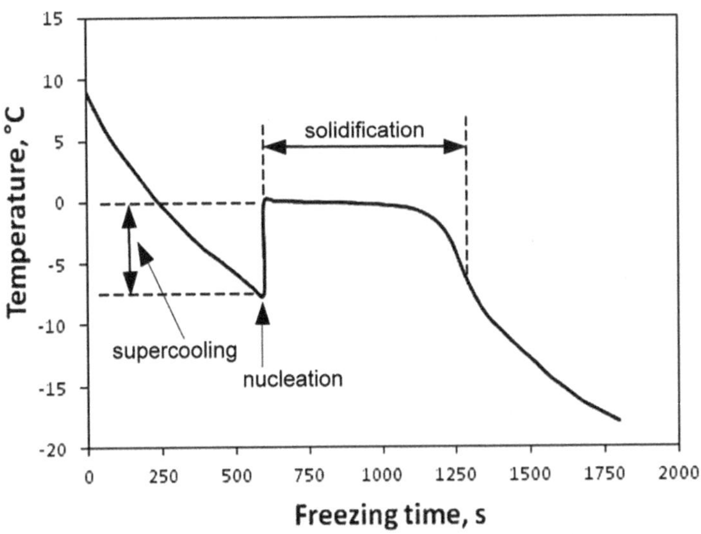

Fig. 5.4 Cell injuries respond to cooling rate. (Adapted from [22] with permission from Taylor & Francis)

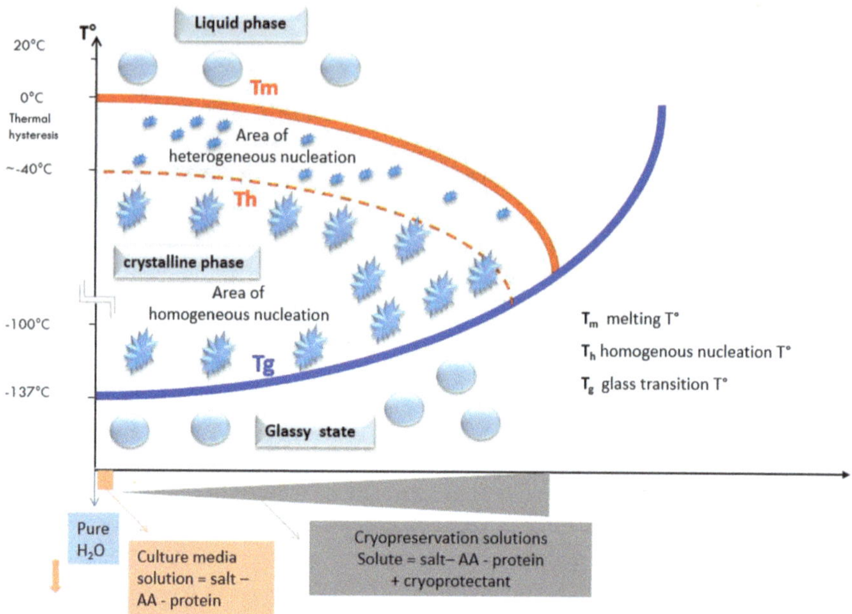

Fig. 5.5 Phase diagram of different solutions during cooling/warming. T_m: melting temperature; T_h: homogeneous nucleation temperature (when T_h is reached, the system ends up crystallizing uniformly even in the absence of any crystallization seeds, transforming into a stable crystalline solid); T_g: glass transition temperature, AA: amino acid. (Adapted from [24])

high CPA concentrations. Rapid cooling is constrained by the thermal properties of devices, specimen thickness, solution volume, and related factors [3]. Xiaoming He et al. demonstrated that quartz micro-capillaries can be utilized to achieve ultra-fast cooling at a relatively low CPA concentration (2 M) [25]. Preferential vitrification of water in small microcapsules was further achieved by combining quartz micro-capillaries with cell microencapsulation at low CPA concentrations. Calorimetric data showed reduced ice formations (smaller heat flux peaks during ice thawing) in the presence of microcapsules and with increasing CPA concentration (Fig. 5.6). Moreover, encapsulated cells displayed significantly higher post-thaw viability compared to non-encapsulated controls (Fig. 5.7) [26]. Vitrification has also been used for organ cryopreservation including liver, kidney, and heart, particularly with the

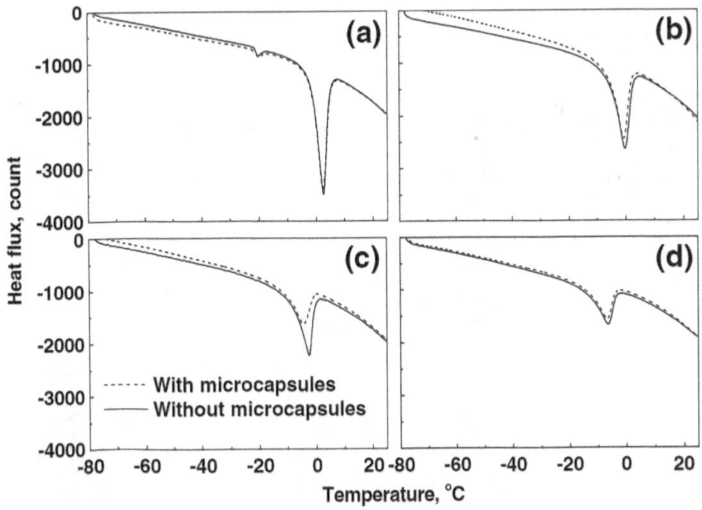

Fig. 5.6 Calorimetric data showing the heat flux peak of ice thawing for samples with (solid line) and without (dashed line) microcapsules in the presence of 0 (**a**), 5 (**b**), 10 (**c**), and 15% (v/v) (**d**) cryoprotectant—dimethylsulfoxide (DMSO). (Adapted from [26] with permission from Springer Nature)

recent breakthrough in combing vitrification with nanowarming (i.e., warming through alternating magnetic fields that heat nanoparticles within organ vasculature) [18, 27]. For example, a recent study demonstrated that vitrified kidneys can be cryogenically stored up to 100 days and successfully recovered by nanowarming, enabling transplantation and restoration of life-sustaining renal function in nephrectomized male rats (Fig. 5.8). In these studies, vitrification was achieved using a controlled-rate freezer (−40 °C/min), followed by a 25-min annealing step once the chamber reached −122 °C (just above the T_g). After annealing, cooling was continued more slowly at −5 °C/min from −122 °C to −150 °C [28, 29].

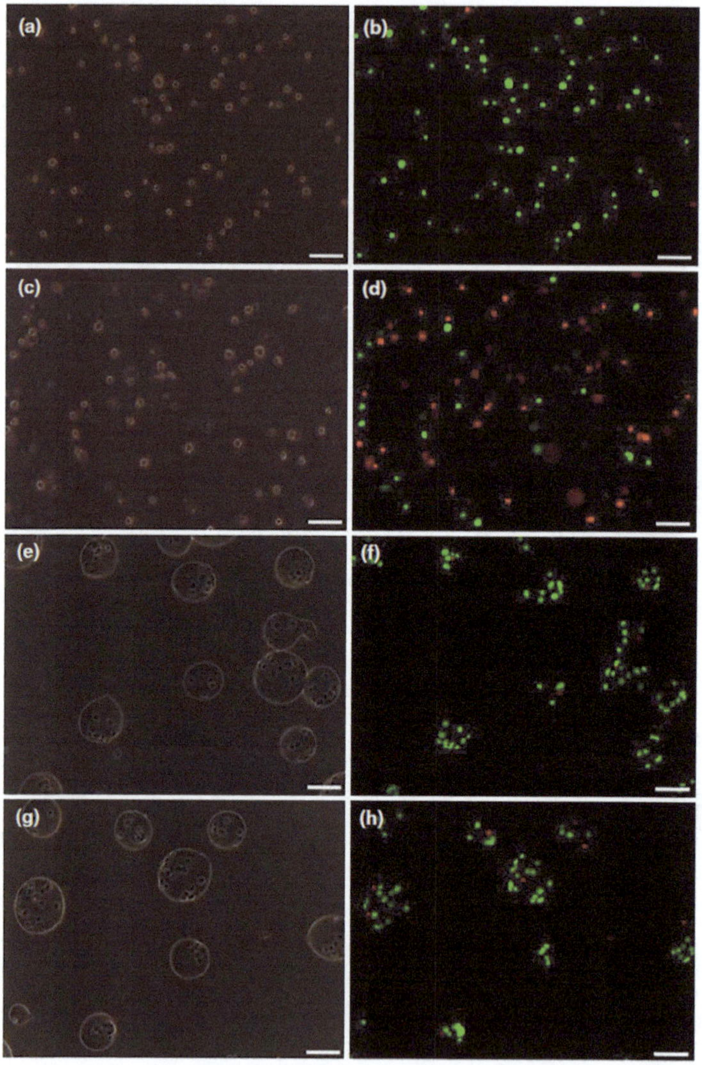

Fig. 5.7 Typical phase contrast and fluorescence images of control (non-microencapsulated, (**a–d**) and microencapsulated (**e–h**) cells. Panels (**a**) and (**b**) show control cells before cryopreservation, (**c**) and (**d**) show control cells after cryopreservation, (**e**) and (**f**) show microencapsulated cells before cryopreservation, and (**g**) and (**h**) show microencapsulated cells after cryopreservation. In the fluorescence images, live and dead cells are stained green and red, respectively. Scale bars: 100 μm. (Adapted from [26] with permission from Springer Nature)

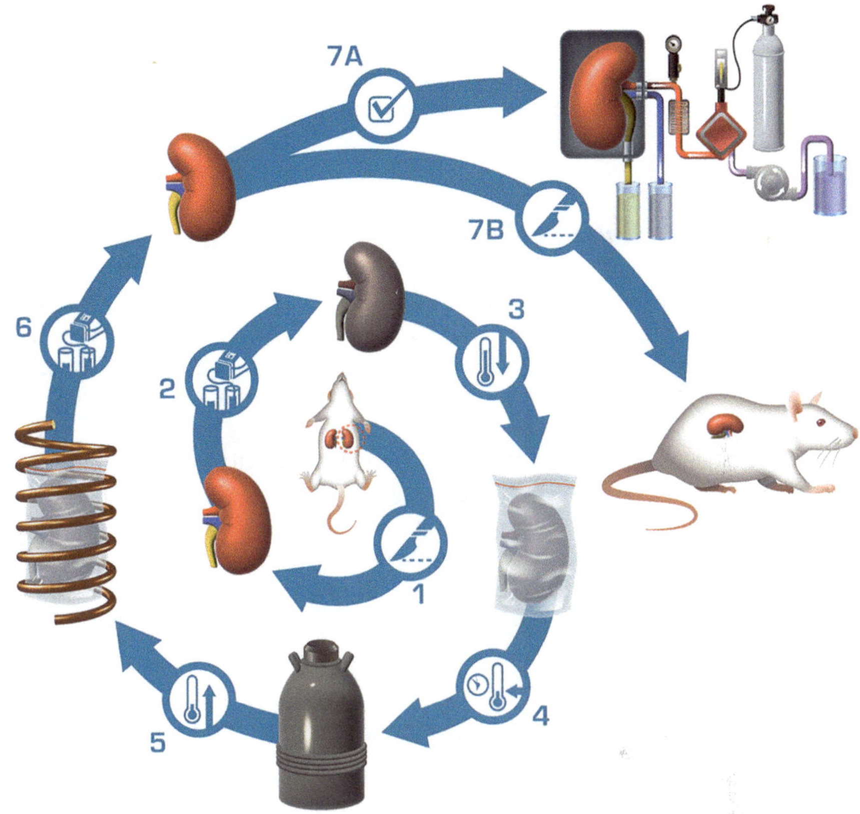

Fig. 5.8 Sequential steps of the vitrification and nanowarming procedure: (1) kidney recovery from the donor; (2) loading of cryoprotective agents (CPA) and iron oxide nanoparticles (IONPs); (3) rapid cooling to a vitrified state; (4) storage at −150 °C until needed; (5) rapid and uniform warming using a radiofrequency alternating electromagnetic field; (6) unloading of CPA and IONPs; and finally, organs are ready for (7A) assessment by normothermic machine perfusion or (7B) transplantation. (Adapted from [29])

5.3 Warming Post-Cryopreservation

Warming is another critical step affecting the cryopreservation outcomes [30, 31]. The term "warming" is more inclusive, as thawing generally refers only to frozen samples. One of the major challenges during warming is recrystallization and devitrification [30]. Recrystallization of ice in liquid solution is believed to occur *via* the Ostwald ripening process, in which large ice crystals grow at the expense of small ones. To avoid devitrification, the warming rate must exceed the critical warming rate (CWR). The CWR is typically several orders of magnitude higher than critical cooling rate (CCR; the cooling must exceed the CCR to achieve vitrification). The addition of CPAs can significantly decrease the CWR [32]. Therefore, successful warming requires both a sufficiently high warming rate and the use of CPAs.

Typically, cryovials are warmed rapidly in a 37 °C water bath for 90–120 seconds. For large samples, novel approaches such as nanowarming (described in Sect. 5.2) have been developed. Electric and laser fields have also been used for warming [15]. It is also essential to remove CPAs promptly after warming, as they can cause osmotic injuries and/or toxicity [33].

5.4 Freeze-Drying of Cells

Cryopreserved cells have to be maintained at low temperature, resulting in high storage and transportation cost, as well as risks of temperature fluctuations and cross-contamination. Successful freeze-drying of cells can eliminate many of these challenges [34]. Freeze-drying is a low-temperature, vacuum-dehydration process. During freeze-drying, liquid water is first converted to ice by lowering the sample temperature at ambient pressure. During primary drying, water is removed *via* sublimation as the pressure is decreased below the triple point of the system (where three different phases coexist), while a controlled amount of heat is supplied. The remaining unfrozen water, or "bound water", is removed by desorption during secondary drying by slowly raising the temperature under vacuum, yielding a lyophilized product with long-term stability at the refrigerated or ambient temperatures (Fig. 5.9) [35, 36].

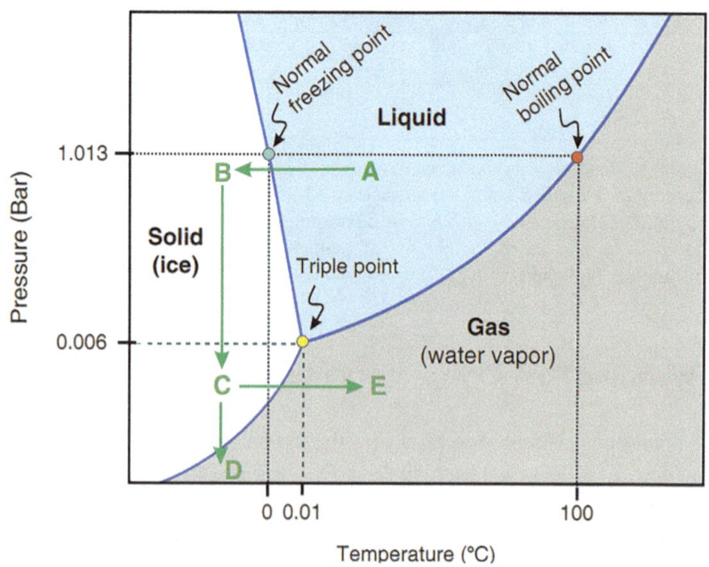

Fig. 5.9 Hypothetical freeze-drying paths superimposed on a pressure-temperature (P-T) phase diagram (not to scale). Freeze-drying involves first converting specimens from a liquid phase to solid (ice) by freezing (path A → B), then lowering pressure below the triple point by applying vacuum (path B → C), and finally converting the solid to gas (water vapor) either by further lowering the pressure below the sublimation line (path C → D) or controlled heating of specimens under vacuum (path C → E). (Adapted from [36] with permission from Springer Nature)

Unfortunately, freeze-drying of mammalian cells has had very limited success. While freeze-drying results in immotile and membrane-compromised sperm, intracytoplasmic sperm injections can be used to initiate fertilization, as the process preserves chromosomal integrity and oocyte-activating factors for long periods depending on storage temperature [36]. Healthy mice have been produced from freeze-dried spermatozoa stored at room temperature [37]. Both freezing and drying impose severe stress on biomolecules and cellular structures. In particular, the water removal surrounding biomolecules may lead to irreversible structural changes in the phase state and organization of cellular membranes, as well as protein aggregation. Increased levels of reactive oxygen species (ROS) can also cause damage, with lipids in membranes being particularly susceptible to free radical attack [38]. Overall, the freeze-drying process affects the chemical, physical, and mechanical environments of biomaterials and cells (Fig. 5.10) [39]. Rehydration of freeze-dried cells can also cause cell death, such as due to osmotic imbalance [34]. In addition to CPAs, protective agents are also needed to protect biomaterials and cells, especially during drying [40]. Lyoprotectants are stabilizers that protect biomaterials and cells during freeze-drying and subsequent storage [41]. Certain sugars, such as trehalose, can act as both cryoprotectants and lyoprotectants. Intracellular delivery of trehalose has been shown to improve outcomes for both cryopreservation and freeze-drying [42, 43]. A single agent is unlikely to be sufficient to protect biomaterials and

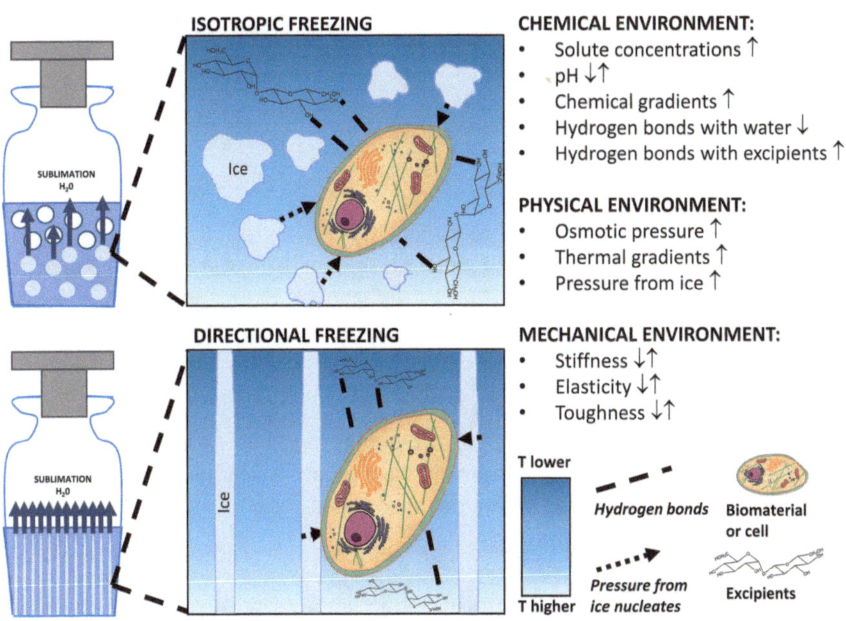

Fig. 5.10 Changes in chemical, physical, and mechanical environments surrounding biomaterials and cells during freeze-drying. Isotropic freezing: cooling rate is controlled, not thermal gradient; Directional freezing: controlled linear thermal gradients are applied during freezing. (Adapted from [39])

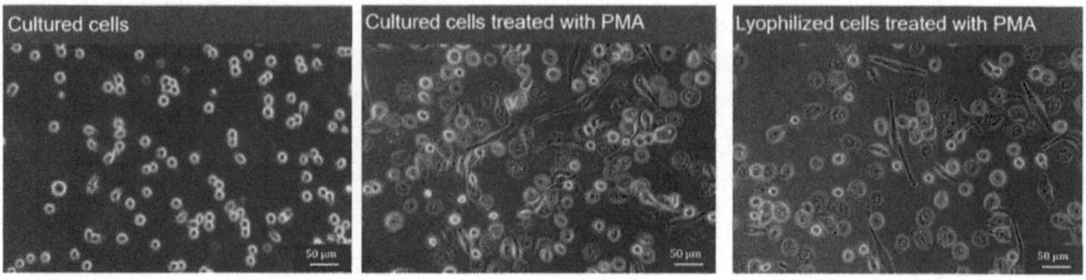

Fig. 5.11 Lyophilized THP-1 cells are able to differentiate after rehydration. Morphological changes post-PMA treatment were observed. Left panel: untreated cultured cells without PMA treatment, showing round, non-adherent morphology. Middle panel: cultured THP-1 cells treated with PMA. Right panel: lyophilized THP-1 cells in medium containing 1 M trehalose + inulin, rehydrated and then treated with PMA. PMA: phorbol 12-myristate 13-acetate. (Adapted from [45])

cells throughout the entire freeze-drying process. For example, replica exchange molecular dynamic simulations indicate that different protectants provide varying degrees of protection depending on the conditions [44]. A recent study demonstrated improved viability of a human monocytic cell line (THP-1) post lyophilization when using a lyophilization medium containing inulin in combination with trehalose (Fig. 5.11) [45].

Review and Practice Questions

1. What is cryopreservation and why is it critical for cell culture and tissue engineering?
2. What is the two-factor hypothesis?
3. What are the two classes of cryoprotectants based on their cell membrane permeability, and what are their respective protective mechanisms during cryopreservation?
4. What is ice seeding, and what are its benefits?
5. What is vitrification, and how is vitrification of cells, tissues, and organs generally achieved?
6. Read the article entitled "Preferential vitrification of water in small alginate microcapsules significantly augments cell cryopreservation by vitrification" and answer the following questions:

   (a) What does the peak area difference in Figure 2C indicate?
   (b) According to the authors, how did the alginate-based hydrogel microsphere improve the cryopreservation outcomes?

7. What cellular injuries can occur during the warming process after cryopreservation?
8. Use the pressure-temperature phase diagram to describe the freeze-drying process.
9. Why is freeze-drying of mammalian cells challenging, and what are the mechanisms of cell injury during the drying process?

References

1. Jaiswal, A.N. and A. Vagga, *Cryopreservation: A Review Article.* Cureus, 2022. **14**(11):e31564.
2. Baust, J.G., D. Gao, and J.M. Baust, *Cryopreservation.* Organogenesis, 2009. **5**(3):90–96.
3. Amini, M. and J.D. Benson, *Technologies for Vitrification Based Cryopreservation.* Bioengineering, 2023. **10**(5):508.
4. Dobruskin, M., G. Toner, and R. Kander, *Cryopreservation practices in clinical and pre-clinical iPSC-based cell therapies: Current challenges and future directions.* Biotechnology Progress, 2025. **41**(4):e70031.
5. Fahy, G.M. and B. Wowk, *Principles of Cryopreservation by Vitrification*, in *Cryopreservation and Freeze-Drying Protocols*, W.F. Wolkers and H. Oldenhof, Editors. 2015, Springer.
6. Whaley, D., et al., *Cryopreservation: An Overview of Principles and Cell-Specific Considerations.* Cell Transplantation, 2021. **30**:0963689721999617.
7. Mazur, P., S.P. Leibo, and E.H.Y. Chu, *A two-factor hypothesis of freezing injury: Evidence from Chinese hamster tissue-culture cells.* Experimental Cell Research, 1972. **71**(2):345-355.
8. Schnebly, R.A., *A Two-Factor Hypothesis of Freezing Injury: Evidence from Chinese Hamster Tissue-Culture Cells" (1972), by Peter Mazur, Stanley Leibo, and Ernest Chu*, in *Embryo Project Encyclopedia* 2021.
9. Fahy, G.M., et al., *Vitrification as an approach to cryopreservation.* Cryobiology, 1984. **21**(4):407-426.
10. Rall, W.F. and G.M. Fahy, *Ice-free cryopreservation of mouse embryos at -196 degrees C by vitrification.* Nature, 1985. **313**(6003):573–575.
11. Elliott, G.D., S. Wang, and B.J. Fuller, *Cryoprotectants: A review of the actions and applications of cryoprotective solutes that modulate cell recovery from ultra-low temperatures.* Cryobiology, 2017. **76**:74-91.
12. Matsumura, K., et al., *Molecular mechanisms of cell cryopreservation with polyampholytes studied by solid-state NMR.* Communications Materials, 2021. **2**(1):15.
13. Anastas, Z.M., et al., *Effect of Bovine Serum Albumin (BSA) Concentration on Cryopreservation of Booroolong Frog Sperm with Evaluation of Post-Thaw Motility in Caffeine.* Veterinary Sciences, 2025. **12**(1):30.
14. Yong, K.W., et al., *Review of non-permeating cryoprotectants as supplements for vitrification of mammalian tissues.* Cryobiology, 2020. **96**:1-11.
15. Kato, Y., et al., *Synthetic zwitterions as efficient non-permeable cryoprotectants.* Communications Chemistry, 2021. **4**(1):151.
16. Chang, T. and G. Zhao, *Ice Inhibition for Cryopreservation: Materials, Strategies, and Challenges.* Advanced Science, 2021. **8**(6):2002425.
17. He, X., *Preservation of Embryonic Stem Cells*, in *Methodological Advances in the Culture, Manipulation and Utilization of Embryonic Stem Cells for Basic and Practical Applications*, C.S. Atwood, Editor. 2011, IntechOpen.
18. Liu, Y., L. He, and G. Zhao, *Multiscale heat and mass transfer in biomaterial cryopreservation.* International Journal of Heat and Mass Transfer, 2025. **252**:127414.
19. Arav, A., *Cryopreservation by Directional Freezing and Vitrification Focusing on Large Tissues and Organs.* Cells, 2022. **11**(7):1072.
20. FOWLER, A. and M. TONER, *Cryo-Injury and Biopreservation.* Annals of the New York Academy of Sciences, 2006. **1066**(1):119-135.
21. Dan, N., et al., *Impact of controlled ice nucleation on intracellular dehydration, ice formation and their implications on T cell freeze–thaw viability.* International Journal of Pharmaceutics, 2024. **665**:124694.
22. Woo, M.W. and A.S. Mujumdar, *Effects of Electric and Magnetic Field on Freezing and Possible Relevance in Freeze Drying.* Drying Technology, 2010. **28**(4):433-443.
23. Meneghel, J., P. Kilbride, and G.J. Morris, *Cryopreservation as a Key Element in the Successful Delivery of Cell-Based Therapies—A Review.* Frontiers in Medicine, 2020. **7**: 592242.

24. Vanderzwalmen, P., et al., *The Evolution of the Cryopreservation Techniques in Reproductive Medicine—Exploring the Character of the Vitrified State Intra- and Extracellularly to Better Understand Cell Survival after Cryopreservation.* Reproductive Medicine, 2020. **1**(2):142-157.
25. He, X., et al., *Vitrification by ultra-fast cooling at a low concentration of cryoprotectants in a quartz micro-capillary: A study using murine embryonic stem cells.* Cryobiology, 2008. **56**(3):223-232.
26. Zhang, W., et al., *Preferential vitrification of water in small alginate microcapsules significantly augments cell cryopreservation by vitrification.* Biomedical Microdevices, 2010. **12**(1):89-96.
27. Chen, J., et al., *Cryopreservation of tissues and organs: present, bottlenecks, and future.* Frontiers in Veterinary Science, 2023. **10**:1201794.
28. Sharma, A., et al., *Vitrification and Nanowarming of Kidneys.* Advanced Science, 2021. **8**(19):2101691.
29. Han, Z., et al., *Vitrification and nanowarming enable long-term organ cryopreservation and life-sustaining kidney transplantation in a rat model.* Nature Communications, 2023. **14**(1):3407.
30. Bojic, S., et al., *Winter is coming: the future of cryopreservation.* BMC Biology, 2021. **19**(1):56.
31. Linkova, D.D., Y.P. Rubtsova, and M.N. Egorikhina, *Cryostorage of Mesenchymal Stem Cells and Biomedical Cell-Based Products.* Cells, 2022. **11**(17):2691.
32. Lin, M., H. Cao, and J. Li, *Control strategies of ice nucleation, growth, and recrystallization for cryopreservation.* Acta Biomaterialia, 2023. **155**:35-56.
33. Gao, D. and J.K. Critser, *Mechanisms of Cryoinjury in Living Cells.* ILAR Journal, 2000. **41**(4):187-196.
34. Rockinger, U., M. Funk, and G. Winter, *Current Approaches of Preservation of Cells During (freeze-) Drying.* Journal of Pharmaceutical Sciences, 2021. **110**(8):2873-2893.
35. Weng, L., *Technologies and Applications Toward Preservation of Cells in a Dry State for Therapies.* Biopreservation and Biobanking, 2021. **19**(4):332-341.
36. Keskintepe, L. and A. Eroglu, *Freeze-Drying of Mammalian Sperm*, in *Cryopreservation and Freeze-Drying Protocols*, W.F. Wolkers and H. Oldenhof, Editors. 2015, Springer.
37. Wakayama, S., D. Ito, and T. Wakayama, *Production of Healthy Mice from Freeze-Dried Spermatozoa Preserved at Room Temperature*, in *Spermatology: Methods and Protocols*, M. Álvarez-Rodríguez, Editor. 2025, Springer.
38. Zhang, M., et al., *Freeze-drying of mammalian cells using trehalose: preservation of DNA integrity.* Scientific Reports, 2017. **7**(1):6198.
39. Merivaara, A., et al., *Preservation of biomaterials and cells by freeze-drying: Change of paradigm.* Journal of Controlled Release, 2021. **336**:480-498.
40. Karunnanithy, V., et al., *Effectiveness of Lyoprotectants in Protein Stabilization During Lyophilization.* Pharmaceutics, 2024. **16**(10):1346.
41. Sundaramurthi, P. and R. Suryanarayanan, *Trehalose Crystallization During Freeze-Drying: Implications On Lyoprotection.* The Journal of Physical Chemistry Letters, 2010. **1**(2):510-514.
42. Stewart, S. and X. He, *Intracellular Delivery of Trehalose for Cell Banking.* Langmuir, 2019. **35**(23):7414-7422.
43. He, H., et al., *Intracellular trehalose improves the survival of human red blood cells by freeze-drying.* Frontiers of Energy and Power Engineering in China, 2007. **1**(1):120-124.
44. Li, D., et al., *Investigating the stabilisation of IFN-α2a by replica exchange molecular dynamics simulation.* Journal of Molecular Modeling, 2022. **28**(8):232.
45. Meiss, A., et al., *Combination of trehalose and inulin efficacy as lyoprotectant for the preservation of human monocytic cell line.* Cryobiology, 2025. **120**:105287.

Cell Engineering 6

Learning Objectives

- Compare different gene delivery approaches—mechanisms, advantages, and limitations
- Discuss the working mechanisms of CRISPR/Cas genome editing and CAR-T therapy
- Examine the biomedical applications of cell encapsulation, emphasizing design considerations
- Evaluate various cell surface modification strategies

6.1 Introduction to Cell Engineering

Cell engineering is critical for various biomedical applications, such as establishing continuous cell lines, developing cellular therapies and engineered tissues. As shown in Fig. 6.1, current cell engineering strategies include genetic engineering, preconditioning, biomaterial encapsulation, surface modification, and cell assembly. Preconditioning of cells encompasses the *ex vivo* treatment with chemical or physical cues to enhance cellular properties against unamiable environments. For example, treatment with inflammatory cytokines promotes communication between mesenchymal stem cells and the immune system, thereby enabling escape from immune-cell-mediated cytotoxicity and improving cellular survival [1]. Genetic engineering, biomaterial encapsulation, and surface modification approaches are discussed in detail in this chapter. Cell assembly is discussed in detail in Chaps. 13 (cell sheets) and 14 (organoids).

W. Zhang, *Cell Culture and Tissue Engineering*,
https://doi.org/10.1007/978-3-032-32096-4_6

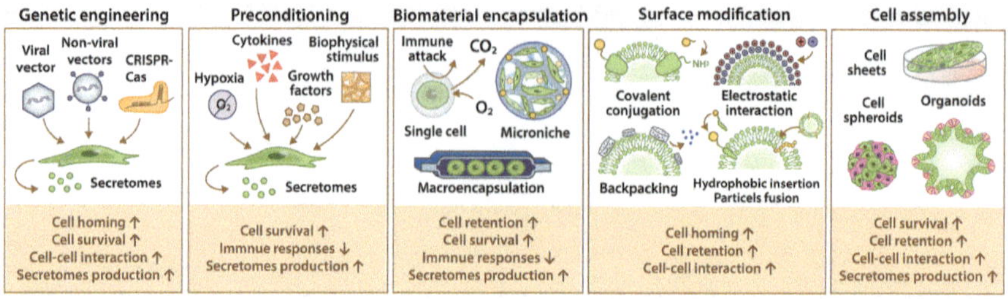

Fig. 6.1 A myriad of cell engineering approaches. (Adapted from [1] with permission from Elsevier)

Table 6.1 Comparison of gene delivery methods

	Methods	Advantages	Disadvantages
Biological	Viral	High efficiency	Biosafety concerns; potential immune response; potential genotoxicity
Chemical	Calcium phosphate coprecipitation	Low cost; easy to perform	Efficiency is highly pH-dependent; low consistency
	Cationic lipids	Relatively high efficiency	Potential cytotoxicity; may not be compatible with some media
	Cationic polymers	High design flexibility	Relatively low efficiency; may not be compatible with some media
Physical	Microinjection	Precise dosing of injected material; selective delivery	Not scalable
	Electroporation	Easy to perform; high efficiency	May cause cell damage

6.2 Gene Delivery and Editing

Gene delivery is the process of transferring genetic material into cells [2]. The genetic materials can be plasmid DNA (pDNA), messenger RNA (mRNA), and RNA interference (RNAi) molecules (e.g., small interfering RNA (siRNA) and micro RNA (miRNA)). pDNA is delivered and then transcribed and translated by host cell machinery. mRNA is delivered and then translated into proteins. RNAi molecules are delivered to target specific mRNA sequences, thereby downregulating host mRNA translation [3]. Gene delivery methods can be mechanical (e.g., electroporation), chemical (e.g., cationic lipids/polymers and nanocarriers), or biological (e.g., viral vectors). Different methods are compared in Table 6.1. Gene delivery using non-viral methods is often referred to as "transfection," while delivery using whole virus or viral vectors is termed as "transduction" [4]. Depending on experimental goals, the type of genetic materials, delivery method, and cell type, transfection can be transient or stable. Transient transfection produces temporary effects on cells, while stable transfection results in permanent genetic changes that are typically passed on to progeny cells [5].

Biologic gene delivery methods typically involve viral vectors, such as those derived from adenoviruses, adeno-associated viruses, and retroviruses. A viral vector generally contains three components: the protein capsid and/or envelope, the transgene of interest, and the regulatory cassette [6]. Although viral vectors offer high specificity and efficiency, safety concerns include potential interference with the host genome, induction of inflammatory system, toxin production, and insertional mutagenesis [5, 7]. In particular, adenoviruses tend to trigger immune responses *in vivo* [5].

Calcium phosphate, cationic lipids/polymers, and nanocarriers are widely used chemical approaches for gene delivery. Positively charged Calcium phosphate forms complexes with negatively charged DNA in a form of precipitates, which then enter cell *via* endocytosis or phagocytosis [5]. Cationic lipids and polymers are commonly used for gene delivery due to their positive charges enabling the formation of polyplexes with genetic material through electrostatic interactions. Uptake of polyplexes into cells generally occurs *via* adsorptive endocytosis. Representative cationic lipids include *N*-[1-(2,3-dioleyloxy) propyl]-*N*, *N*,*N*-trimethylammonium chloride (DOTMA), 1,2-dioleoyloxy-3-trimethylammoniopropane chloride (DOTAP), and 1,2-dimyristyloxypropyl-3-dimethyl-hydroxyethyl ammonium bromide (DMRIE) (Fig. 6.2). Despite their relatively high efficiency, cationic lipids can cause cytotoxicity by altering the biosynthesis and metabolism of cholesterol, lipids, and steroids, interaction with enzymes such as protein kinase C, membrane destabilization and oxidative stress [8]. Cationic polymers such as polyethyleneimine (PEI), chitosan, and poly-l-lysine are attractive alternatives to lipids, offering higher stability and greater design flexibility [9].

Physical methods are frequently employed due to their simplicity. Microinjection and electroporation are two of the most common approaches. Microinjection is widely used for *in vitro* fertilization and generating transgenic animals. But the major challenge lies in the inability to process large numbers of cells efficiently. Electroporation applies an electric field to temporarily disrupt cell membrane permeability, allowing genetic material to enter cells; however, this technique can cause significant cell damage [5].

Gene delivery using a nanocarrier has become increasingly popular. Advantages of using nanoscale carriers include the ability to encapsulate genetic materials, high design flexibility, and potential targetability through carrier engineering. Nanoscale carriers are predominantly lipid- or polymer-based (Fig. 6.3). Liposomes and lipid nanoparticles are the two most widely used lipid-based carriers, while polymeric nanocarriers encompass a broad range of structures such as nanocapsules and nanospheres. Cellular uptake is strongly influenced by the physicochemical properties of nanocarriers, including size and shape, surface charge, hydrophobicity, and surface functionality. Uptake generally occurs *via* endocytosis. After endocytic vesicles pinch off the plasma membrane, they fuse with early endosomes. Endosomes then direct further trafficking into different cellular destinations. If the carriers and/or genetic material are not released from endosomes, degradation may occur when late

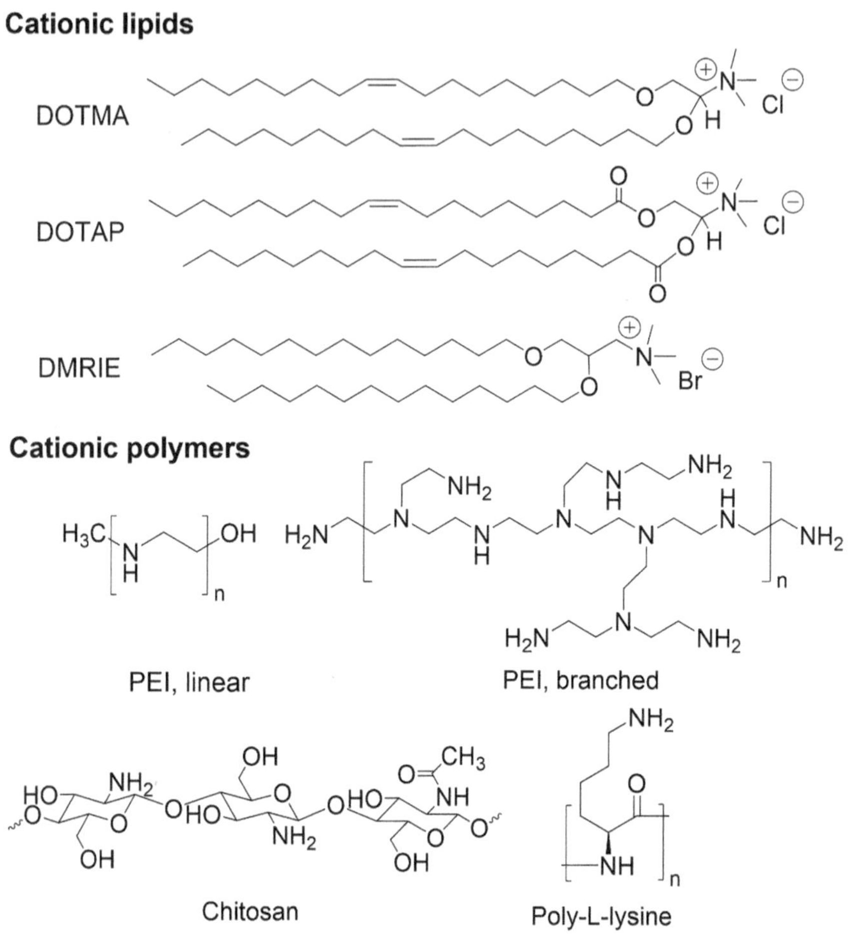

Fig. 6.2 Structures of commonly used cationic lipids and polymers for gene delivery. DOTMA: *N*-[1-(2,3-dioleyloxy) propyl]-*N*, *N*,*N*-trimethylammonium chloride, DOTAP: 1,2-dioleoyloxy-3-trimethylammoniopropane chloride, DMRIE:1,2-dimyristyloxypropyl-3-dimethyl-hydroxyethyl ammonium bromide, PEI: polyethyleneimine

endosomes fuse with lysosomes. Alternatively, release may occur when late endosomes fuse back with the plasma membrane [10, 11]. Therefore, successful gene delivery requires careful consideration of not only transport across the plasma membrane but also the subsequent release of genetic material into the cytosol or nucleus. Designing carriers responsive to specific stimuli can help overcome this challenge. For example, cationic polymer-based carriers exhibit endosomal escape properties explained by the "proton sponge" hypothesis. In this mechanism, a cationic polymer binds protons within endosomes, leading to the translocation of protons into the endosomal compartment. This process is coupled with the influx of chloride ions, resulting in increased ionic concentration and water influx. The resulting osmotic pressure ultimately causes endosomal rupture (Fig. 6.4) [12–14]. Other proposed

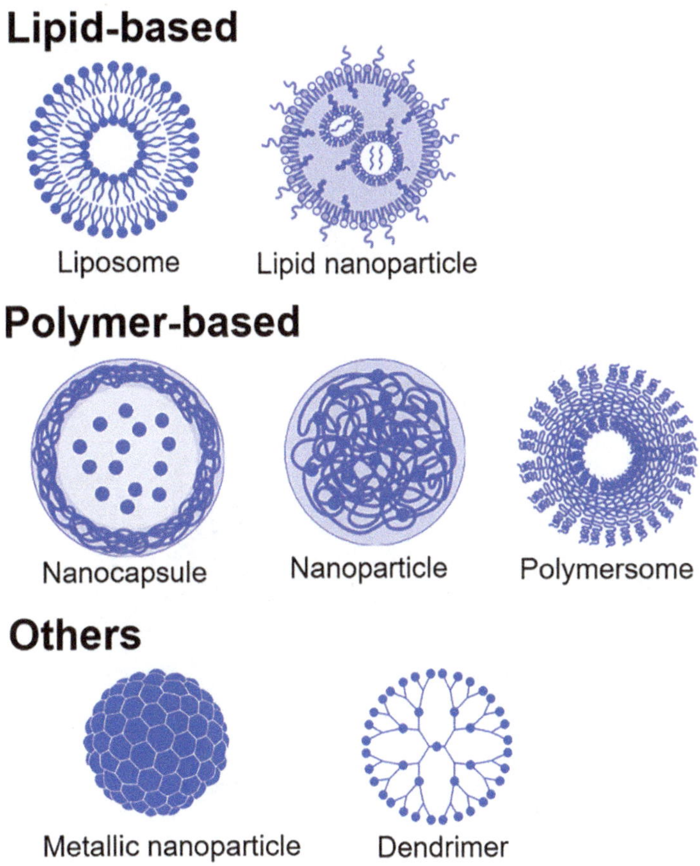

Fig. 6.3 Different types of nanoscale carriers for gene delivery

mechanisms of endosomal escape include membrane destabilization, nanocarrier swelling, and membrane fusion (Fig. 6.4) [14].

Genome editing refers to processes by which the genome sequence is altered by adding (insertions), replacing (substitutions), or removing (deletions) DNA base pairs. The term *genome editing* is preferred over *gene editing* because the modifications are not limited to genes themselves but may also target regulatory regions that control gene expression [15, 16]. Three technologies—clustered regularly interspaced short palindromic repeats (CRISPR)-CRISPR-associated protein 9 (Cas9), transcription activator-like effector nucleases (TALENs), and zinc-finger nucleases (ZFNs)—have driven a revolution in genome editing [17, 18]. Among these, the CRISPR/Cas9 system has rapidly evolved from its initial discovery as part of an adaptive immune system in bacteria to a versatile two-component genome editing tool applicable to mammalian tissues (Fig. 6.5) [19]. CRISPR/Cas9 technology offers several advantages: it is cost effective, relatively simple to perform, and has strong potential for multiplexed genome editing [20].

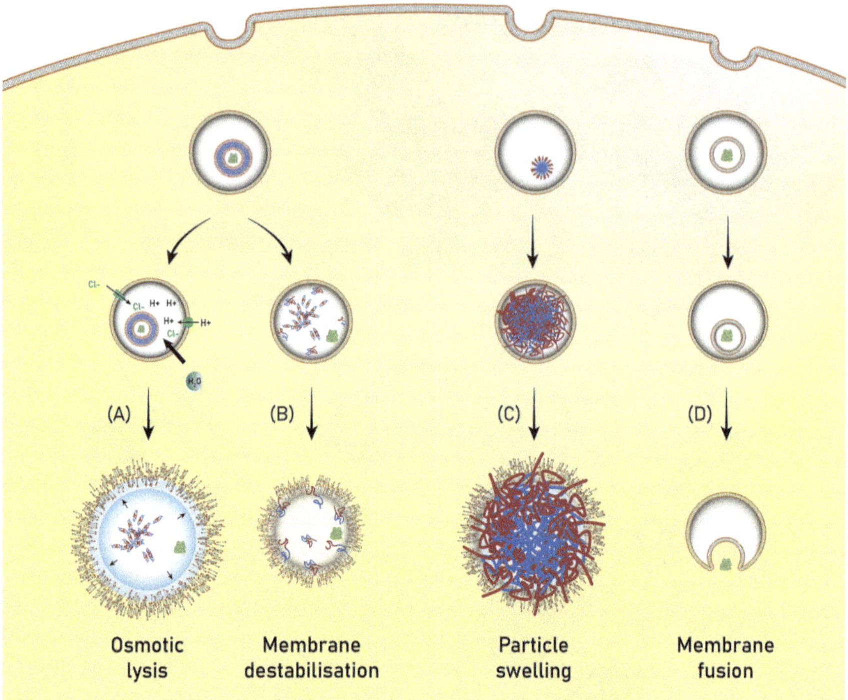

Fig. 6.4 Schematic representation of proposed endosomal escape mechanisms of nanocarriers such as polymersomes, liposomes and nanoparticles. (**a**): osmotic membrane rupture *via* the "proton sponge" effect. (**b**): disassembly of pH/reduction-responsive polymersomes that destabilize the endosomal/lysosomal membrane. (**c**): swelling of pH-responsive nanocarriers inducing mechanical stress on the membrane. Swelling may also contribute to osmotic rupture/"proton sponge" effect. (**d**): fusion of the nanocarrier with the membrane. (Adapted from [14])

Chimeric antigen receptor (CAR)-T cell therapy has made remarkable breakthroughs in treating cancers, particularly hematologic malignancies. However, its broader application in cancer treatment is hindered by multiple challenges. Consequently, there is hence a rapidly growing interest in applying CRISPR, unprecedentedly efficient tools for genomic and cellular reprogramming, to engineer CAR-T cells [21–23]. In particular, CRISPR shows strong potential for enhancing the efficacy of CAR-T cells while mitigating their adverse events [22]. CARs are synthetic modular proteins designed to redirect immune cells toward specific target antigens. First-generation CARs consist of a single-chain variable fragment (scFv) for antigen recognition, a hinge and transmembrane domain for signal transduction, and an intracellular CD3ζ domain for activation. Second- and third-generation CARs incorporate one and two costimulatory domains, respectively, which enhance proliferative potential and antitumor effects. Fourth-generation CARs, known as TRUCKs (T cells redirected for universal cytokine killing) or

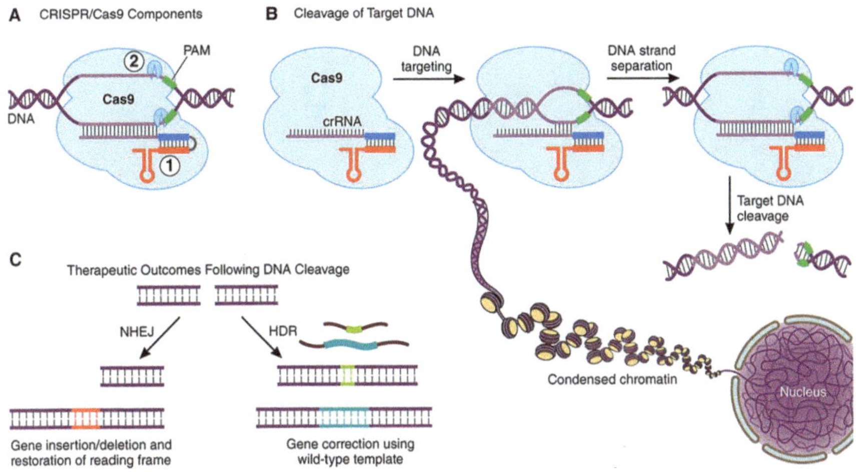

Fig. 6.5 The nuts and bolts of the CRISPR (Clustered Regularly Interspaced Palindromic Repeat)/Cas9. (**a**) the CRISPR/Cas9 system consists of two essential components: (1) a ribonucleic acid guide sequence that targets host deoxyribonucleic acid (DNA) containing a protospacer adjacent motif (PAM) and (2) the Cas9 endonuclease enzyme. (**b**) DNA cleavage involves target recognition within the complex genomic milieu, followed by DNA strand separation and precise enzymatic scission of the DNA backbone. (**c**) after DNA cleavage, non-homologous end joining (NHEJ) may restore an out-of-frame transcript without requiring a DNA template, whereas the homology-directed repair (HDR) pathway can use an error-free template to correct a gene mutation. CrRNA: CRISPR RNA. (Adapted from [19] with permission from Wolters Kluwer Health, Inc.)

"armored" CAR, are further engineered to secrete cytokines that enhance functionality and therapeutic potential (Fig. 6.6a) [21]. A typical CAR-T therapy begins with collection of a patient's T cells, which are then engineered into CAR-T cells. Following expansion, the CAR-T cells are reinfused into the patient to target and kill cancer cells (Fig. 6.6b) [21, 24]. Recently, *in vivo* CAR-T cell generation for cancer and autoimmune diseases have also been reported [25]. In an innovative approach, T cells and bacteria were engineered to collaborate in tumor eradication. A nonpathogenic strain of *Escherichia coli* (a tumor-colonizing, probiotic bacterium) was engineered and administered to deliver synthetic antigens to the tumor microenvironment, effectively "tagging" the tumor. The CAR-T cells were programmed (i.e., probiotic-guided CAR-T cells) to recognize these tags and kill the tumor cells. This novel platform broadens the scope of CAR-T cell therapy to include difficult-to-target tumors [26, 27].

Since 2017, the U.S. Food and Drug Administration has approved several CAR-T cell therapies (Table 6.2) [28, 29]. Currently, hundreds of autologous and allogenic CAR-T and CAR-natural killer (NK) products are in clinical trials worldwide [25]. In 2024, the FDA approved the first cellular therapy for a solid tumor (skin cancer melanoma) -Lifileucel (Amtagvi), which utilizes tumor-infiltrating lymphocytes

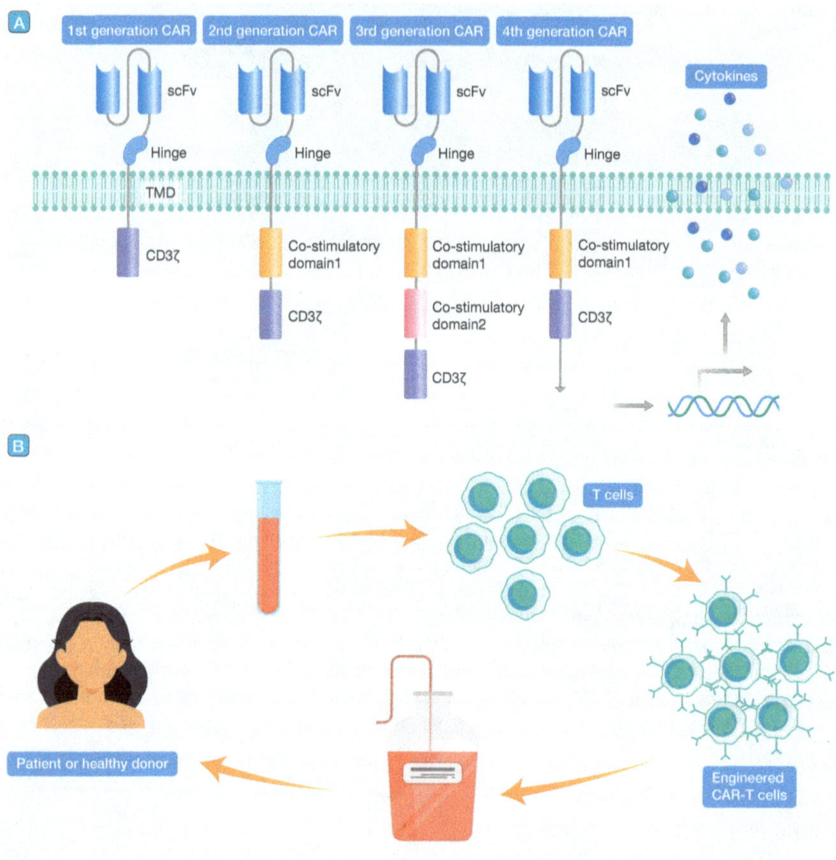

Fig. 6.6 Overview of CAR and CAR-T cell therapy. (**a**) structural illustration of the four generations of CARs. (**b**) schematic overview of the CAR-T cell therapy procedure. TMD: transmembrane domain, scFv: single-chain variable fragment, CAR: chimeric antigen receptor. (Adapted from [21] with permission from Elsevier)

collected from the patient's tumor rather than circulating blood (as in CAR-T) [30]. In the same year, the FDA also approved the first T-cell receptor (TCR) therapy, TECELRA (afamitresgene autoleucel), for metastatic synovial sarcoma. A key distinction between CAR-T cell therapy and TCR therapy lies in the receptors employed: CAR-T utilizes a synthetic receptor, whereas TCR therapy relies on a modified form of a naturally occurring receptor [31].

Table 6.2 FDA-approved CAR-T cell therapies

Product and trade name	Cell type	Target antigen	Targeted disease(s)
YESCARTA (axicabtagene ciloleucel)	Autologous T cell	CD19	Large B-cell lymphoma; follicular lymphoma
KYMRIAH (tisagenlecleucel)	Autologous T cell	CD19	B-cell precursor acute lymphoblastic leukemia; large B-cell lymphoma; follicular lymphoma
TECARTUS (brexucabtagene autoleucel)	Autologous T cell	CD19	B-cell precursor acute lymphoblastic leukemia; mantle cell lymphoma
BREYANZI (lisocabtagene maraleucel)	Autologous T cell	CD19	Large B-cell lymphoma; chronic lymphocytic leukemia or small lymphocytic lymphoma; follicular lymphoma; mantle cell lymphoma
ABECMA (idecabtagene vicleucel)	Autologous T cell	B-cell maturation antigen	Multiple myeloma
CARVYKTI (ciltacabtagene autoleucel)	Autologous T cell	B-cell maturation antigen	Multiple myeloma
AUCATZYL (obecabtagene autoleucel)	Autologous T cell	CD19	B-cell precursor acute lymphoblastic leukemia

Note: This table is subjected to change and is generated based on [28, 29]

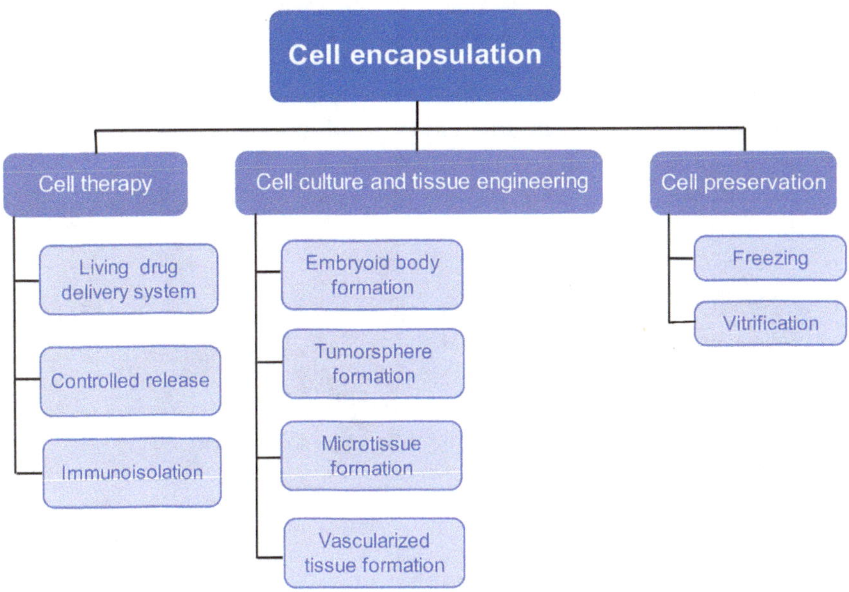

Fig. 6.7 Biomedical applications of cell encapsulation

6.3 Cell Encapsulation

Cell encapsulation refers to the entrapment of live cells within a semi-permeable matrix or device [32]. It has been applied in various biomedical applications (Fig. 6.7). Based on system sizes, cell encapsulation can be categorized into cell macroencapsulation and cell microencapsulation. One major challenge of macroencapsulation is its low surface area/volume ratio, which requires large amounts of nutrients and oxygen or vascularization to support encapsulated cells. Recently, the FDA approved an encapsulated cell therapy for adults with idiopathic macular telangiectasia type 2—ENCELTO. A single-dose implant containing 200,000 to 440,000 allogenic retinal pigment epithelial cells expressing recombinant human ciliary neurotrophic factor (rhCNTF). ENCELTO is an opaque semi-permeable capsule with a width of 1.2 ± 0.1 mm, length of 6.1 ± 0.4 mm, and internal diameter of 0.88 ± 0.02 mm [33] (Fig. 6.8).

Cell microencapsulation has shown great promise in biomedical applications such as cell and gene therapy, cell culture and tissue engineering (e.g., embryoid body and tumorsphere formation; see Chap. 4), and cell cryopreservation (see Chap. 5). First introduced in 1980 by Franklin Lim and Anthony M. Sun, embedding of islets in alginate microspheres for diabetes treatment was demonstrated [34]. For cell therapies, microcapsules/microspheres protect encapsulated cells from both the host immune system and mechanical stresses, while allowing free diffusion of nutrients and metabolic wastes necessary for cells to survive. Moreover, microencapsulation systems can be engineered to enable controlled/sustained release of therapeutic agents produced by the encapsulated cells (Fig. 6.9) [35]. Clinical applications include treatment for type 1 diabetes as well liver, cardiac and neural tissue repair [36]. Both allogenic and xenogeneic cells have been incorporated into microencapsulation systems. However, in xenotransplantation, simple microencapsulation systems are often insufficient to avoid immune rejection (Fig. 6.10)

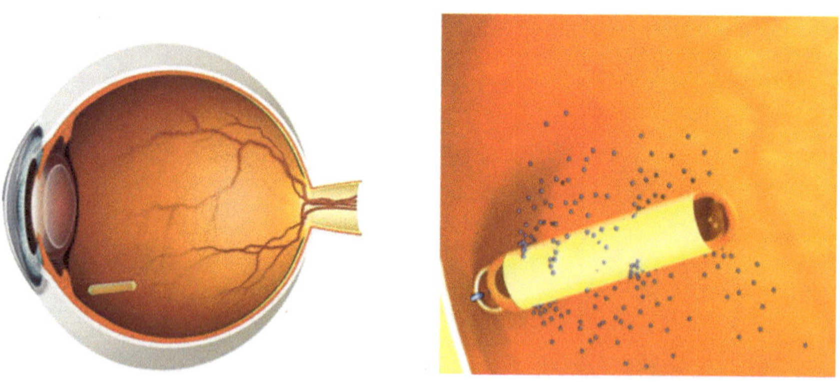

Fig. 6.8 Neurotech Pharmaceuticals' Encapsulated Cell Thearpy (ECT) implant provides constant delivery of thereapeutic proteins to the eye. (Available from: https://www.neurotechpharmaceuticals.com/science-technology/, reprinted with permission from Neurotech Pharmaceuticals Inc.)

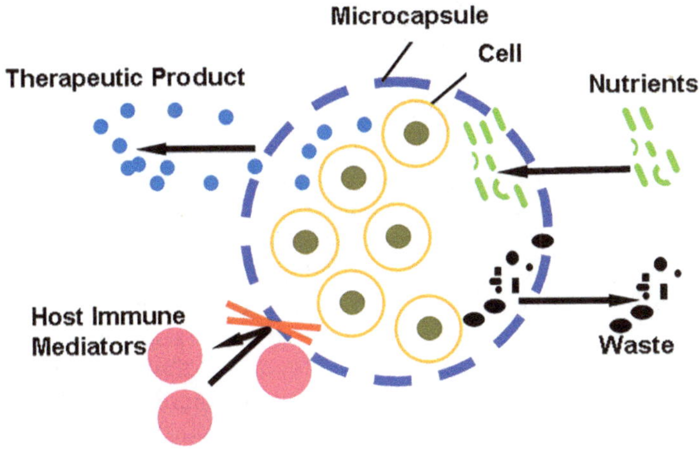

Fig. 6.9 Schematic illustration of the concept of cell microencapsulation as a living drug delivery system. (Adapted from [35])

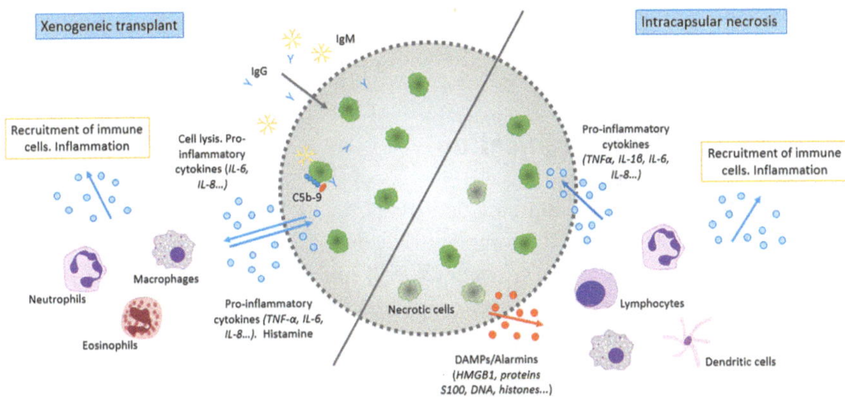

Fig. 6.10 Pro-inflammatory molecule secretion from the microcapsules activates the immune response in different scenarios. Xenogeneic transplant: some antibodies can enter the capsule or contact the encapsulated cells located in the most superficial layers of the implant. These antibodies recognize sequences absent in the host species, leading to the activation of complement pathways. This activation causes direct cell lysis and the release of molecules that promote inflammation and the recruitment of immune cells, such as neutrophils, to the implant site. Recruited cells release cytokines that can cross the microcapsule membrane and damage encapsulated cells. In advanced phases, fibrotic processes may isolate the implant, compromising nutrients and oxygen supply to the capsule interior. Intracapsular necrosis: necrotic cells release damage-associated molecular patterns (DAMPs) or alarmins, such as high-mobility group box-1 (HMGB1), heat shock proteins (HSPs), S100 proteins, and DNA/RNA fragments to the extracellular fluid. These small molecules can diffuse outside the microcapsules and activate immune cells, such as macrophages, dendritic cells, neutrophils or lymphocytes, *via* pattern recognition receptors (PRRs), such as TLRs. Activated immune cells, in response, secrete pro-inflammatory cytokines, such as IL-1β, IL-6, interleukin-8 (IL-8), or TNF-α, which induce inflammation and recruit more immune cells to the area. These molecules can penetrate microcapsules and damage the encapsulated cells. Persistent exposure may activate the adaptive immune response. (Adapted from [32])

[32]. Both natural and synthetic polymers have been applied in cell microencapsulation, including alginate (most widely used), collagen, agarose, chitosan, cellulose, dextran, gelatin, hyaluronic acid, poly(vinyl alcohol), polyethylene glycol, and poly(lactic-*co*-glycolic acid) [35–38]. Common fabrication methods include extrusion (e.g., electrostatic and vibrational extrusion), emulsion, lithography, and microfluidics [39].

Recent advances in cell microencapsulation include the use of interfacial assembly of metal-phenolic network nanofilms to maintain stem cell stemness in an inflammatory microenvironment [40]. Bone marrow-derived mesenchymal stromal cells have been encapsulated in type I collagen-based microcapsules for intramyocardial delivery in a murine model of myocardial infarction [41]. A pancreatic extracellular matrix (pECM)-based microencapsulation platform for islets significantly improved glucose tolerance and achieved glycemic correction without any immunosuppression (Fig. 6.11) [42]. The combination of cell microencapsulation with microfluidic chips for drug evaluation has also gained increasing interest [43, 44]. For example, primary colorectal cancer cells were encapsulated in hydrogel microcapsules with carboxymethyl cellulose cores and alginate shells to form spheroids. The spheroids were then exposed to dynamic drug gradients with the chip, enabling high-throughput chemotherapy screening (Fig. 6.12) [44].

6.4 Cell Surface Modification

The cell surface is a highly complex and dynamic environment comprising lipids, proteins and carbohydrates. Its composition governs all interactions between the cell and its environment, and consequently, cell functions such as adhesion, migration and cell-cell interactions can be controlled through cell surface modification [45]. Chemical approaches to cell surface modification can be either covalent or non-covalent [45–47], while biological approaches include genetic engineering, enzyme-mediated cell surface functionalization, and metabolic engineering [47] (Fig. 6.13).

Starting with chemical approaches, cationic polymers have been used for temporary single-cell coating to enhance stability under *in vitro* conditions, control kinetics of cell proliferation, and support bioprocessing applications [48, 49]. Modification of the cell surface with single-stranded DNA (ssDNA)-poly(ethylene glycol)

Alg-Is and pECM-Is, 3 days post-encapsulation. (**c**) bright-field micrographs of 7-day cultured Alg-Is and pECM-Is. (**d**) vinculin (green) expression in the encapsulated islets, stained for actin (red, upper panel) or counterstained for nuclei (grey, bottom panel) 14 days post-encapsulation. (**e**) insulin (red) and collagen I (white) expression in 14-day cultured encapsulated islets. (**f**): PDX1 (green) and insulin (red) within 14-day cultured encapsulated islets. (**g**): MAFA (green) and insulin (red) within 14-day cultured encapsulated islets. Scale bars: (**a, c–e**) 20 μm; (**b, f–g**) 50 μm. pECM: porcine pancreatic extracellular matrix, N-Is: non-encapsulated islets, Alg-Is: islets entrapped withing alginate microcapsules, pECM-Is: islets entrapped withing pECM microcapsules. (Adapted from [42] with permission from Elsevier)

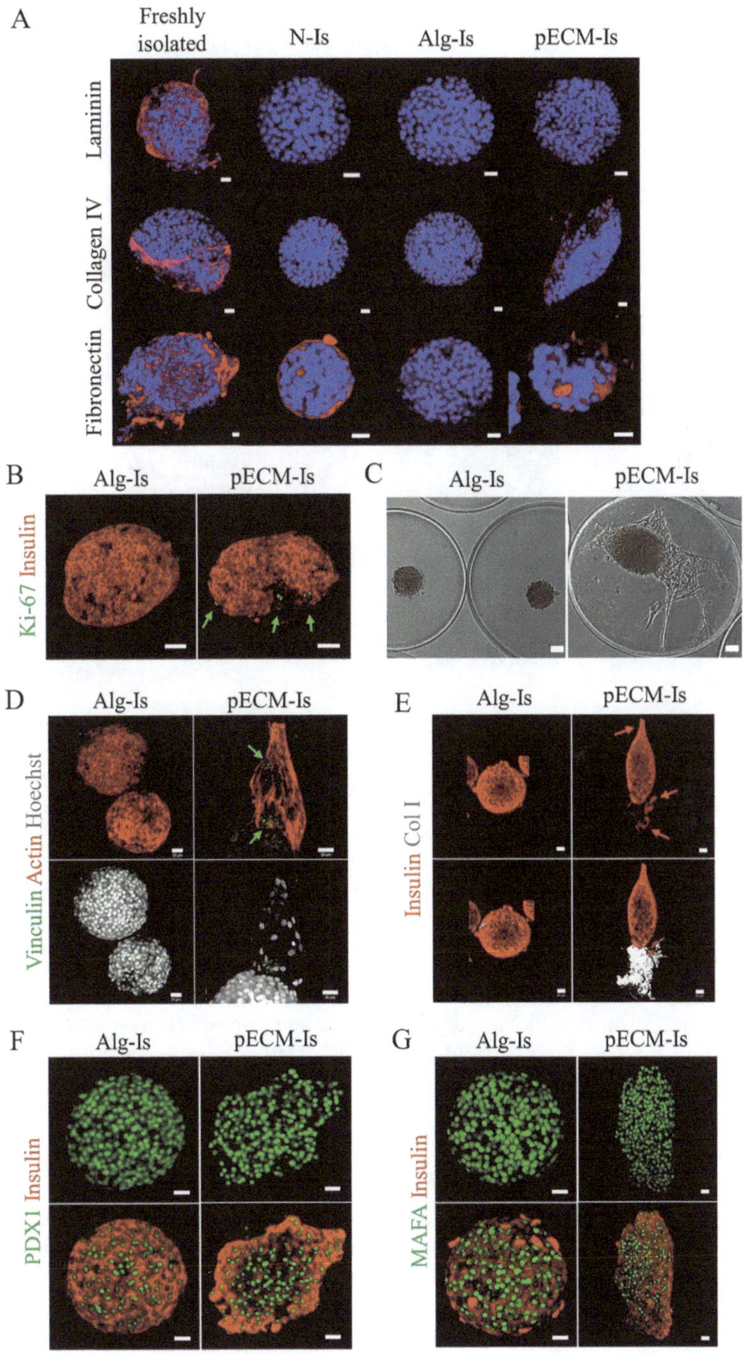

Fig. 6.11 Impact of pECM encapsulation on islet composition and structure. (**a**) representative fluorescence images of laminin, collagen IV, and fibronectin (red) within freshly isolated islets and 14-day cultured N-Is, Alg-Is, and pECM-Is. (**b**) insulin (red) and Ki-67 (green) expression within

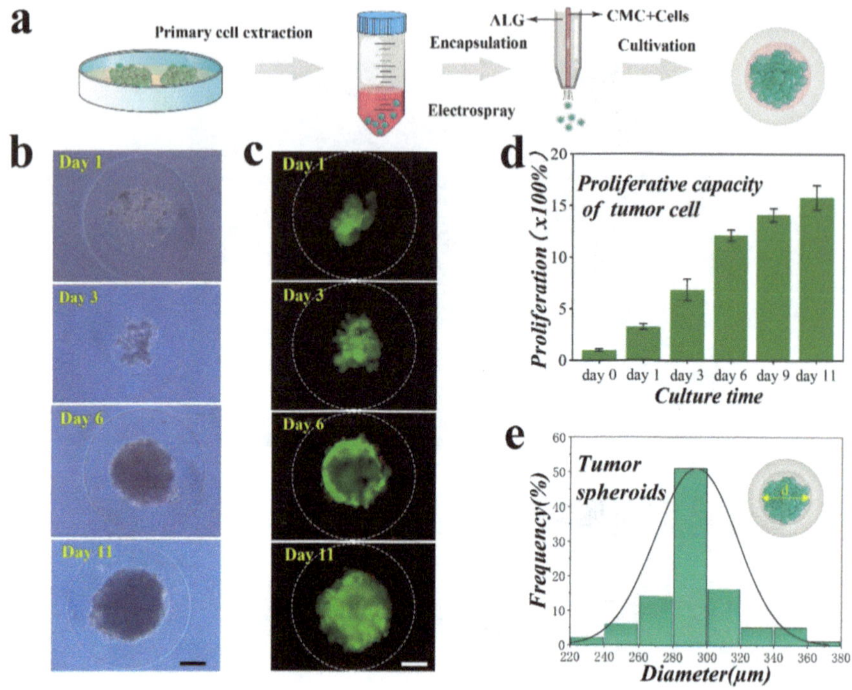

Fig. 6.12 Preparation and characterization of colorectal cancer cell (CRC) hydrogel microcapsules. (**a**) schematic diagram of primary CRC extraction and microfluidic electrospray preparation of tumor microcapsules. (**b**) microscope images of primary CRC microcapsules after incubation for 1, 3, 6, and 11 days. (**c**) fluorescent images of live/dead staining of CRC microcapsules after incubation for 1, 3, 6, and 11 days. (**d**) quantitative analysis of cell proliferation in CRC microcapsules after incubation for 1, 3, 6, 9, and 11 days. (**e**) size distribution of CRC microcapsules at day 11. Scale bars: 100 μm (**b, c**). ALG: alginate, CMC: carboxymethyl cellulose. (Adapted from [44] with permission from Elsevier)

(PEG)-lipid has enabled the analysis of intercellular interactions between different cells by incorporating complementary ssDNA strands on different cells to induce cell attachment through DNA hybridization [50]. Cell-penetrating peptide (CCP)-PEG-lipid has been used to induce cell adhesion (Fig. 6.14) [51, 52], and promote cell fusion [53].

Biologically-driven cell surface modification approaches offer additional versatility. Genetic engineering leverages the cell's biosynthetic machinery to express exogenous proteins on the cell surface. Enzyme-mediated surface modifications use enzymes to recognize and transform naturally occurring or genetically inserted substrates. Moreover, exogenous enzymes catalyze reactions in the presence of specific substrates, enabling *in situ* modification with functional moieties. Metabolic engineering exploits intrinsic cell metabolism and native biosynthetic machinery, such as glycan biosynthesis pathways, to incorporate chemically functionalized

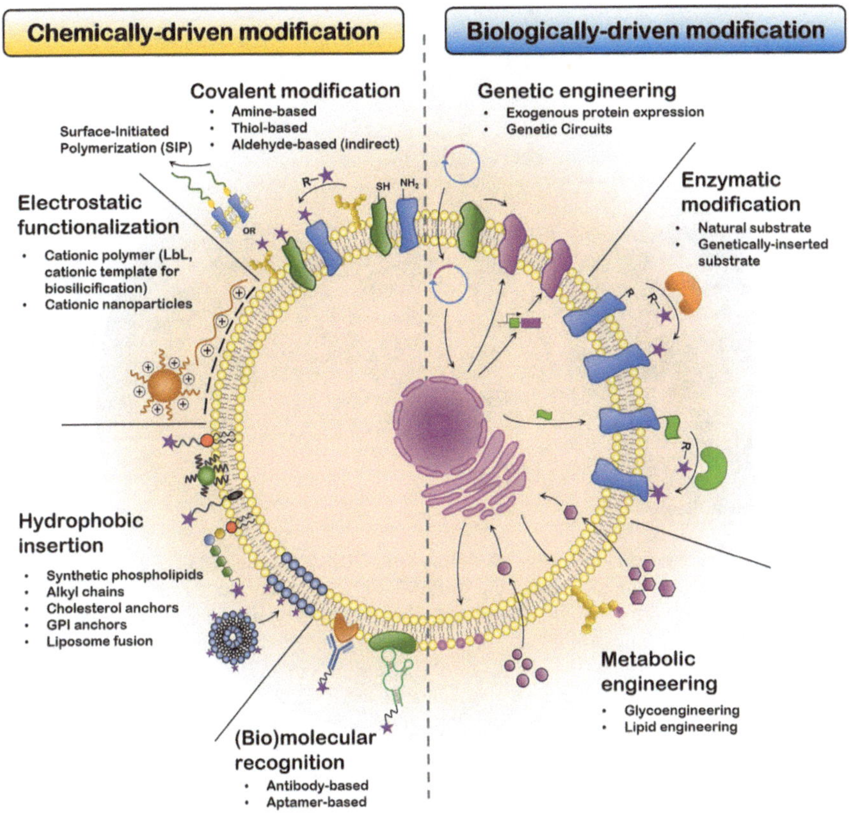

Fig. 6.13 Schematics of currently available cell surface engineering toolboxes, including chemically- and biologically-driven approaches. Chemically-driven approaches exploit the naturally occurring chemical and biophysical features of the cell surface. Biologically-driven approaches use intrinsic biological machinery or bioactive mediators (e.g., enzymes) to install added functionalities in living cells. These toolsets enable the processing of cells into randomly aggregated or spatiotemporally programmed living assemblies. (Adapted from [47])

precursors into biomacromolecules on the cell surface [47]. For example, Shengjie Liu et al. combined metabolic oligosaccharide engineering and click chemistry to construct fibrinolytic cells surfaces. HeLa cells were incubated with the azido-sugar Ac4ManNAz to generate azido-labelled cells, which were then incubated with a polymer *via* the biorthogonal azide-dibenzocyclooctyne (DBCO) "click" reaction, resulting in surface-modified cells (Fig. 6.15) [54]. The advantages of using metabolic oligosaccharide engineering for cell surface modification include cytocompatibility, a range of biorthogonal functional groups for selection, and broad applicability across diverse cell types [55].

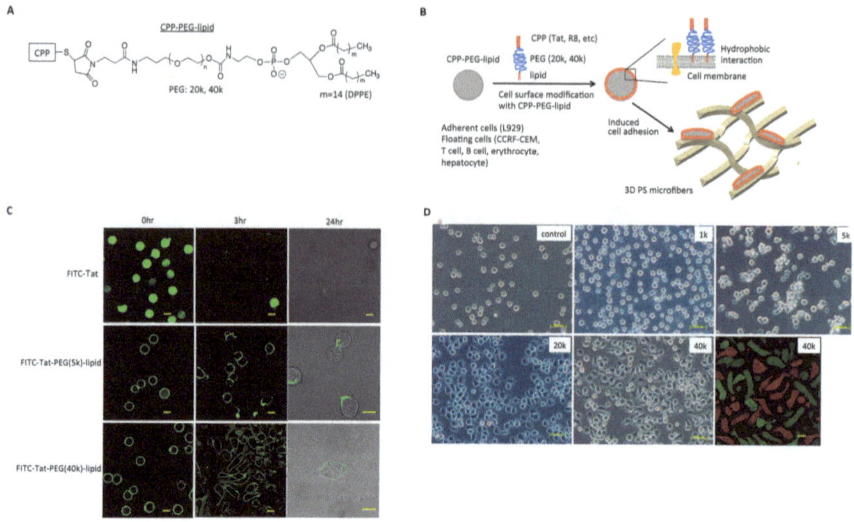

Fig. 6.14 Influence of cell-surface modification with CPP-conjugated PEG-lipids on cell adhesion. (**a**) chemical structure of amphiphilic CPP-PEG-lipids. (**b**) schematic representation of cell-surface modification with CPP-PEG-lipids *via* hydrophobic interactions and the induction of firm cell adhesion to 3D PS microfibers. (**c**) confocal laser-scanning microscopy of CCRF-CEM cells after treatment with FITC-Tat (top), FITC-Tat-PEG(5k)-lipids (middle), and FITC-Tat-PEG(40k)-lipids (bottom). Scale bars: 10 μm. (**d**): influence of the PEG chain length on cell adhesion of CCRF-CEM cells modified with CPP-PEG-lipids. Phase contrast images of CCRF-CEM cells 3 h after modification with Tat-PEG-lipids with 1 kDa, 5 kDa, 20 kDa, or 40 kDa PEG. Nonmodified CCRF-CEM cells serve as the control. For Tat-PEG(40k)-lipids, cells were fluorescently labelled with Celltracker Green and Orange, mixed, and treated with Tat-PEG(40k)-lipids for confocal laser-scanning microscope observation on glass-bottom dishes. Scale bars: 50 μm (phase contrast) and 10 μm (confocal). CPP: cell-penetrating peptide, PEG: poly(ethylene glycol), Tat: transactivator of transcription peptide (a representative CPP), PS: polystyrene, FITC: fluorescein isothiocyanate. (Adapted with permission from [51]. Copyright© 2017 American Chemical Society)

Review and Practice Questions

1. What are the different cell engineering approaches? Provide an example of each approach.
2. What is gene delivery?
3. Compare different gene delivery approaches.
4. Why is gene delivery using nanocarriers becoming increasingly popular? And what are some mechanisms of endosomal escape for nanocarriers?
5. What is genome editing? What is the working mechanism of the CRISPR/Cas system in genome editing?
6. What is the typical CAR-T therapy process? Compare CAR-T therapy with TCR (T-cell receptor) therapy.

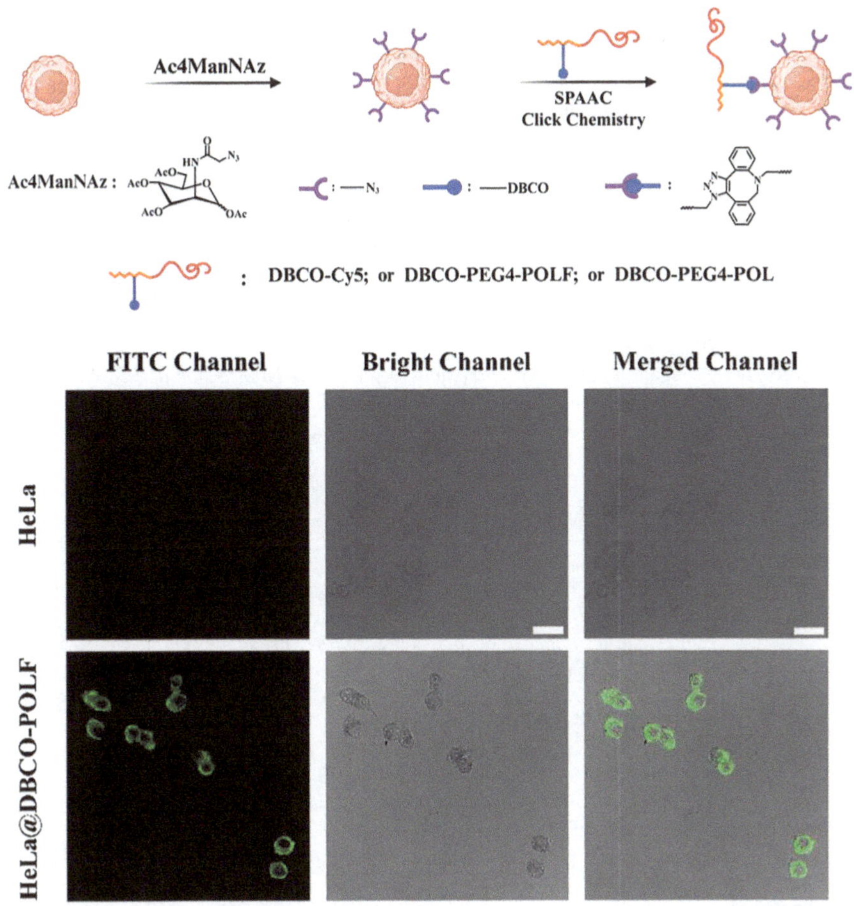

Fig. 6.15 Top: scheme of cell surface modification using DBCO-labeled copolymer. Bottom: confocal microscopic images of HeLa cells and HeLa@DBCO-POLF cells after 3 days of incubation with Ac4ManNAz and 15 min staining with DBCO-PEG4-POLF. scale bar: 50 μm. DBCO: dibenzocyclooctyne, PEG4: 4-unit ethylene glycol spacer arm, POL: poly(oligoethylene glycol methacrylate)-*co*-(Lys), POLF: POL- fluorescein isothiocyanate. (Adapted with permission from [54]. Copyright© 2024 American Chemical Society)

7. Compare cell macroencapsulation with cell microencapsulation.
8. What are some biomedical applications of cell encapsulation?
9. Conduct a literature review and write a short summary of a recent biomedical application of cell microencapsulation not covered in this chapter.
10. What are the different cell surface modification approaches?
11. Explain the process and advantages of metabolic oligosaccharide engineering for cell surface modification.

References

1. Liang, K. and Y. Du, *Cell engineering techniques improve pharmacology of cellular therapeutics.* Biomaterials and Biosystems, 2021. **2**:100016.
2. Shahlaei, M., S.M. Asl, and M. Saeidifar, *Chapter 5—Nanotechnology in gene delivery for neural regenerative medicine*, in *Neural Regenerative Nanomedicine*, M. Razavi, Editor. 2020, Academic.
3. Puhl, D.L., A.R. D'Amato, and R.J. Gilbert, *Challenges of gene delivery to the central nervous system and the growing use of biomaterial vectors.* Brain Research Bulletin, 2019. **150**:216-230.
4. Capes-Davis, A., et al., *22. Genetic Modification and Immortalization.* Eighth edition. ed. Freshney's culture of animal cells: a manual of basic technique and specialized applications. 2021, Wiley Blackwell.
5. Fus-Kujawa, A., et al., *An Overview of Methods and Tools for Transfection of Eukaryotic Cells in vitro.* Frontiers in Bioengineering and Biotechnology, 2021. **9**:701031.
6. Bulcha, J.T., et al., *Viral vector platforms within the gene therapy landscape.* Signal Transduction and Targeted Therapy, 2021. **6**(1):53.
7. Nayerossadat, N., T. Maedeh, and P.A. Ali, *Viral and nonviral delivery systems for gene delivery.* Advanced Biomedical Research, 2012. **1**:27.
8. S, P.R., et al., *Permanently Charged Cationic Lipids—Evolution from Excipients to Therapeutic Lipids.* Small Science, 2024. **4**(7):2300270.
9. Liu, Z., et al., *Hydrophobic modifications of cationic polymers for gene delivery.* Progress in Polymer Science, 2010. **35**(9):1144-1162.
10. Behzadi, S., et al., *Cellular uptake of nanoparticles: journey inside the cell.* Chemical Society Reviews, 2017. **46**(14):4218-4244.
11. Ahmad, A., et al., *Endolysosomal trapping of therapeutics and endosomal escape strategies.* Drug Discovery Today, 2024. **29**(8):104070.
12. Vermeulen, L.M.P., et al., *The proton sponge hypothesis: Fable or fact?* European Journal of Pharmaceutics and Biopharmaceutics, 2018. **129**:184-190.
13. Casper, J., et al., *Polyethylenimine (PEI) in gene therapy: Current status and clinical applications.* Journal of Controlled Release, 2023. **362**:667-691.
14. Gouveia, M.G., et al., *Polymersome-based protein drug delivery—quo vadis?* Chemical Society Reviews, 2023. **52**(2):728-778.
15. National Academies of Sciences, Engineering, and Medicine, *Human Genome Editing: Science, Ethics, and Governance.* 2017, The National Academies Press.
16. Jayachandran, M., Z. Fei, and S. Qu, *Genetic advancements in obesity management and CRISPR–Cas9-based gene editing system.* Molecular and Cellular Biochemistry, 2023. **478**(3):491-501.
17. Gaj, T., et al., *Genome-Editing Technologies: Principles and Applications.* Cold Spring Harbor Perspectives in Biology, 2016. **8**(12): a023754.
18. Almeida-Porada, G. and C.D. Porada, *Chapter 57—Prenatal Cell- and Gene-Based Therapies for Regenerative Medicine*, in *Principles of Regenerative Medicine (Third Edition)*, A. Atala, et al., Editors. 2019, Academic.
19. Munshi, N.V., *CRISPR (Clustered Regularly Interspaced Palindromic Repeat)/Cas9 System.* Circulation, 2016. **134**(11):777-779.
20. Mani, I., *Chapter Twelve—Genome editing in cardiovascular diseases*, in *Progress in Molecular Biology and Translational Science*, V. Singh, Editor. 2021, Academic.
21. Yan, L., et al., *Boosting CAR-T cell therapy with CRISPR technology.* hLife, 2024. **2**(8):380-396.
22. Lei, T., et al., *Leveraging CRISPR gene editing technology to optimize the efficacy, safety and accessibility of CAR T-cell therapy.* Leukemia, 2024. **38**(12):2517-2543.
23. Dimitri, A., F. Herbst, and J.A. Fraietta, *Engineering the next-generation of CAR T-cells with CRISPR-Cas9 gene editing.* Molecular Cancer, 2022. **21**(1):78.
24. Couzin-Frankel, J., *The counterattack.* Science, 2024. **386**(6723):719-723.

25. Hunter, T.L., et al., *In vivo CAR T cell generation to treat cancer and autoimmune disease.* Science, 2025. **388**(6753):1311-1317.
26. Vincent, R.L., et al., *Probiotic-guided CAR-T cells for solid tumor targeting.* Science, 2023. **382**(6667):211-218.
27. Bressler, E.M. and W.W. Wong, *Engineered bacteria guide T cells to tumors.* Science, 2023. **382**(6667):154-155.
28. Bhaskar, S.T., et al., *Overview of approved CAR-T products and utility in clinical practice.* Clinical Hematology International, 2024. **6**(4):93-99.
29. U.S. Food & Drug Administration. *Approved Cellular and Gene Therapy Products*. Available from: https://www.fda.gov/vaccines-blood-biologics/cellular-gene-therapy-products/approved-cellular-and-gene-therapy-products.
30. National Cancer Institute. *First Cancer TIL Therapy Gets FDA Approval for Advanced Melanoma.* Available from: https://www.cancer.gov/news-events/cancer-currents-blog/2024/fda-amtagvi-til-therapy-melanoma.
31. National Cancer Institute. *FDA Approves Engineered Cell Therapy for Advanced Synovial Sarcoma.* Available from: https://www.cancer.gov/news-events/cancer-currents-blog/2024/fda-tecelra-synovial-sarcoma-mage-a4.
32. Lopez-Mendez, T.B., et al., *Cell microencapsulation technologies for sustained drug delivery: Latest advances in efficacy and biosafety.* Journal of Controlled Release, 2021. **335**:619-636.
33. *Package Insert—ENCELTO.* Available from: https://www.fda.gov/media/185726/download?attachment.
34. Lim, F. and A.M. Sun, *Microencapsulated Islets as Bioartificial Endocrine Pancreas.* Science, 1980. **210**(4472):908-910.
35. Zhang, W. and X. He, *Microencapsulating and Banking Living Cells for Cell-Based Medicine.* Journal of Healthcare Engineering, 2011. **2**(4):875406.
36. Marikar, S.N., et al., *Microencapsulation-based cell therapies.* Cellular and Molecular Life Sciences, 2022. **79**(7):351.
37. Abbas, S.E.M., et al., *Mesenchymal Stem/Stromal Cells Microencapsulation for Cell Therapy.* Cells, 2025. **14**(3):149.
38. pezhman, a., *Biomaterial in Microencapsulation: How Microencapsulation is Changing the Medicine World*, in *Biomaterials in Microencapsulation*, A. Sharma, Editor. 2024, IntechOpen.
39. Lee, S.-Y., et al., *Polysaccharide-Based Hydrogels for Microencapsulation of Stem Cells in Regenerative Medicine.* Frontiers in Bioengineering and Biotechnology, 2021. **9**:735090.
40. Zhao, Y., et al., *Stem cell microencapsulation maintains stemness in inflammatory microenvironment.* International Journal of Oral Science, 2022. **14**(1):48.
41. Jacques, E., et al., *Collagen-Based Microcapsules As Therapeutic Materials for Stem Cell Therapies in Infarcted Myocardium.* ACS Biomaterials Science & Engineering, 2020. **6**(8):4614-4622.
42. Krishtul, S., et al., *ECM-based bioactive microencapsulation significantly improves islet function and graft performance.* Acta Biomaterialia, 2023. **171**:249-260.
43. Song, T., et al., *Primary Human Pancreatic Cancer Cells Cultivation in Microfluidic Hydrogel Microcapsules for Drug Evaluation.* Advanced Science, 2023. **10**(12):2206004.
44. Pan, C., et al., *Primary colorectal cancer cells cultured in hydrogel microcapsules-integrated microfluidic chip for chemotherapy evaluation.* Chemical Engineering Journal, 2025. **519**:164173.
45. Custódio, C.A. and J.F. Mano, *Cell surface engineering to control cellular interactions.* ChemNanoMat, 2016. **2**(5):376-384.
46. Lee, D.Y., et al., *Cell surface engineering and application in cell delivery to heart diseases.* Journal of Biological Engineering, 2018. **12**(1):28.
47. Almeida-Pinto, J., et al., *Cell Surface Engineering Tools for Programming Living Assemblies.* Advanced Science, 2023. **10**(34):2304040.
48. Hong, D. and S.H. Yang, *Cationic Polymers for Coating Living Cells.* Macromolecular Research, 2018. **26**(13):1185-1192.

49. Ribeiro, R.D.C., et al., *Temporary Single-Cell Coating for Bioprocessing with a Cationic Polymer.* ACS Applied Materials & Interfaces, 2017. **9**(15):12967-12974.
50. Teramura, Y., *Cell surface modification with ssDNA–PEG–lipid for analysing intercellular interactions between different cells.* Biomaterials, 2015. **48**:119-128.
51. Teramura, Y., et al., *Cell Adhesion Induced Using Surface Modification with Cell-Penetrating Peptide-Conjugated Poly(ethylene glycol)-Lipid: A New Cell Glue for 3D Cell-Based Structures.* ACS Applied Materials & Interfaces, 2017. **9**(1):244-254.
52. Noiri, M., et al., *Exogenous Cell Surface Modification with Cell Penetrating Peptide-Conjugated Lipids Causes Spontaneous Cell Adhesion.* ACS Applied Bio Materials, 2021. **4**(5):4598-4606.
53. Sato, Y., U.-i. Chung, and Y. Teramura, *Induction of intercellular interaction and cell fusion by cell-penetrating peptide-conjugated lipids.* Journal of Materials Chemistry B, 2025. **13**(39):12546-12556.
54. Liu, S., et al., *Integrating Metabolic Oligosaccharide Engineering and SPAAC Click Chemistry for Constructing Fibrinolytic Cell Surfaces.* ACS Applied Materials & Interfaces, 2024. **16**(28):35874-35886.
55. Tomás, R.M.F. and M.I. Gibson, *100th Anniversary of Macromolecular Science Viewpoint: Re-Engineering Cellular Interfaces with Synthetic Macromolecules Using Metabolic Glycan Labeling.* ACS Macro Letters, 2020. **9**(7):991-1003.

Part II

Tissue Engineering Basics

Introduction to Tissue Engineering

7

Learning Objectives

- Define tissue engineering and regenerative medicine
- Explain and evaluate the importance of tissue engineering from multiple perspectives
- Discuss and analyze the historical development of tissue engineering, highlighting its key milestones and events

7.1 What Is Tissue Engineering?

Mary Shelley's *Frankenstein; or, The Modern Prometheus* was published in 1818 (Fig. 7.1a). One source of inspiration for her work was Luigi Galavni's experiment which demonstrated that a spark could induce muscle contractions in a dead frog (Fig. 7.1b) [1]. Two centuries later, *Science* journal revisited this iconic story in Volume 359, Issue 6372, publishing a series of featured articles reflecting on its scientific and cultural impact [2]. One article posed a provocative question: "*But what if Shelly had written her book today? What technologies might give rise to her iconic creature?*" [3]. The author of this book immediately pointed to the advances in tissue engineering as a possible answer. As shown in Fig. 7.2, one of the most significant achievements in tissue engineering is the development of lab-grown organs.

So, what is tissue engineering? The most widely cited definition comes from Robert Langer and Joseph P. Vacanti: "*Tissue engineering is an interdisciplinary field that applies the principles of engineering and the life sciences toward the development of biological substitutes that restore, maintain, or improve tissue function*" [4]. Various government agencies have also provided definitions that generally reflect their specific missions and focus areas [5]. A more recent definition from the National Institute of Biomedical Imaging and Bioengineering (NIBIB) states that

W. Zhang, *Cell Culture and Tissue Engineering*,
https://doi.org/10.1007/978-3-032-32096-4_7

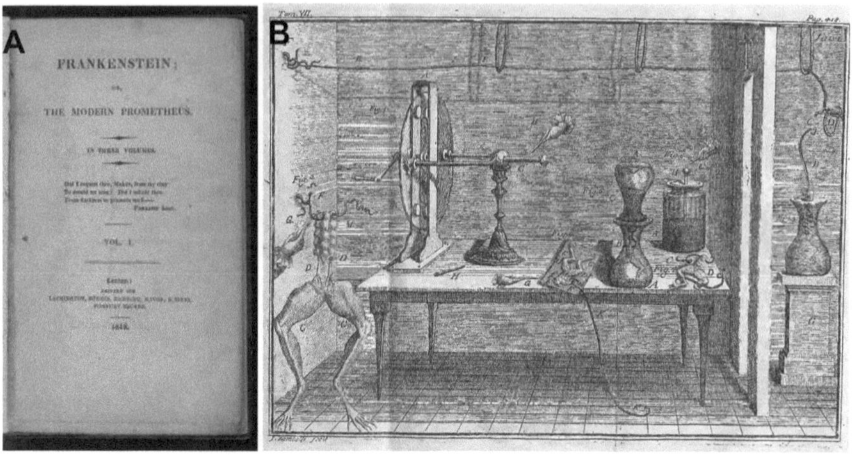

Fig. 7.1 A page view from the 1818 edition of the book *Frankenstein; or, The Modern Prometheus* (**a**; Credit: Library of Congress, Rare Book and Special Collections Division; https://www.loc.gov/resource/rbc0001.2018gen51218v1/?sp=9&r=-1.387,0.834,3.773,1.669,0) and an illustration of Luigi Galvani's Frog's Leg Experiment (**b**; adapted from *De viribus electricitatis in motu musculari commentaries*; https://archive.org/details/AloysiiGalvaniD00Galv))

the goal of tissue engineering is "*to assemble fully functional constructs by combining scaffolds, cells, and biologically active molecules, that restore, maintain, or improve damaged tissue or a whole organ*" [6].

Another term—regenerative medicine—has also been widely adopted and is often used interchangeably with tissue engineering. Together, the fields are frequently referred to by the abbreviation TERM (Tissue Engineering and Regenerative Medicine). However, regenerative medicine is a broad field that encompasses tissue engineering but also incorporates the concept of self-healing—where the body leverages its own systems, sometimes with help from added biological material from outside the body, to recreate cells or rebuild organs [6]. In essence, regenerative medicine is grounded in the understanding of morphogenesis and natural, inherent self-repair mechanisms, and, as such, regenerative medicine typically involves the use of stem and progenitor cells [7].

Giving the evolving nature of the tissue engineering field, it is critical to establish a working definition. One such definition describes tissue engineering as "*a multidisciplinary field bringing together experts from engineering, life sciences and medicine, utilizing the building blocks of cells, biomaterials, bioactive molecules, and bioreactors for the development of three-dimensional artificial tissues and organs which can be used to augment, repair and/or replace damaged and/or diseased tissues/organs*" [5]. This definition emphasizes both the multidisciplinary nature of the field and its core components—building blocks. As illustrated in Fig. 7.3, the four fundamental building blocks of tissue engineering are cells, biomaterials (for scaffolding), bioactive molecules, and bioreactions.

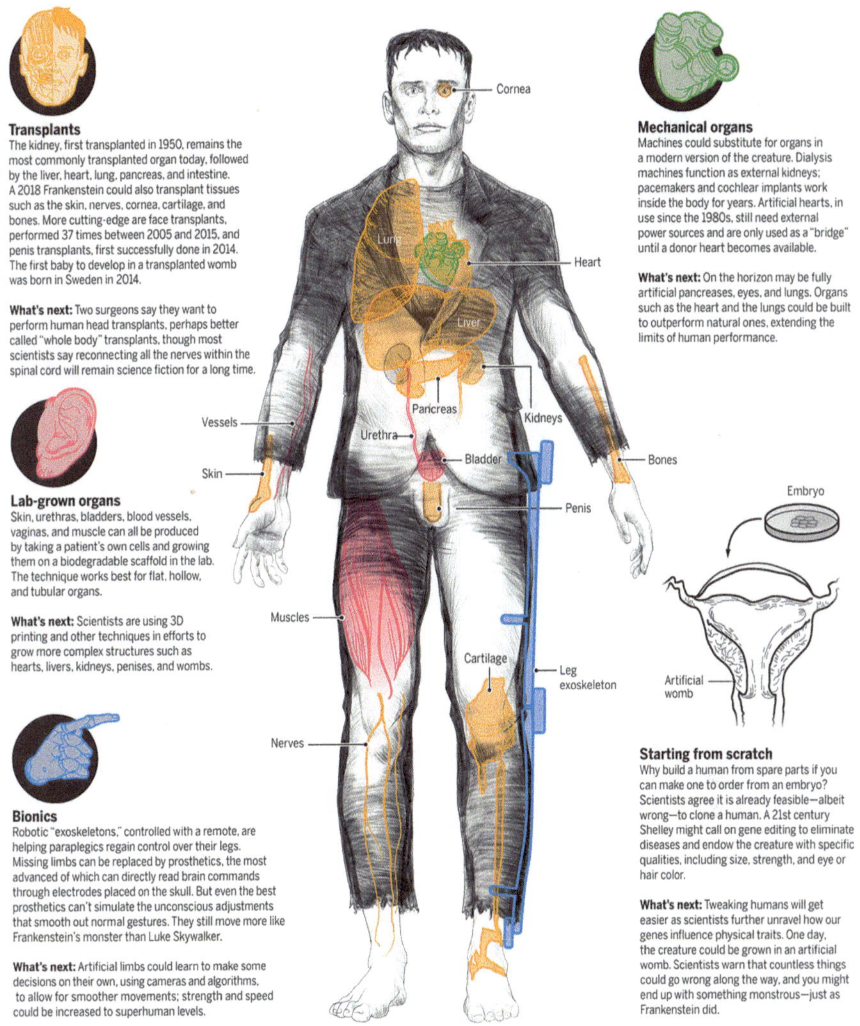

Fig. 7.2 Creating a modern monster. (*Adapted from* [3] *with the permission from the American Association for the Advancement of Science*)

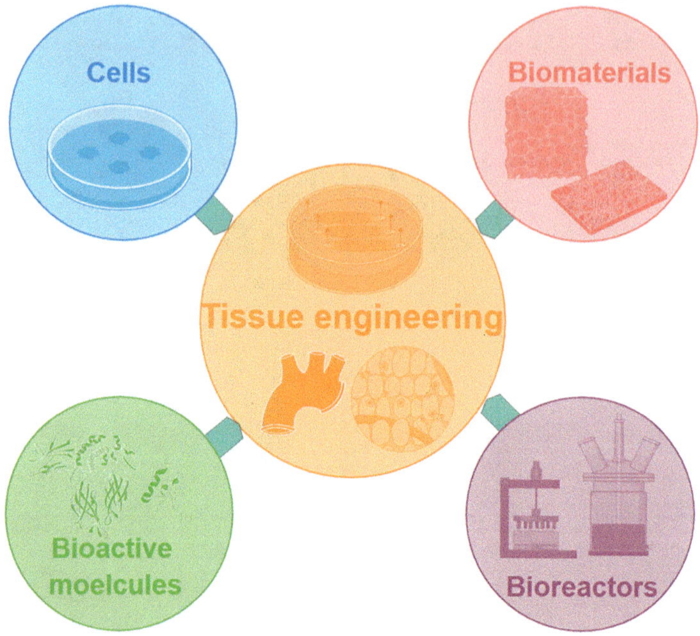

Fig. 7.3 Four building blocks of tissue engineering

Fig. 7.4 Organ donation and transplant waitlist statistics. (*Data from:* https://data.hrsa.gov/topics/health-systems/organ-donation)

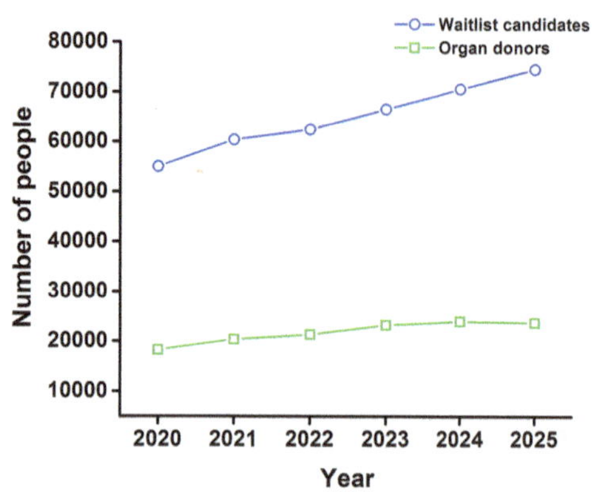

7.2 Why Tissue Engineering?

According to organdonor.gov, a new person is added to the transplant waiting list every 8 min, and 13 people die each day waiting for an organ transplant [8]. Moreover, as shown in Fig. 7.4, the gap between the numbers of patients on the waiting list and available organ donors continues to widen. Even for those fortunate enough to receive a transplant, lifelong use of immunosuppressant drugs is required—often accompanied by severe side effects [9, 10].

Tissue engineering offers a promising alternative to help bridge this gap [11]. Beyond organ replacement, tissue engineering shows significant potential in developing human tissue chips and organoids, which can replace animal models for drug screening and disease research [12, 13]. With advances in stem cell research, particularly induced pluripotent stem cells (iPSCs), it is possible to engineer patient-specific tissues/organs. This progress supports the growing field of personalized medicine or precision medicine, which aims to tailor medical treatment based on a patient's unique clinical, genetic, and environmental characteristics [14].

7.3 Historical Perspectives of Tissue Engineering

The ideas of generating new tissues and restoring body parts or organs have existed in the human imagination since the dawn of history [7]. These concepts are reflected in ancient legends, such as Prometheus and liver regeneration (Fig. 7.5a) and the miraculous transplantation of a limb by the Saints Cosmos and Damien (Fig. 7.5b). In the myth of Prometheus, he was punished by Zeus for stealing fire and giving it to mankind. Chained to a rock in the Caucasus Mountains, Prometheus is tormented daily by an eagle that ate part of his liver—only for it to regenerate overnight,

Fig. 7.5 Prometheus depicted in a sculpture by Nicolas-Sébastien Adam (**a**; *from Wikimedia Commons*; https://commons.wikimedia.org/w/index.php?title=File:Prometheus_Adam_Louvre_MR1745_edit_atoma.jpg&oldid=928522446.) and Saints Cosmas and Damian performing a miraculous cure by transplantation of a leg (**b**; *Credit: Wellcome Library, London;* https://commons.wikimedia.org/w/index.php?title=File:A_verger%27s_dream_Wellcome_L0064583.jpg&oldid=967542944)

forcing him to endure this cycle of eternity. The story has long inspired both artists and scientists. In medicine, it has become a powerful symbol of the liver's regenerative capacity, provoking the question: "*Did the ancient Greeks know about the liver's amazing capacity for self-repair?*" [15, 16]. Regarding the miracle of the leg, the saints are said to have amputated the leg of Deacon Justin and successfully transplanted the leg of a recently deceased Black African man. This event is considered the first pictorial representation of a transplant in history. Interestingly, one article noted inconsistencies in these depictions—specifically regarding which leg (left or right) was transplanted and which part of the leg (above or below the knee). Some paintings even portray the transplantation of the "wrong" leg [17].

In the 1580s, Gaspare Tagliacozzi (1545–1599), a professor in surgery and anatomy at the University of Bologna in Italy, was the first medical doctor to practice and write about plastic surgery (first described the nose reconstruction from the forearm flap), and he was also the first to teach the subject. As such, he is widely considered as the founder of the plastic surgery [7, 18]. Advancements in anesthesia and infection prevention during the nineteenth century accelerated the development of surgical procedures. These breakthroughs enabled the first applications of living tissues and organs to recover from malfunction [7]. In the nineteenth century, the Western world began adopting skin grafting and it has since been used in chronic ulcers treatment, burn care, and reconstructive surgeries [19]. In 1981, Eugene Bell demonstrated a way to repair skin wounds with artificial skin made from a person's or an animal's own cells [20] and is considered as the father of tissue engineering by some [21]. He was also one of the undisputed pioneers in the commercialization of tissue engineering and regenerative medicine and was one of founding team members of the first biotech company—Organogenesis [22].

The field of tissue engineering was born in the mid to late 1980s, largely established and popularized by Joseph Vacanti and Robert Langer. Their pioneering work involved generating functional tissue equivalents by seeding viable cells onto synthetic biocompatible/biodegradable polymeric scaffolds mimicking native tissue structures [9, 23]. Tissue engineering gained widespread public attention with the airing of a BBC broadcast that featured the images of now-infamous "mouse with human ear" (Fig. 7.6) [23]. This work, from the laboratory of Charles Vacanti, showed a tissue engineering approach for engineering cartilage tissue using

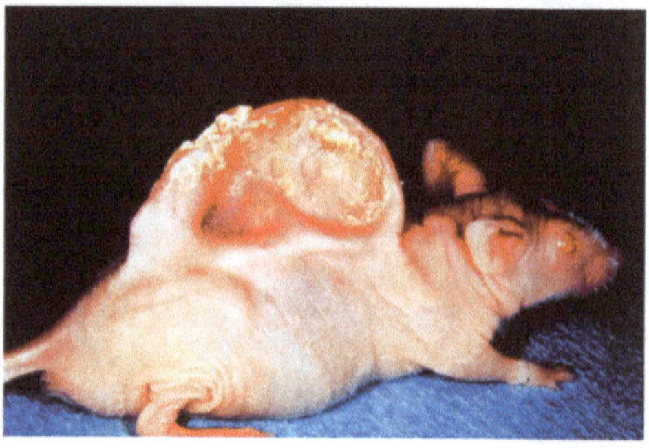

Fig. 7.6 Photograph of infamous mouse with the human ear, depicting new tissue-engineered cartilage generated in the shape of a human ear. (*Adapted from* [23] *with permission from John Wiley and Sons*)

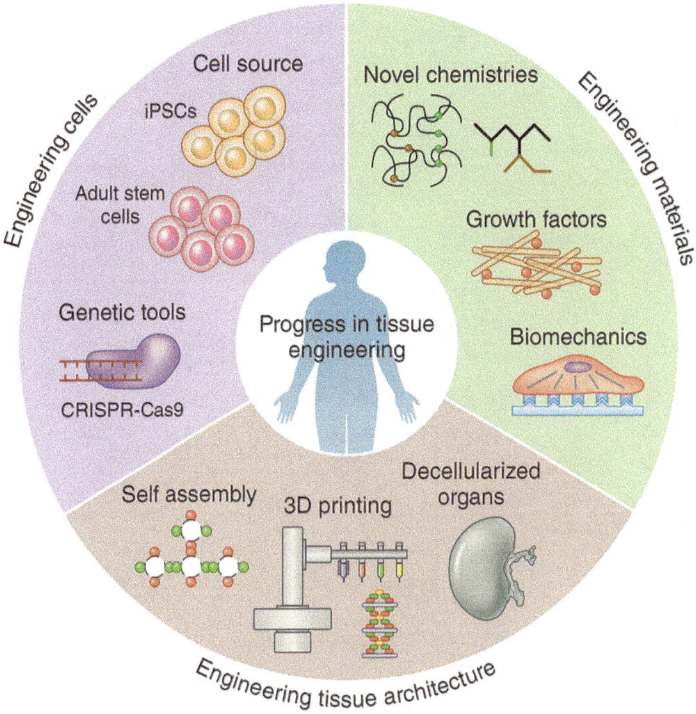

Fig. 7.7 Summary of tissue engineering progress from the early 2000s to mid-2010s. (*Adapted from* [28] *with permission from Springer Nature*)

chondrocytes seeded onto a synthetic biodegradable polymer fashioned in the shape of an ear and then implanted on the back of mice (known as the "Vacanti mouse") [9, 24].

Since then, the field of tissue engineering has expanded rapidly. Key milestones include the first implantation of engineered human bladders in patients [25], decellularization and recellularization of a whole heart [26], and the implantation of bioengineered lungs in a large animal model [27]. Major advances related to tissue engineering include innovations in bioprinting, mechanobiology, induced pluripotent stem cells (iPSCs), decellularization, immunomodulation, biorthogonal chemistries, and clustered regularly interspaced short palindromic repeats (CRISPER) gene-editing technology [28]. A summary of the progress in tissue engineering from the early 2000s to the mid-2010s is presented in Fig. 7.7.

Emerging topics in tissue engineering include the fabrication of vascularized tissues and organs, or tissues with potential of vascularization upon transplantation [29]; the development of human organoids for complex disease modeling and drug discovery [30, 31]; and the fabrication of organs-on-chips that more accurately mimic human physiology [32, 33].

Looking ahead, several major challenges still remain. These include achieving adequate vascularization (formation of blood vessels) and innervation (process by

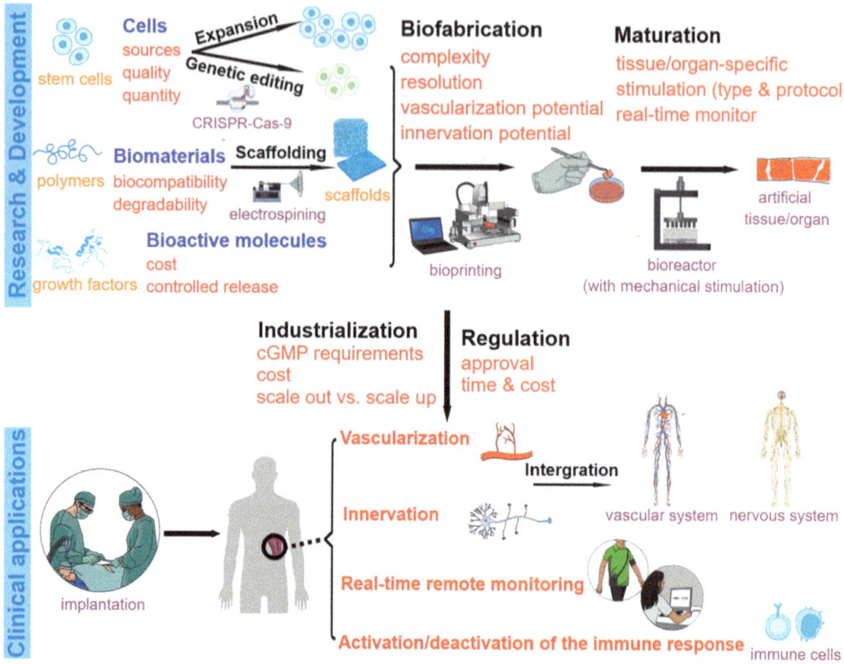

Fig. 7.8 Challenges facing tissue engineering—from research and development to clinical applications. Challenge-related text is highlighted in red. CRISPR: clustered regularly interspaced short palindromic repeats, cGMP: current good manufacturing practice

which nerves grow and form connections), ensuring controlled scaffold degradation, and overcoming regulatory and cost barriers [34–37]. Additionally, after implantation, critical considerations include the activation/deactivation of the immune response, external and remote control of tissue response, and real-time monitoring of tissue function (Fig. 7.8) [35].

7.4 Resources for Studying Tissue Engineering

The tissue engineering field is a rapidly growing and continuously evolving field, making it necessary to consult resources outside this textbook. The Tissue Engineering and Regenerative Medicine International Society (TERMIS) is a global organization dedicated to advancing tissue engineering & regenerative medicine worldwide to generate knowledge with a view to improving patient outcomes globally (https://termis.org/). TERMIS's official journal is *Tissue Engineering, Parts A, B, & C* (https://home.liebertpub.com/publications/tissue-engineering-parts-a-b-and-c/595). The society also collaborates with Springer to publish an updatable book series—Tissue Engineering and Regeneration (https://termis.org/springer-book-series). Many regional societies and professional organizations also support communities in this field. For example, the American Institute of Chemical Engineers (AIChE) hosts the Regenerative Engineering Society (https://www.aiche.org/resociety). Besides *Tissue Engineering, Parts A, B, & C*, there are numerous journals focused specifically on tissue engineering (Table 7.1). There are many additional resources available for studying tissue engineering that are not specifically mentioned here.

Table 7.1 Scientific journals related to tissue engineering and regenerative medicine

Title	Website
Artificial Organs	https://onlinelibrary.wiley.com/journal/15251594
Bioprinting	https://www.sciencedirect.com/journal/bioprinting
Cell Transplantation	https://us.sagepub.com/en-us/nam/cell-transplantation/journal203416
International Journal of Artificial Organs	https://journals.sagepub.com/home/JAO
International Journal of Bioprinting	https://accscience.com/journal/IJB
Journal of Artificial Organs	https://link.springer.com/journal/10047
Journal of Biomaterials and Tissue Engineering	http://www.aspbs.com/jbt.html
Journal of Stem Cell & Regenerative Medicine	https://www.pubstemcell.com/site/pubstemcell
Journal of Tissue Engineering	https://journals.sagepub.com/home/tej
Journal of Tissue Engineering and Regenerative Medicine	https://onlinelibrary.wiley.com/journal/jterm
npj Regenerative Medicine	https://www.nature.com/npjregenmed/
Organoid	https://j-organoid.org/
Organoids	https://www.mdpi.com/journal/organoids
Regenerative Engineering and Translational Medicine	https://link.springer.com/journal/40883
Regenerative Medicine	https://www.tandfonline.com/journals/irme20
Regenerative Therapy	https://www.sciencedirect.com/journal/regenerative-therapy
Tissue Engineering and Regenerative Medicine	https://link.springer.com/journal/13770
Tissue Engineering Part A: Research Advances	https://home.liebertpub.com/publications/tissue-engineering-part-a/263
Tissue Engineering Part B: Reviews	https://home.liebertpub.com/publications/tissue-engineering-part-b-reviews/262
Tissue Engineering Part C: Methods	https://home.liebertpub.com/publications/tissue-engineering-part-c-methods/261

Note: This is by no means an exhaustive list and is subject to updates and additions

Review and Practice Questions

1. What is tissue engineering? And what is regenerative medicine?
2. Why is tissue engineering important in biomedical engineering? Discuss its significance from multiple perspectives.
3. What are some of the current challenges facing tissue engineering?
4. Can we trust AI tools?

   (a) Identify an AI tool (e.g., ChatGPT).
   (b) Ask the AI tool to generate an essay with references using at least two keywords—one must be tissue engineering.
   (c) Verify all references provided by ChatGPT to ensure they are real and credible.
   (d) Carefully read the essay and identify at least two sections that contain inaccurate, misleading, or unsubstantiated information. Revise those sections using reliable references, such as peer-reviewed journal articles, and cite them appropriately.

5. Googling tissue engineering: One interesting way to estimate a topic's relative prominence is by counting the number of hits (results) using an internet search engine like Google© [38]. Perform your own topic searches using the terms—"Tissue Engineering" and "Regenerative Medicine", respectively, and graph (use the time range for the past 5 years; an example is shown below). Briefly describe any observable trends.

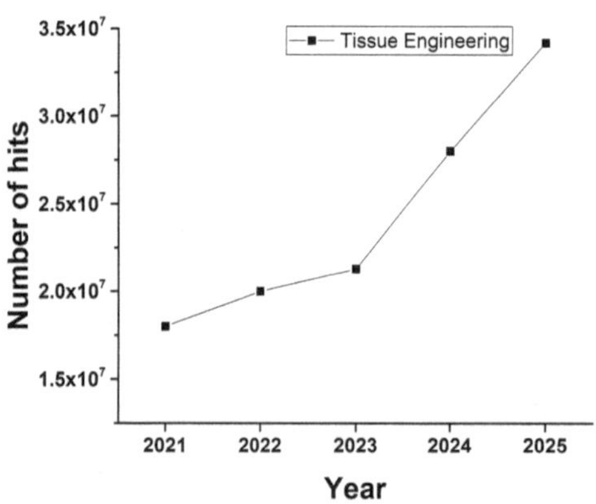

References

1. Cohen, J., *How a horror story haunts science.* Science, 2018. **359**(6372):148–150.
2. van den Belt, H., *Frankenstein lives on.* Science, 2018. **359**(6372):137–137.
3. Shultz, D., *Creating a modern monster.* Science, 2018. **359**(6372):151–151.
4. Langer, R. and J.P. Vacanti, *Tissue Engineering.* Science, 1993. **260**(5110): 920–926.

5. Birla, R., *Introduction to Tissue Engineering*, in *Introduction to Tissue Engineering: Applications and Challenges*. 2014, John Wiley & Sons.
6. National Institute of Biomedical Imaging and Bioengineering. *Tissue Engineering and Regenerative Medicine*. Available from: https://www.nibib.nih.gov/sites/default/files/Tissue%20Engineering%20Fact%20Sheet%20508.pdf.
7. Mhanna, R. and A. Hasan, *Introduction to Tissue Engineering*, in *Tissue Engineering for Artificial Organs*. 2017, Wiley-VCH.
8. Health Resources & Services Administration. *Organ Donation Statistics*. Available from: https://www.organdonor.gov/learn/organ-donation-statistics.
9. Sohn, S., et al., *Whole Organ Engineering: Approaches, Challenges, and Future Directions.* Applied Sciences, 2020. **10**(12):4277.
10. Augustine, R., et al., *Stem cell-based approaches in cardiac tissue engineering: controlling the microenvironment for autologous cells.* Biomedicine & Pharmacotherapy, 2021. **138**: 111425.
11. Zieman, J., et al., *Development of Gelatin-Coated Hydrogel Microspheres for Novel Bioink Design: A Crosslinker Study.* Pharmaceutics, 2023. **15**(1):90.
12. Mandrycky, C., K. Phong, and Y. Zheng, *Tissue engineering toward organ-specific regeneration and disease modeling.* MRS Communications, 2017. **7**(3):332–347.
13. Tebon, P.J., et al., *Drug screening at single-organoid resolution via bioprinting and interferometry.* Nature Communications, 2023. **14**(1):3168.
14. Park, S., et al., *Engineering considerations of iPSC-based personalized medicine*. Biomaterials Research, 2023. **27**(1):67.
15. Power, C. and J.E. Rasko, *Whither prometheus' liver? Greek myth and the science of regeneration*. Ann Intern Med, 2008. **149**(6):421–6.
16. van Gulik, T.M., M.M. van Gulik, and H.H. Koning, *[Prometheus and liver regeneration: the dissection of a myth]*. Ned Tijdschr Geneeskd, 2018. **162**:D2882.
17. Maggioni, F. and G. *Maggioni, A Closer Look at Depictions of Cosmas and Damian*. American Journal of Transplantation, 2014. **14**(2):494–495.
18. Ménard, S., *An Unknown Renaissance Portrait of Tagliacozzi (1545–1599), the Founder of Plastic Surgery*. Plastic and Reconstructive Surgery – Global Open, 2019. **7**(1):e2006.
19. Andrew, K. and R. Hage, *History of skin grafting: Ode to Dr. Otto Lanz*. Translational Research in Anatomy, 2024. **37**:100318.
20. Bell, E., et al., *Living Tissue Formed in Vitro and Accepted as Skin-Equivalent Tissue of Full Thickness*. Science, 1981. **211**(4486):1052–1054.
21. *Eugene Bell, 'father of tissue engineering,' dies at 88*. Available from: https://news.mit.edu/2007/obit-bell-0712.
22. Kemp, P., *Obituary: Professor Eugene Bell*. Regenerative Medicine, 2007. **2**(5):861–861.
23. Vacanti, C.A., *The history of tissue engineering*. Journal of Cellular and Molecular Medicine, 2006. **10**(3):569–576.
24. Cao, Y., et al., *Transplantation of Chondrocytes Utilizing a Polymer-Cell Construct to Produce Tissue-Engineered Cartilage in the Shape of a Human Ear*. Plastic and Reconstructive Surgery, 1997. **100**(2):297–302.
25. Atala, A., et al., *Tissue-engineered autologous bladders for patients needing cystoplasty*. The Lancet, 2006. **367**(9518):1241–1246.
26. Ott, H.C., et al., *Perfusion-decellularized matrix: using nature's platform to engineer a bioartificial heart*. Nature Medicine, 2008. **14**(2):213–221.
27. Zhou, H., et al., *Bioengineering Human Lung Grafts on Porcine Matrix*. Annals of Surgery, 2018. **267**(3):590–598.
28. Khademhosseini, A. and R. Langer, *A decade of progress in tissue engineering*. Nature Protocols, 2016. **11**(10):1775–1781.
29. Janson, K.D., et al., *Strategies for the vascular patterning of engineered tissues for organ repair*. Nature Biomedical Engineering, 2025. **9**(7):1007–1025.
30. Pocock, M.W., et al., *Maturation of human cardiac organoids enables complex disease modeling and drug discovery*. Nature Cardiovascular Research, 2025. **4**(7):821–840.

31. Saiki, N., et al., *Self-organization of sinusoidal vessels in pluripotent stem cell-derived human liver bud organoids*. Nature Biomedical Engineering, 2025. **9**(11):1869–1885.
32. Leung, C.M., et al., *A guide to the organ-on-a-chip*. Nature Reviews Methods Primers, 2022. **2**(1):33.
33. Zhao, Y., et al., *Organs-on-a-chip: a union of tissue engineering and microfabrication*. Trends in Biotechnology, 2023. **41**(3):410–424.
34. Meretsky, C.R., et al., *Advances in Tissue Engineering and Its Future in Regenerative Medicine Compared to Traditional Reconstructive Techniques: A Comparative Analysis*. Cureus, 2024. **16**(9):e68872.
35. Uquillas, J.A., L. Moroni, and J. de Boer, *Chapter 1 - An introduction to tissue engineering; the topic and the book, in Tissue Engineering (Third Edition)*, J. De Boer, et al., Editors. 2023, Academic Press.
36. Das, S., et al., *Innervation: the missing link for biofabricated tissues and organs*. npj Regenerative Medicine, 2020. **5**(1):11.
37. Chen, S., et al., *Vascularization and Innervation for Bone Tissue Engineering*. Accounts of Materials Research, 2024. **5**(9):1121–1133.
38. Rogers, B., J. Adams, and S. Pennathur, *Big Picture and Principles of the Small World*, in *Nanotechnology: Understanding Small Systems*. 2014, CRC Press.

Cells for Tissue Engineering

8

Learning Objectives

- Evaluate the importance of organ formation process during embryogenesis in the context of tissue engineering
- Discuss the significance of epithelial-mesenchymal transition (EMT)
- Assess the potential of MSCs and iPSCs as promising cell sources for tissue engineering
- Compare the methods available for isolating MSCs
- Analyze and design manufacturing processes for MSCs and iPSCs

8.1 Introduction to Cells for Tissue Engineering

Cells are the fundamental units of life and constitute one of the building blocks of tissue engineering [1]. A tissue is a collection of cell types arising from a single progenitor or stem cell (e.g., intestinal epithelium). Organs are typically composed of several tissue layers and arise from multiple developmental origins (e.g., the stomach consists of gastric epithelium, smooth muscle and connective tissues, blood vessels, nerve fibers, and cells of the blood and immune system). Most organs contain both an epithelial component (usually the functional part) and mesenchymal component (which provides mechanical support, growth factors, and other functions) (Fig. 8.1) [2]. Epithelial cells adhere tightly to one another at their lateral surfaces and to an organized extracellular matrix (ECM) at their basal domain, whereas mesenchymal cells are individual cells with a bipolar morphology that are embedded within a 3D ECM. The conversion of epithelial cells into mesenchymal cells, known as epithelial-mesenchymal transition (EMT), is central to many aspects of embryonic morphogenesis, adult tissue repair and fibrosis, and various disease states [3, 4] (Fig. 8.2). Moreover, growing evidence indicates that EMT transition transcription factors (EMT-TFs), ZEB, SNAI, and TWIST, play important

W. Zhang, *Cell Culture and Tissue Engineering*,
https://doi.org/10.1007/978-3-032-32096-4_8

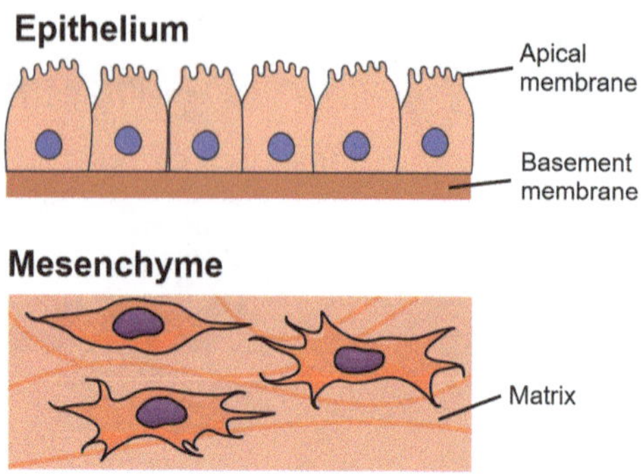

Fig. 8.1 Most organs are composed of epithelial and mesenchymal components

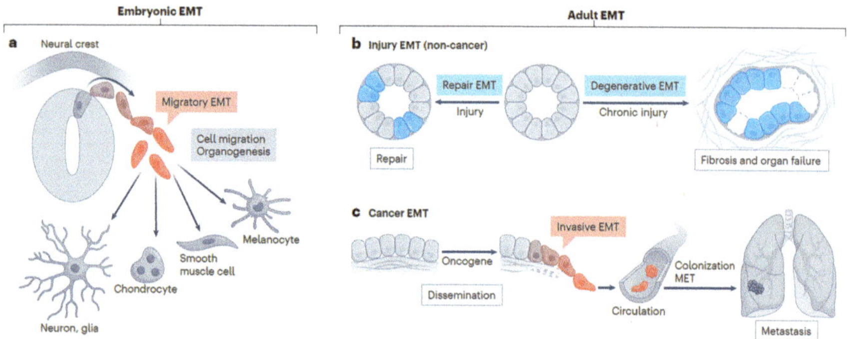

Fig. 8.2 Epithelial–mesenchymal transition (EMT) processes in development and disease. (**a**) EMT promotes cell delamination and migration during embryonic development in various tissues; (**b**) In the adult (non-cancer), EMT is repressed in homeostasis; (**c**) EMT is activated in cancer cells to promote cell invasion and dissemination, leading to metastasis. (*Adapted from* [4] *with permission from Springer Nature*)

regulatory roles in wound healing and mesenchymal tissue regeneration, such as osteogenesis and neurogenesis [5].

Embryogenesis, the development of a fertilized oocyte into a multicellular organism, builds embryonic structure and increases organismal complexity [6]. Most adult tissues are capable of self-repair and their intrinsic repair mechanisms recapitulate many of the processes involved in organ formation during embryogenesis. Hence, it is critical to mimic these processes in tissue engineering. Lessons to be learned include: the origins of cells that contribute to tissue formation; growth factors and other stimuli that drive differentiation; regulatory mechanisms of differentiation pathways; cell-cell interactions; the role of cell migration in the formation of specific tissue structures; and the selection of markers to monitor differentiation and tissue maturation [7].

8.2 Cells for Tissue Engineering

There are multiple sources of cells for tissue engineering, including stem cells, human-derived cells, animal-derived cells, and cell lines. For cells, they can be either autologous (from the patient) or allogenic (from human donors) (Fig. 8.3) [1].

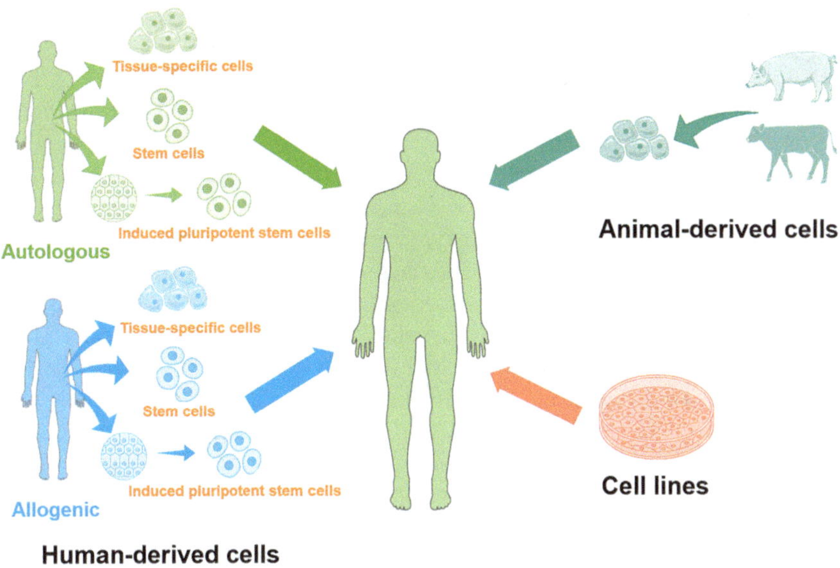

Fig. 8.3 Cell sources for tissue engineering

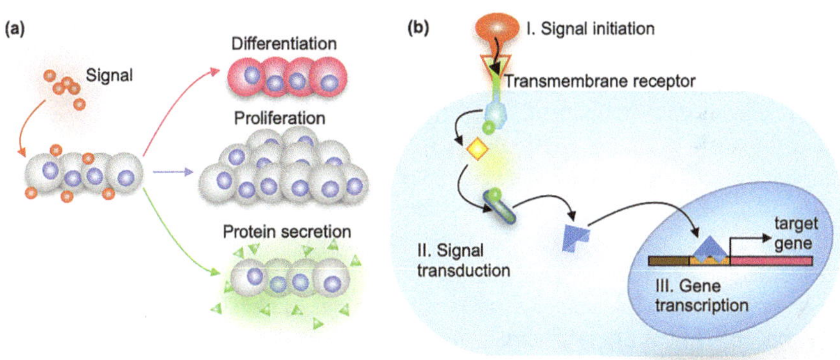

Fig. 8.4 (**a**) Cellular signaling leads to changes in the responding cell. (**b**) Three distinct steps of cellular signaling. (*Adapted from* [10] *with permission from Elsevier*)

At research and preclinical stages, it is common to use cell lines and animal-derived cells. However, stem cells and human-derived cells, particularly autologous cells, are preferred for translational and clinical applications. Unfortunately, tissue-specific cells are often difficult to harvest and/or expand *in vitro* [8]. Stem cells are the most promising sources for tissue engineering and regenerative medicine [9]. A key reason for their use is their ability to differentiate into desired cell types (Chap. 4). Cells respond to their environment and communicate with each other through cellular signaling, which typically involves three steps: signal initiation, signal transduction, and gene activation (Fig. 8.4). Cell signaling leads to changes in the responding cell, including cell differentiation (Fig. 8.4) [10]. Changes in gene expression drive cell differentiation. It is generally accepted that genes are activated during differentiation through interaction between tissue- and developmental-stage-specific enhancers and their cognate promoters. Recent studies suggest that enhancers predominantly control gene expression during differentiation by recruitment and/or initiation of transcription [11].

Mesenchymal stem/stromal cells (MSCs)-based therapy has emerged as the most prevalent approach in cell therapy. MSCs can be found in various neonatal and adult tissues. They can be easily expanded *in vitro* and exhibit low immunogenicity, making them promising candidates for allogenic uses. In addition to differentiation and replacement of damaged cells, MSCs demonstrate extensive paracrine activity and can interact with immune cells [12]. Recently, the FDA approved the first MSC-based therapy, Ryoncil (remestemcel-L-rknd), for the treatment of steroid-refractory acute graft-versus-host disease using allogenic bone marrow-derived MSCs [13]. Increasing evidence suggests that MSCs originate as perivascular cells and, rather than differentiating into different tissues *in vivo*, they primarily instruct tissue-specific progenitor cells responsible for regenerating injured tissue. Human MSCs (hMSCs) used in clinical settings can be broadly categorized into two types: adult MSCs (from bone marrow, adipose tissue, peripheral blood, and dental pulp) and neonatal tissue-derived MSCs (from placenta, amnion, and umbilical cord). Bone marrow-derived MSCs (BM-MSCs) remain the most widely used in clinical trials (Fig. 8.5) [14].

There are various approaches to isolate BM-MSCs: direct planting of whole marrow, density gradient centrifugation, and advanced cell-sorting methods (such as flow cytometry-based methods) [15, 16]. Direct planting selects plastic-adhesive cells; however, this approach cannot fully distinguish MSCs from other bone marrow cells, such as endothelial cells [16]. In density gradient centrifugation, the mononuclear cell fraction is collected and plated in culture medium (Fig. 8.6) [15]. MSCs must be characterized and manufactured in accordance with Current Good Manufacturing Practice (CGMP) regulations (Chap. 18). Key parameters for release control are shown in Table 8.1 [16, 17].

For large-scale MSC manufacturing for clinical applications, culture medium selection is critical. In particular, serum sources must be CGMP-compliant.

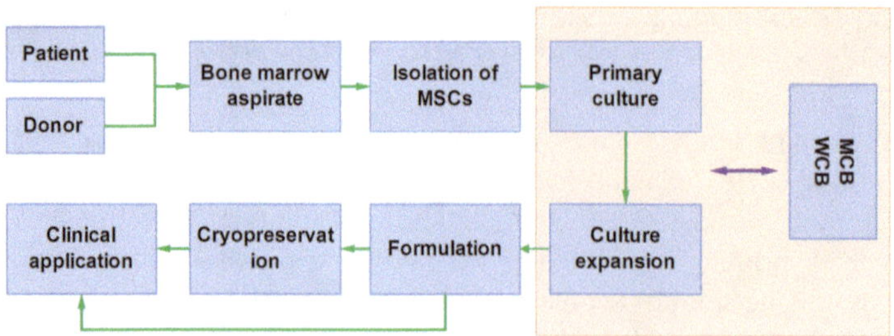

Fig. 8.5 A typical process of clinical applications of bone-marrow derived mesenchymal stem/stromal cells (BM-MSCs), starting from isolation. MCB: master cell bank, WCB: working cell bank

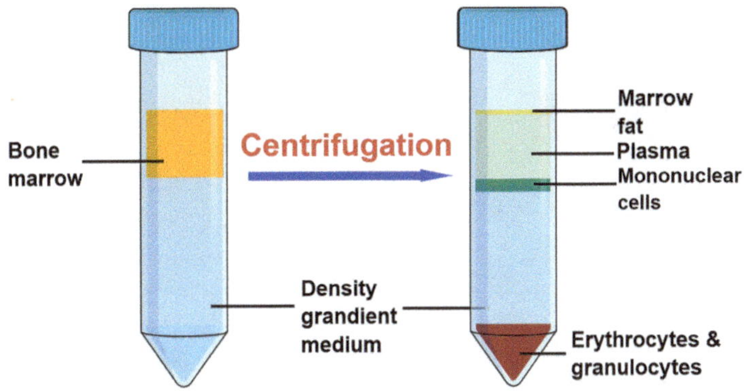

Fig. 8.6 Isolation of mesenchymal stem cells from bone marrow using density gradient centrifugation

Table 8.1 The release control parameters for MSCs

Parameters	Requirements
Cell number	5–10 × 10^8 cells
Viability	>90%
Sterility	Positive
Immune phenotype	CD105, CD73, and CD90 (>95% total cells). CD34, CD45, CD14 or CD11b, CD79α or CD19 and HLA-DR (≤2%)
Microbial	Negative
Endotoxin	Negative
Mycoplasma test	<50 CFU/mL

Note: The information is from [16]

Xenogenic-free and serum-free media are increasingly popular alternatives. Beyond multilayer culture flasks, automated and closed cell culture systems are becoming essential. Several commercial platforms have been developed for this purpose (Fig. 8.7) [18].

While the number of embryonic stem cell (ESC)-based trials has recently declined, the number of trials using induced pluripotent stem cells (iPSCs) has increased, owing to advances in reprogramming and differentiation technologies that have made personalized medicine more accessible [12]. To overcome the limitations of 2D iPSC culture, 3D culture systems show great potential, especially for scale-up and automated culture processes. For example, a closed culture system based on atelocollagen (a low-immunogenic collagen derivative obtained by removing the N- and C-terminal telopeptide components) microcarriers (3D culture) has been developed for autologous iPSCs (Fig. 8.8) [19].

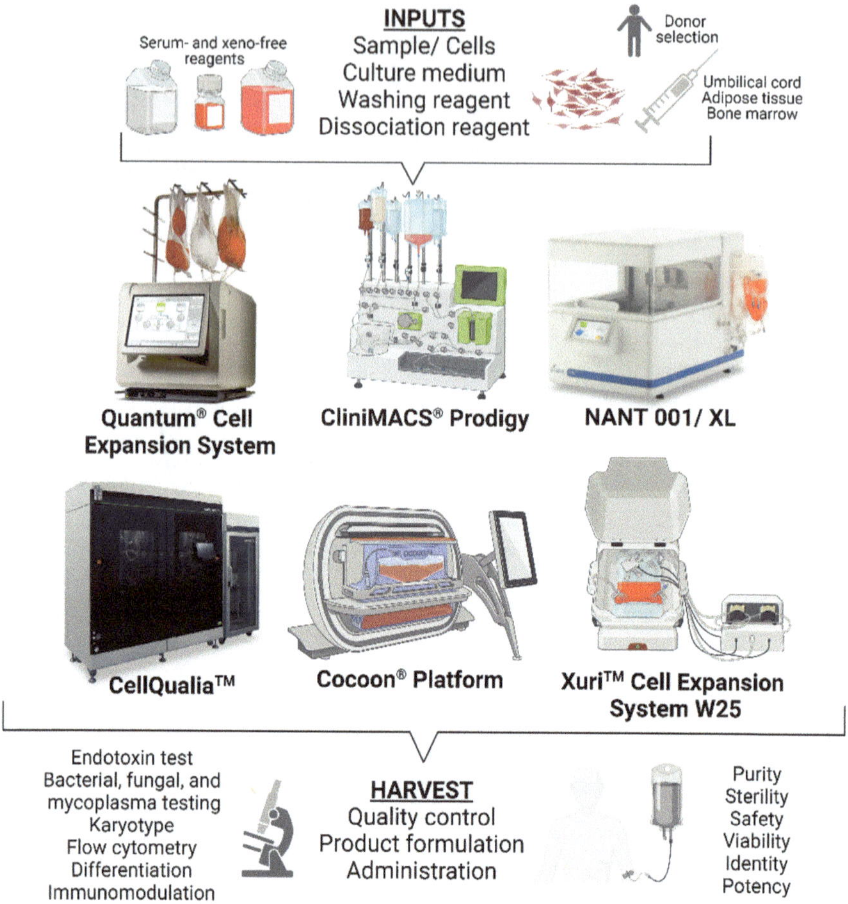

Fig. 8.7 Overview of bioprocessing strategies for large-scale production of clinical-grade mesenchymal stem/stromal cells (MSCs). MSCs expansion platforms, such as Quantum® Cell Expansion System, CliniMACS Prodigy®, NANT001/XL, CellQualia™, Cocoon® Platform, and Xuri™ Cell Expansion System W25, require sample or isolated cells, culture media, washing reagent, dissociation reagent, and optional supplements. Downstream processes including cell harvest, quality control (purity, identity, potency, and safety of final product), cryopreservation, and product formulation are also essential for clinical-grade MSC manufacturing. (Original image was created with BioRender.com. *Adapted from* [18])

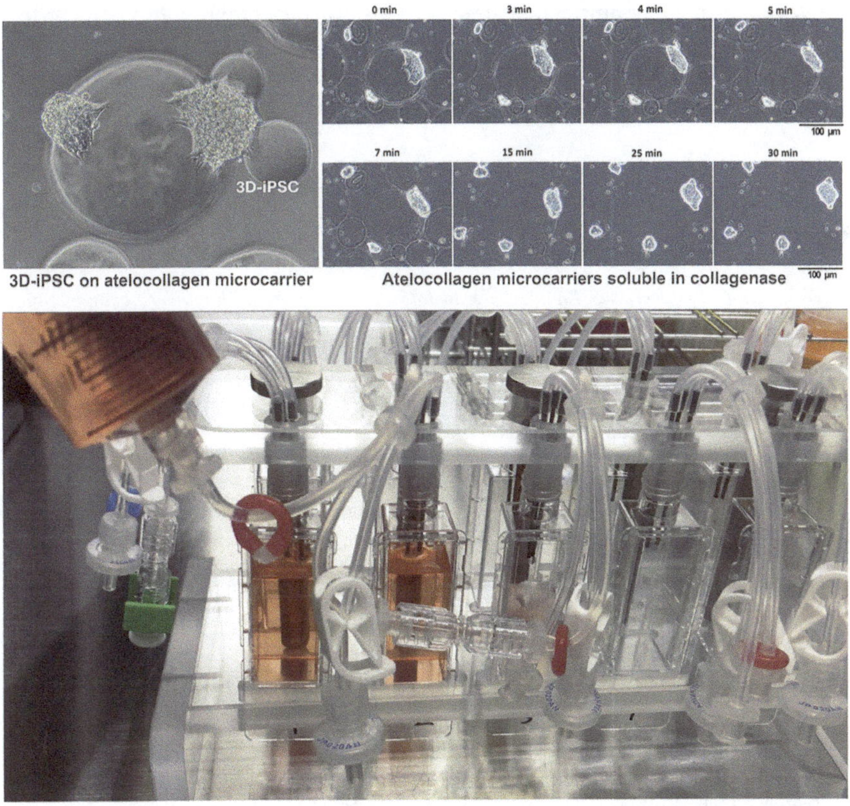

Fig. 8.8 A closed culture system using atelocollagen for autologous induced pluripotent stem cells (iPSCs). Top: 3D culture of iPSCs on atelocollagen microcarriers. Bottom: a prototype plastic culture vessel constructed to mimic a closed-bag system. (*Adapted from* [19])

Review and Practice Questions

1. What is epithelial-mesenchymal transition (EMT), and what is its significance in embryogenesis?
2. Why is it critical to mimic organ formation processes during embryogenesis in tissue engineering?
3. What are the main cell sources for tissue engineering?
4. What are the benefits of using stem cells in tissue engineering?
5. What are the benefits of using mesenchymal stem/stromal cells (MSCs) in cellular therapy?
6. What is the typical process for the clinical application of bone marrow-derived mesenchymal stem/stromal cells (BM-MSCs) starting from isolation?
7. What is the mechanism of density gradient centrifugation for isolating BM-MSCs?
8. Various commercial systems exist for large-scale production of clinical-grade MSCs. Conduct literature research and describe the working mechanisms of one such system.

9. Why are induced pluripotent stem cells (iPSCs) becoming more popular than embryonic stem cells (ESCs) in clinical applications?
10. What is atelocollagen, and why is it preferred in tissue engineering for clinical applications.

References

1. Birla, R., *Cells for Tissue Engineering*, in *Introduction to Tissue Engineering: Applications and Challenges*. 2014, John Wiley & Sons.
2. Slack, J.M.W., *Chapter 7 - Molecular Biology of the Cell*, in *Principles of Tissue Engineering (Fourth Edition)*, R. Lanza, R. Langer, and J. Vacanti, Editors. 2014, Academic Press.
3. Ahlstrom, J.D., *Chapter 8 - Molecular Organization of Cells*, in *Principles of Tissue Engineering (Fourth Edition)*, R. Lanza, R. Langer, and J. Vacanti, Editors. 2014, Academic Press.
4. Youssef, K.K. and M.A. Nieto, *Epithelial–mesenchymal transition in tissue repair and degeneration.* Nature Reviews. Molecular Cell Biology, 2024. **25**(9):720–739.
5. Wei, Z., et al., *Epithelial-to-mesenchymal transition transcription factors: New strategies for mesenchymal tissue regeneration.* Cytokine & Growth Factor Reviews, 2025. **83**:99–124.
6. Ghosh, S., et al., *Developmental energetics: Energy expenditure, budgets and metabolism during animal embryogenesis.* Seminars in Cell & Developmental Biology, 2023. **138**:83–93.
7. Karperien, M., et al., *Chapter 3 - Tissue formation during embryogenesis*, in *Tissue Engineering (Third Edition)*, J. De Boer, et al., Editors. 2023, Academic Press.
8. Yin, P.T., E. Han, and K.B. Lee, *Engineering Stem Cells for Biomedical Applications.* Advanced Healthcare Materials, 2016. **5**(1):10–55.
9. Ude, C.C., et al., *Application of stem cells in tissue engineering for defense medicine.* Military Medical Research, 2018. **5**(1):7.
10. LaPointe, V. and K.A. Kilian, *Chapter 4 - Cellular signaling*, in *Tissue Engineering (Third Edition)*, J. De Boer, et al., Editors. 2023, Academic Press.
11. Larke, M.S.C., et al., *Enhancers predominantly regulate gene expression during differentiation via transcription initiation.* Molecular Cell, 2021. **81**(5):983–997.
12. Hoang, V.T., et al., *Tissue Engineering and Regenerative Medicine: Perspectives and Challenges.* MedComm, 2025. **6**(5):e70192.
13. U.S. Food & Drug Administration. *FDA Approves First Mesenchymal Stromal Cell Therapy to Treat Steroid-refractory Acute Graft-versus-host Disease*. Available from: https://www.fda.gov/news-events/press-announcements/fda-approves-first-mesenchymal-stromal-cell-therapy-treat-steroid-refractory-acute-graft-versus-host.
14. Rodríguez-Fuentes, D.E., et al., *Mesenchymal Stem Cells Current Clinical Applications: A Systematic Review.* Archives of Medical Research, 2021. **52**(1):93–101.
15. Alm, J.J., H. Qian, and K. Le Blanc, *Chapter 13 - Clinical Grade Production of Mesenchymal Stromal Cells*, in *Tissue Engineering (Second Edition)*, C.A.V. Blitterswijk and J. De Boer, Editors. 2014, Academic Press.
16. Chu, D.-T., et al., *An Update on the Progress of Isolation, Culture, Storage, and Clinical Application of Human Bone Marrow Mesenchymal Stem/Stromal Cells.* International Journal of Molecular Sciences, 2020. **21**(3):708.
17. Papantoniou, I., et al., *Chapter 21 - Product and Process Design: Toward Industrial TE Manufacturing*, in *Tissue Engineering (Second Edition)*, C.A.V. Blitterswijk and J. De Boer, Editors. 2014, Academic Press.
18. Strecanska, M., et al., *Automated Manufacturing Processes and Platforms for Large-scale Production of Clinical-grade Mesenchymal Stem/Stromal Cells.* Stem Cell Reviews and Reports, 2025. **21**(2):372–389.
19. Nakashima, Y. and M. Tsukahara, *Development of a cell therapy set prototype for autologous iPS cells.* Cytotherapy, 2025. **27**(11):1351–1361.

Bioscaffolds for Tissue Engineering 9

Learning Objectives

- Explain the functions and components of the extracellular matrix
- Compare different classes of biomaterials used for bioscaffolding
- Design and evaluate bioscaffolds for specific tissue engineering applications
- Select and apply appropriate bioscaffold fabrication techniques for specific tissue engineering applications

9.1 Extracellular Matrix (ECM) and Bioscaffolds

A scaffold in tissue engineering is defined as a three-dimensional (3D) biomaterial before cells have been added [1]. In tissue engineering, bioscaffolds are produced to mimic the native extracellular matrix (ECM) and generate organotypic models [2]. ECMs are complex, well-organized 3D architectural networks with critical structural and functional roles in regulating cellular processes. ECMs guide tissue morphogenesis, development, and homeostasis through regulation of cellular physiology, growth, survival, differentiation, and adhesion [3]. The major building blocks of the ECM are proteins/glycoproteins—such as collagen, fibronectin, and laminin which serve as structural molecules, and glycosaminoglycans and growth factor which serve functional roles. These molecules provide mechanical properties required for proper tissue function and facilitate signal transduction between adjacent cells and between cells and the ECM [4]. The ECM can be divided into two categories: the basement membrane (in contact with epithelium and endothelium) and the interstitial matrix. These differ in molecular composition, architecture, localization, and the nature of cell-ECM interactions and functions (Fig. 9.1) [2, 5]. Mature ECM is also capable of undergoing dynamic remodeling in response to environmental stimuli [6].

W. Zhang, *Cell Culture and Tissue Engineering*,
https://doi.org/10.1007/978-3-032-32096-4_9

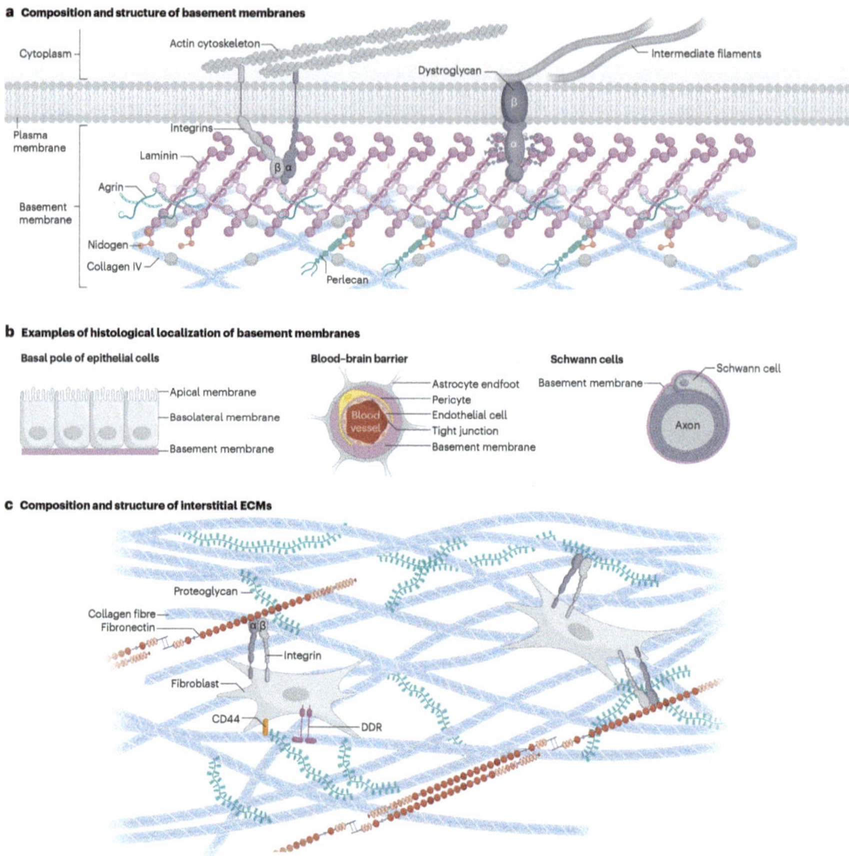

Fig. 9.1 Structural diversity of the extracellular matrix (ECM). (**a**) The basement membrane is a thin sheet-like assembly of ECM proteins primarily composed of a self-assembled lattice of laminin trimers, a network of collagen IV, and the proteoglycans agrin and perlecan. Additional proteins contribute to the linkage between these components (e.g., nidogen). Cells express ECM receptors such as integrins or the laminin receptor dystroglycan on their surface. These receptors form a bridge between the ECM and the cellular cytoskeleton. (**b**) Examples of anatomical locations of basement membranes. (**c**) The interstitial ECM is a 3D meshwork found in connective tissues and primarily made of fibrillar ECM proteins (collagens I, III, V, IX; fibronectin) and glycosaminoglycans and proteoglycans. Cells in connective tissues (e.g., fibroblasts) are surrounded by the interstitial ECM and interact with proteins of interstitial ECMs through receptors such as integrins, the tyrosin kinase discoidin domain receptors (DDRs), or the hyaluronic acid receptor CD44. (*Adapted from* [5] *with permission from Springer Nature*)

Collagens are generally composed of highly organized triple-helical polypeptide chains. There are many collagen types (28 identified), with type I being the most abundant in the human body [7]. They consist of homo- or hetero-trimeric α-chains and function as supportive tissue materials, providing elasticity and stability (Fig. 9.2) [3, 6]. Laminins are glycoproteins composed of α, β, and γ chains,

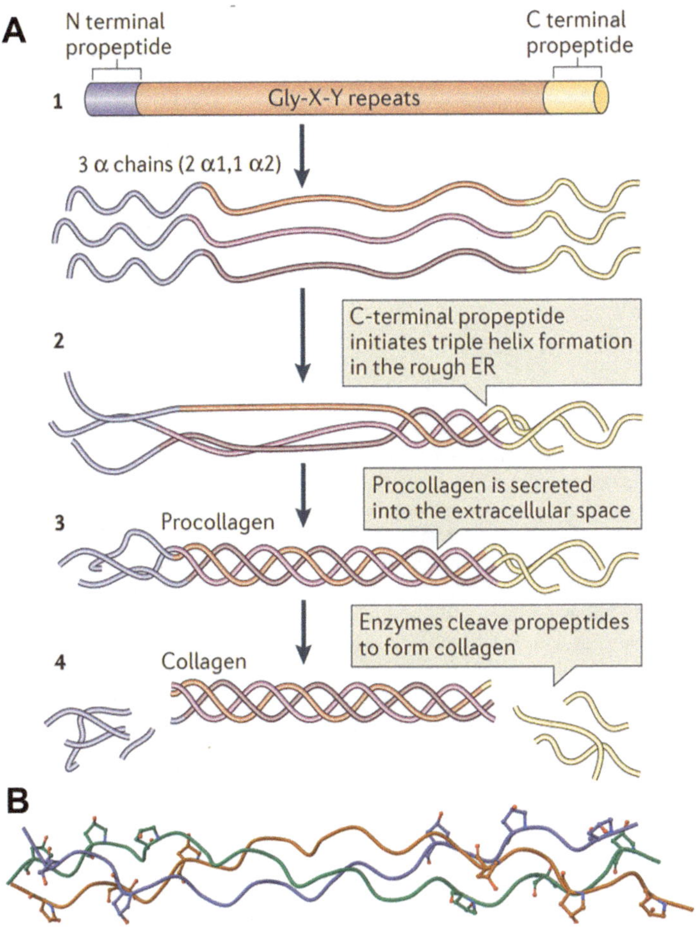

Fig. 9.2 (**a**) Collagen structure: the standard fibrillar collagen molecule is characterized by amino- and carboxy-terminal propeptide sequences flanking a series of Gly-X-Y repeats (where X and Y represent any amino acids but are frequently proline and hydroxyproline) (step 1). These form the central triple helical structure of procollagen and collagen. Three α-chains (the illustration shows two α1-chains and one α2-chain, which is representative of type I collagen) are assembled intracellularly into a triple helix following initiation of this process by the C-terminal domain (step 2). Procollagen is secreted by cells into the extracellular space (step 3) and converted into collagen by the removal of the N- and C-propeptides *via* metalloproteinase enzymes (step 4). ER: endoplasmic reticulum. (*Adapted from* [6] *with permission from Springer Nature*). (**b**) Structure of a triple-helix region of human collagen I. (*Created with PDB ID 7CWK (PDB* https://doi.org/10.2210/pdb7CWK/pdb*) and RCSB PDB*)

which associate *via* a central triple-helical coiled-coil domain to form cruciform (cross-shaped), Y-shaped (three arms), or rod-shaped assembled heterotrimers (Fig. 9.3). Five α-chains, four β-chains, and three γ-chains have been identified [6]. Laminins are primarily located in the basal lamina. In general, the C-terminal ends interact with cell surface receptors, relaying biochemical and mechanical signals

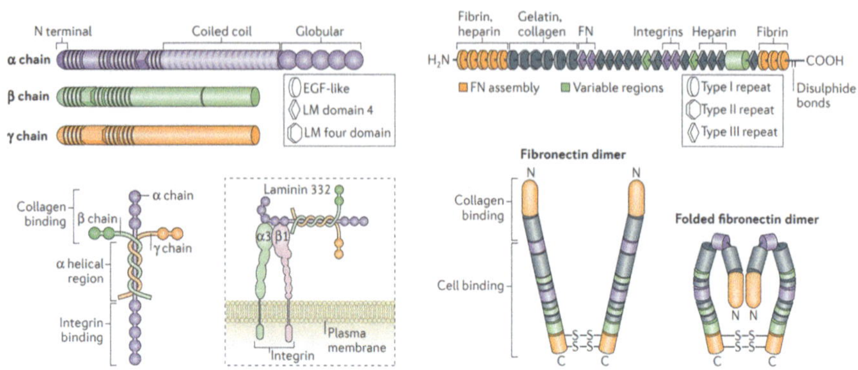

Fig. 9.3 Laminins are formed by the incorporation of α, β and γ chains into a cruciform, Y-shaped, or rod-like structures (left). These chains are characterized by different domains, as shown. The domain structures depicted represent only one isoform for each chain type, but major differences between the basic composition of the chains (such as the lack of globular regions in β and γ chains) are also present in the other isoforms. Laminins interact with cell surface receptors, such as integrins, primarily through globular domains in the chain. Fibronectin domain structure and the domains to which extracellular matrix molecules and cell surface receptors bind are indicated (right). The fibronectin molecule forms a dimer through disulphide (disulfide) bonds at its C terminus. The folded fibronectin molecule forms *via* ionic interactions between type III domains of neighboring molecules and is deformed by mechanical force to reveal cryptic binding sites for other fibronectin molecules and cell surface receptors when interacting with cells. EGF: epidermal growth factor, LM: laminin, FN: fibronectin. (*Adapted from* [6] *with permission from Springer Nature*)

between intracellular and extracellular molecular networks, while the N-terminal ends interact with other ECM molecules [8]. Overall, laminins perform various adhesive and stimulatory functions as well as serve as reservoirs for growth factors [9]. Fibronectin, a large glycoprotein, assembles *via* cell-mediated processes into fibrillar structures. Each subunit contains three types of repeating modules (types I, II and III), which contain binding motifs that facilitate interaction with cell surface receptors, such as integrins. Fibronectin dimers form *via* a pair of disulfide bonds at the C-terminus (Fig. 9.3) [6]. Although fibronectin originates from a single gene, as many as 20 different human variants have been identified. The RGD (arginyl-glycyl-aspartic acid) motif binds several integrins. Fibronectin fibrils possess unique mechanical properties that influence mechanotransduction signals. Binding of soluble growth factors to fibronectin fibrils modulates signal transduction, while binding of other ECM proteins facilitates ECM maturation and tissue specificity [10].

Proteoglycans play major regulatory roles in signal transduction and various cellular processes. They are upstream of many signaling cascades and modulate intracellular phosphorylation events and distinct signaling pathways [11]. Proteoglycans are characterized by a core protein covalently linked to glycosaminoglycans (GAGs). For example, aggrecan, composed of a protein core with three globular regions, is a predominant proteoglycan in hyaline cartilage and forms aggregates containing a central hyaluronic acid filament to which multiple aggrecans are attached (Fig. 9.4) [12]. GAGs are long, negatively charged, and linear chains of disaccharide repeats, such as heparan sulfate, chondroitin sulfate, dermatan sulfate, hyaluronic acid (hyaluronan), and keratan sulfate (Fig. 9.4). The primary biological function of proteoglycans is derived from their GAG chains, which bind water to provide hydration and compressive resistance [6]. GAGs participate in numerous physiological processes, including binding, activating, and sequestering a variety of protein ligands, such as growth factors and cytokines. Their binding specificity is encoded in their primary structures but ultimately depends on how functional groups are presented in 3D space [13].

Two major advantages of using ECM as bioscaffolds are the presence of all the attendant growth factors in the relevant amounts and the preservation of 3D

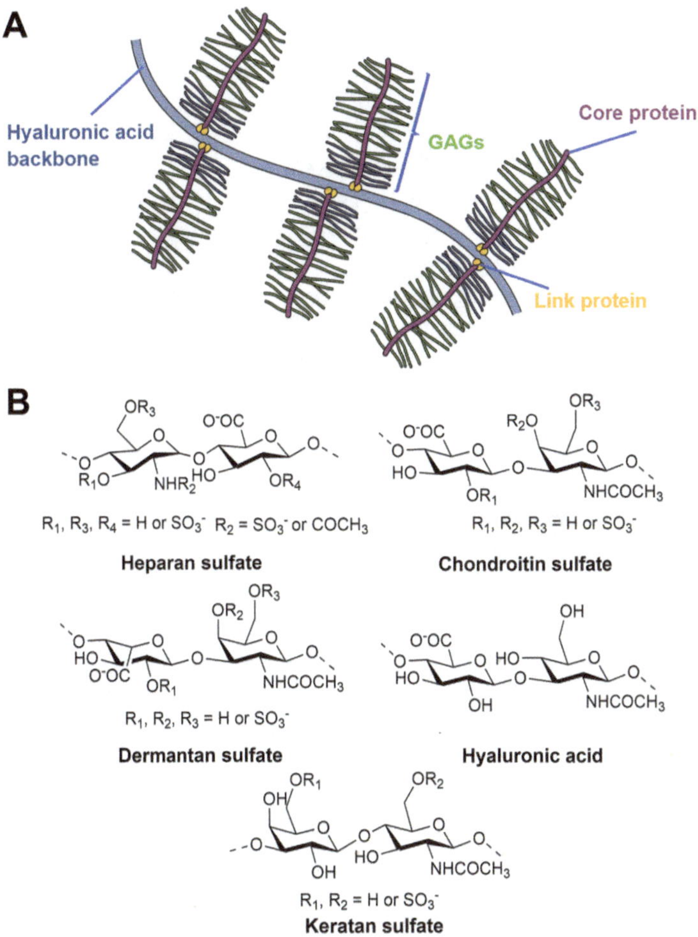

Fig. 9.4 (**a**) Structure of aggrecan. Aggrecan aggregates contain glycosaminoglycans (GAGs)—chondroitin sulfate and keratan sulfate. (**b**) Principal constituent disaccharide units of GAGs

ultrastructure that exits in native tissue [4]. To obtain ECM for tissue engineering applications, decellularization followed by sterilization is commonly used (further discussed in Chap. 16). Additionally, cell culture-derived ECMs have attracted increasing interest due to their ability to overcome many of the limitations imposed by tissue-derived ECMs, while exhibiting characteristics that enable them to recapitulate, to a great extent, the complex biological environment of native tissue [14].

9.2 Biomaterials for Bioscaffolds Fabrication

The first Biomaterials Symposium was held at Clemson University (considered the international birthplace of the biomaterials field) in 1969 and marked a key milestone in recognizing the need to integrate engineering and medicine for the development of biomedical materials [15, 16]. According to the journal *Biomaterials*, a biomaterial is a substance that has been engineered to take a form which, alone or as part of a complex system, is used to direct, by control of interactions with components of living systems, the course of any therapeutic or diagnostic procedure [17]. The definition from the 2018 Consensus Conference of Biomaterial Definitions states that a biomaterial is a material designed to take a form that can direct, through interactions with living systems, the course of any therapeutic or diagnostic procedure. And it was agreed that "biomaterial" is synonymous with "biomedical material" [18, 19]. From a tissue engineering perspective, a biomaterial is any substance that simulates the extracellular matrix by functionally interacting with isolated cells to support fabrication and maturation of 3D artificial tissue [20]. As shown in Fig. 9.5, biomaterials can be classified in various ways—for example, as polymers, metals, or ceramics/glasses based on composition, and as natural or synthetic based on origin [21]. It is common for a bioscaffold to contain more than one type of biomaterial.

Polymers are commonly used in tissue engineering because most components of the ECM are polymeric. Natural polymers such as collagen including gelatin,

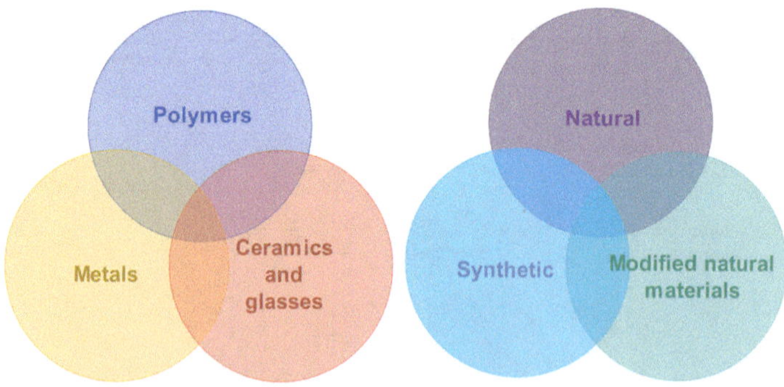

Fig. 9.5 Different types of biomaterials

fibronectin, hyaluronic acid, and laminin are commonly used to fabricate bioscaffolds. Other natural and synthetic polymers include alginate, pectin, chitosan, polyethylene glycol, and poly(lactic-*co*-glycolic) acid (Fig. 9.6) [22–25]. Stimuli-responsive ("smart") polymers that respond to pH, temperature, light, or enzymes have attracted great interest because they enable reversible control over ECM-like properties, such as modulating feedback mechanisms between cells and their microenvironment [26]. Responsiveness can also be exploited during bioscaffold fabrication. However, polymers are generally not used for load-bearing biomedical applications. Thermoresponsive polymers used in tissue engineering include poly(*N*-isopropylacrylamide) (PNIPAM), Pluronics, elastin-like polypeptides (ELP), and poly(*N*-vinylcaprolactam) (PNVCL) [27]. Enzyme-responsive polymers can mimic the dynamics of natural ECM, enabling communication between biomaterials and cells through enzyme-responsive moieties. Common targeting enzymes include proteases, esterases, glucosidases, and oxidoreductases [28]. A notable exception to the load-bearing limitation is ultrahigh-molecular-weight polyethylene, which has been used as an articulating surface in joint replacements [29].

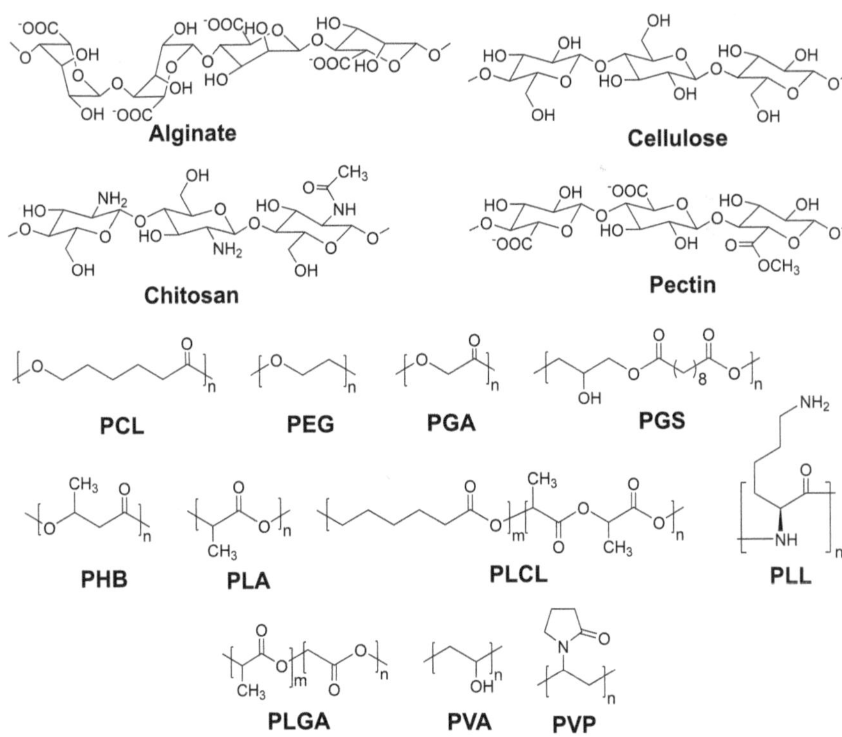

Fig. 9.6 Structure of common polymers in tissue engineering. PCL: polycaprolactone, PEG: polyethylene glycol, PGA: polyglycolic acid, PHB: polyhydroxybutyrate, PGS: poly(glycerol sebacate), PLA: polylactic acid, PLCL: poly(L-lactide-*co*-caprolactone), PLGA: poly(lactide-*co*-glycolide), PLL: poly-L-lysine, PVA: polyvinyl alcohol, PVP: polyvinylpyrrolidone

Alginate, chitosan, and pectin are biodegradable natural polysaccharides. Alginate, mostly derived from brown algae, is one of the most widely used biomaterials in tissue engineering and regenerative medicine. Alginate consists of α-L-guluronic acid (G) and β-D-mannuronic acid blocks and form hydrogels in the presence of divalent ions such as Ca^{2+} due to ionic cross-linking (between cations and carboxyl groups of alginate). Modifications of alginate are routinely implemented to improve stability and cell interactions [30–33]. Pectin, found in plant cell walls, is primarily composed of α-(1-4)-linked D-galacturonic acid with interspersed 1,2-linked L-rhamnose residues. Commercial pectin is often derived from orange peels and apple pomace. Based on the degree of esterification (DE), pectin is classified as high DE and low DE pectin. Very similar to alginate, pectin forms a gel when exposed to divalent cations, low DE pectin in particular. Modifications broaden its tissue engineering applications [33–36]. Chitin, β-(1-4)-poly-*N*-acetyl-*D*-glucosamine, is the second most abundant natural polysaccharide, while chitosan is produced from deacetylation of chitin (with the conversion of a fraction of the *N*-acetyl-*D*-glucosamine to D-glucosamine). Mostly, commercial chitosan products are crustacean-based and vary greatly in their degree of deacetylation and molar mass. Chitosan is an attractive material for bioscaffolds due to its biodegradability, antimicrobial, and immunomodulatory effects. One major challenge of applying chitosan is its low solubility—it only dissolves in weak acid solutions due to the protonation of the amino groups obtained through the deacetylation process. Current research focuses on combining chitosan (or modified chitosan) with other biomaterials and bioactive agents to improve performance [33, 37, 38].

Synthetic polymers offer tunability and manufacturing consistency due to controlled molar mass and purity, although they are often less biocompatible and biodegradable [39, 40]. Polylactic acid (PLA) and polycaprolactone (PCL) are the most commonly used synthetic polymers for tissue engineering applications [25]. PLA is a linear aliphatic polyester and can be produced from renewable resources like corn and sugarcane. PLA exists in three stereochemical forms, poly(L-lactic acid) (PLLA), poly(D-lactic acid) (PDLA), and poly(DL-lactic acid) (PDLLA), and their properties, such as crystallinity and degradability, are different. PLA is biodegradable and has certain mechanical properties, such as for repairing small bone damage in non-load-bearing parts. Another advantage of PLA is its high 3D printing ability, which shows high potential for fabricating complex and personalized scaffolds with precise structural features. However, acidic by-products produced during PLA degradation cause long-term biocompatibility concerns. Moreover, PLA has low cell adhesion. Chemical modification and combination with other biomaterials have been widely explored to relieve the challenges associated with using PLA alone in tissue engineering applications [25, 40, 41]. PCL is another biodegradable aliphatic polymer with high processability (e.g., 3D printing) and strong mechanical properties. However, slow degradability and low cell affinity limit its tissue engineering applications. Similar to PLA, its applicability can be enhanced by chemical modification and combination with other biomaterials [25, 42].

Metals and metal alloys, such as titanium (Ti), magnesium (Mg), cobalt-chromium (Co-Cr), iron (Fe), zinc (Zn), are widely used in bone, cartilage, dental and cardiovascular tissue engineering [43]. Traditionally applied in load-bearing implants, recent work explores their intrinsic bioactivity, immune modulation, and promotion of osseointegration and vascularization [44]. Metallic biomaterials must be selected carefully with regard to mechanical strength, biodegradability, and functionality. For example, Mg-based biomaterials degrade and release beneficial ions but have lower mechanical strength. Conductive metallic biomaterials are useful for pacemakers and neuromuscular devices [43]. A recent review analyzed 67 references and concluded that Ti and Mg alloys are potentially suitable for use in bone tissue engineering [45]. Corrosion can be harmful, but controlled ion release can assist in functionality [43]. For example, different ions affect bone regeneration processes differently (Fig. 9.7) [46]. Surface modification, encapsulation, and other approaches have been employed to improve biological performance [47, 48] (Fig. 9.8).

Additive manufacturing enables fabrication of controlled porous interconnected structures. For example, porous Mg scaffolds fabricated *via* 3D gel-printing promoted cancellous bone osteogenesis and delayed degradation when oxide-coated [49]. Additionally, a triply periodic minimal surface-structured porous magnesium alloy was fabricated using the laser powder bed fusion process. Strontium (Sr)-doped octacalcium phosphate and strontium hydrogen phosphate biphasic

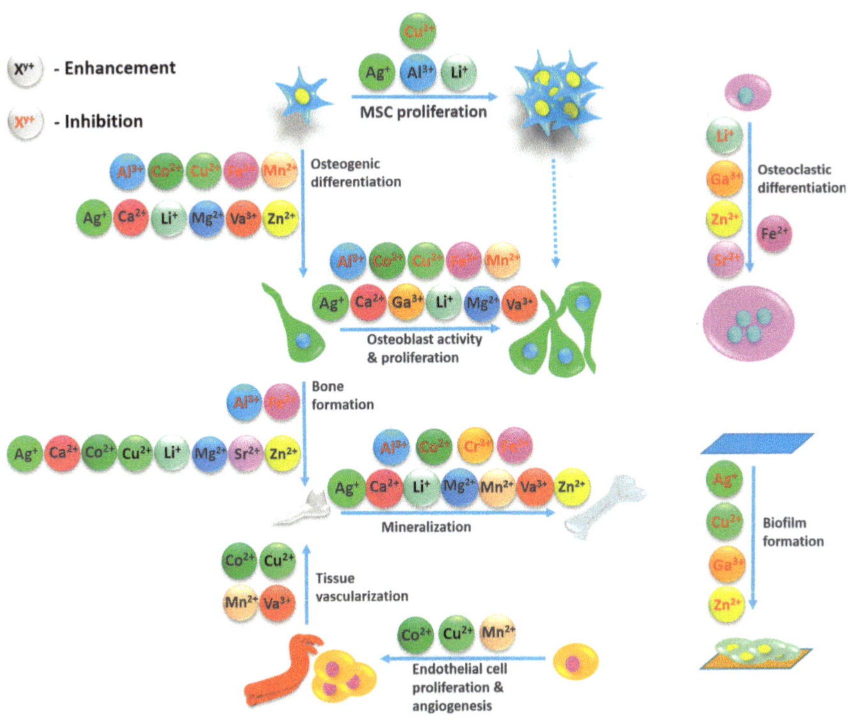

Fig. 9.7 Influence of metal ions on processes involved in bone regeneration. (*Adapted from* [46])

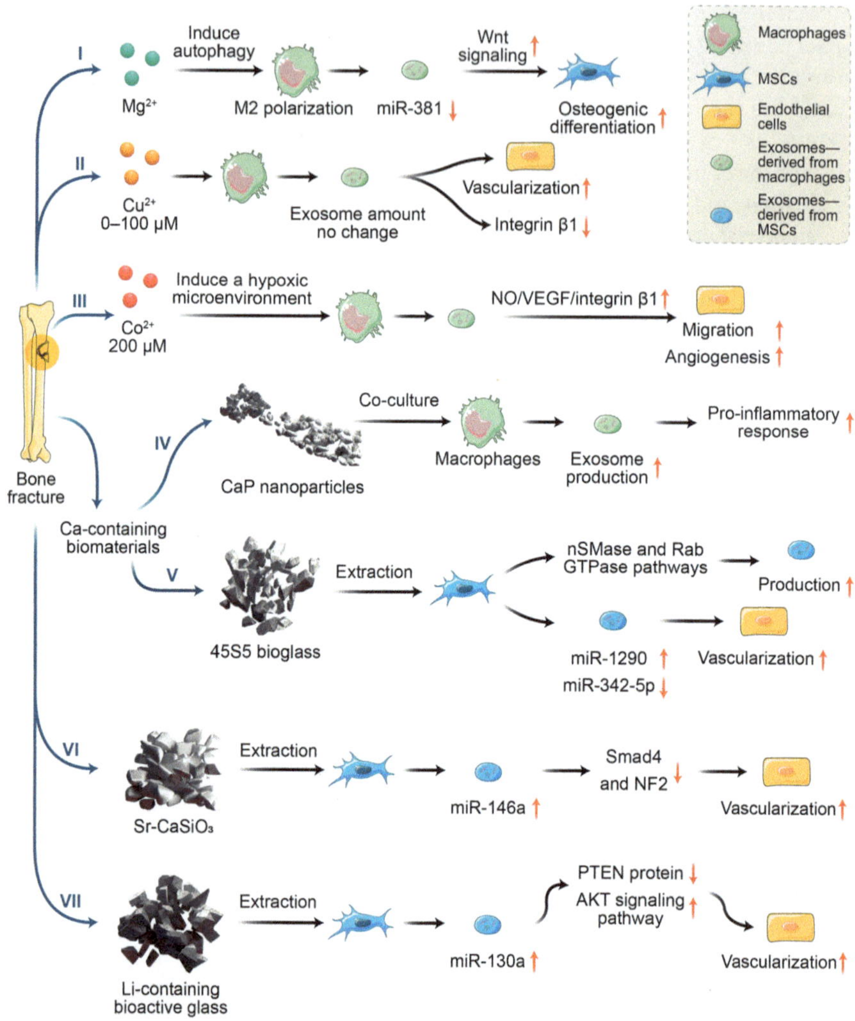

Fig. 9.8 Effects of metallic ions or metallic ion-incorporated biomaterials on angiogenesis and osteogenesis *via* macrophages and MSCs secreting exosomes. (*Adapted from* [48])

composite coatings were then applied to the scaffolds for enhanced bone defect repair [50] (Fig. 9.9).

Ceramics are solid materials composed of inorganic, non-metallic substances. Ceramics exist as both crystalline and amorphous compounds [51]. A glass is an inorganic product of fusion that has cooled to a rigid condition without crystallization [52]. Bioceramics are biocompatible ceramics with apatite-formation ability, allowing bonding between the material and tissues. They are produced by sintering at high temperature [53]. Bioglass is generally a series of specially designed silica-based glasses in which a 3D SiO_2 network is modified by the incorporation of Na_2O,

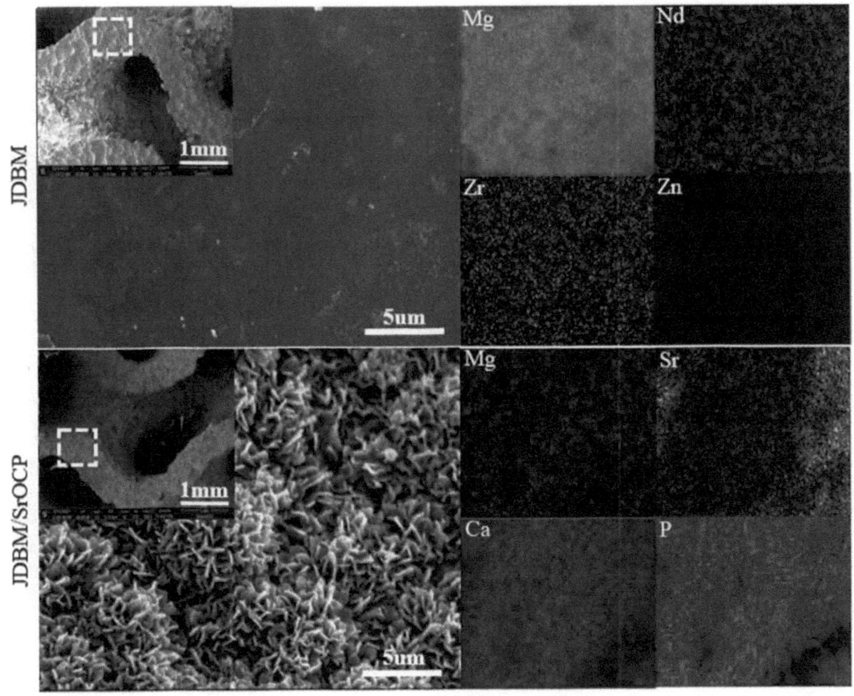

Fig. 9.9 Surface morphology and elemental distribution of triply periodic minimal surface (TPMS)-structured porous magnesium alloy (JDBM) scaffolds and coated JDBM (JDBM/SrOCP) scaffolds examined by scanning electron microscopy (SEM) and energy dispersive spectroscopy (EDS). (*Adapted from* [50] *with permission from Springer Nature*)

CaO, and P_2O_5 [54]. Bioactive ceramics and glasses are mainly developed and used for bone tissue engineering. There is increasing interest in the utilization of glasses in soft tissue engineering. The most commonly used ceramics include hydroxyapatite (HA; $Ca_{10}(PO_4)_6(OH)_2$), beta-tricalcium phosphate (β-TCP; $Ca_3(PO_4)_2$), and biphasic calcium phosphate (a combination of HA and TCP) [55–57]. Pure, highly crystalline synthetic HA bioceramics are considered to be non-degradable, whereas β-TCP bioceramics are considered to be degradable [58]. 45S5, S53P4, and 13-93B3 are commonly used bioglasses. For example, 45S5 contains 45% SiO_2, 24.5% CaO, 24.5% Na_2O, and 6.0% P_2O_5 by weight [59]. Current focuses on tissue engineering application of bioceramics include development of antimicrobial bioceramics, electrospinning of bioceramics, 3D printing of bioceramics, and the combination of bioceramics with other biomaterials [60–63]. For example, a recent study screened 24 different porous structures using 3D printed calcium phosphate (CaP) ceramic scaffolds (Fig. 9.10). The complex relationship between osteoinductive properties and scaffold configurations were then explored through *in vitro* and *in vivo* evaluations, along with machine learning and nonlinear fitting. The results indicated that porosity and specific surface area, rather than pore geometry, significantly affected the bone regenerative ability [64].

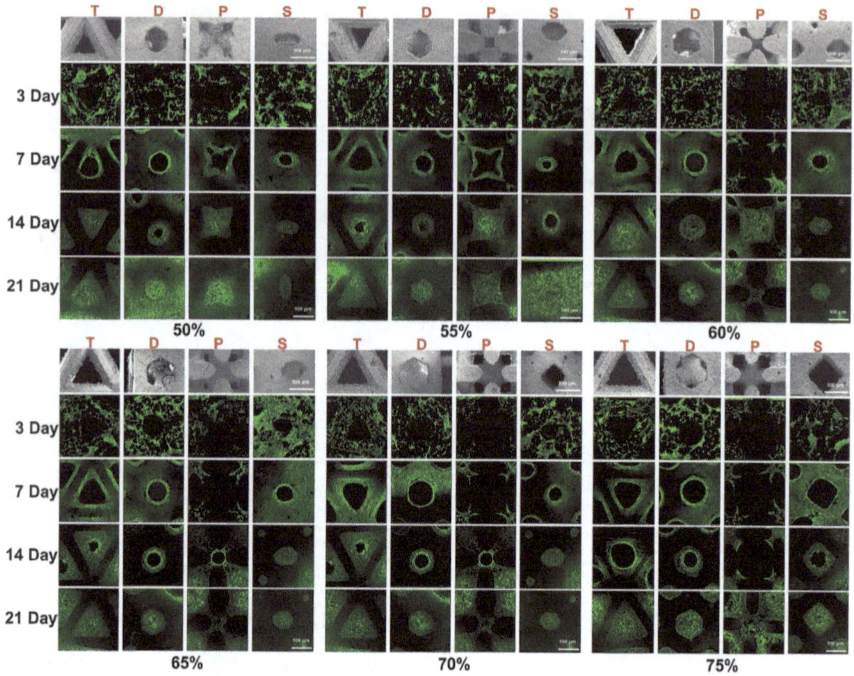

Fig. 9.10 *In vitro* evaluation of CaP Chip. CLSM observations for BMSCs growth on CaP Chip. The scaffolds included triangular (T), diamond (D), square (S), and polyhedral (P) structures, each featuring six porosity levels ranging from 50% to 75%. CLSM: confocal laser scanning microscope. (*Adapted from* [64] *with permission from Elsevier*)

With time, materiomics has emerged as an -omics approach to biomaterials research and development. The linking of mechanisms across multiple scales using a materials science approach to provide structure-process-property relationships characterizes the field of materiomics. Materiomics can be defined as the holistic study of material systems [65]. Materobiology is another emerging concept that examines the biological effects of biomaterial properties on biological functions at different levels (cells, tissues, organs, and the whole organism) [66]. Microarray strategies, such as polymer microarray, have been applied as high-throughput biomaterials screening platforms. Microarray technology is particularly useful for examining many unique and spatially defined identities in parallel—such as unique materials and material-topography combinations [66, 67]. In a polymer microarray, small polymer spots are deposited onto a substrate at designated locations. Parameters to be considered include the base substrate (compatible with biological assay and associated high-throughput readouts), substrate coating (adherent to the polymers but not biomolecules and cells), polymer library and synthesis (presynthesized or *in situ* polymerization and deposition through ink-jet printing or other methods), and the coupled biological assay (relevant readouts can be assessed in an automated fashion) [67]. Laurence Burroughs et al. reported a novel

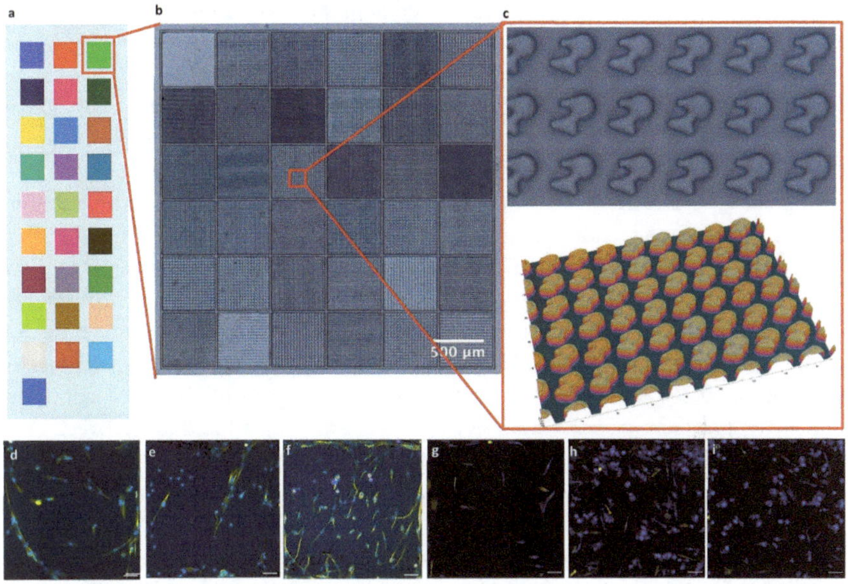

Fig. 9.11 (**a**) Schematic showing ChemoTopoChip layout with colors representing different chemistries; (**b**) interference profilometer image of ChemoTopo unit (30 μm-high walls separate each Topo unit); (**c**) example features from a ChemoTopo unit; hiMSCs (blue = ALP, yellow = α-tubulin); on (**d**) flat TMPMP-co-TEGDA area (**e**) TMPMP-co-TEGDA + Topo 3 area; (**f**) mMAOES + Topo 3 area; macrophages (blue = IL-10, yellow = TNFα) on (**g**) flat TMPMP-co-TEGDA area; (**h**) TMPMP-co-TEGDA + Topo 22 area; (**i**) BzHPEA + Topo 22 area. hiMSCs: human immortalized mesenchymal stem cells, ALP: alkaline phosphatase. (*Adapted from* [68])

combinatorial chemistry-topography screening platform, the ChemoTopoChip, for identifying materials suitable for bone regeneration. The platform contained 1008 microtopography and materials chemistry combinations (36 topographical structures and 28 chemistries) to simultaneously probe their combined effects on cellular response—including differentiation of hMSCs and polarization of macrophages toward an M2 state (pro-healing phenotype). The chemistries were chosen from libraries of (meth)acrylate and (meth)acrylamide monomers to provide maximum chemical diversity for structures and classification of their number H-bond donors, number of H-bond receptors, and log partition coefficient (Fig. 9.11) [68, 69].

9.3 Design and Fabrication of Bioscaffolds

There are a number of key considerations when designing a bioscaffold: biocompatibility, biodegradability, surface properties, mechanical properties, scaffold architecture, and manufacturing technology [1]. To fully explore these key considerations, it is critical to review the cell-material interactions first, as they are the defining features of biomaterials [70]. *In vitro* 2D culture conditions do not fully capture the physicochemical features experienced by cells in the body. Bioinspired 3D geometry that recapitulates as many as aspects of the natural ECM as possible provides combinative biochemical/physical cues to mediate cell response. To take the dynamic mechanostructural changes into consideration, the development of 4D biomimetic matrices that are stimuli-responsive have gained increasing attention (Fig. 9.12) [66].

In general, these are the steps that cells follow during cell-material interactions, in order of occurrence: adhesion, spreading, polarization, and migration [71]. Cell adhesion is a complex process that can be divided into different phases: (1) a cell physically contacts with the substrate in the first phase of the cell-matrix interaction; (2) the cell adheres and starts sensing the substrate through electrostatic interaction, being guided by gravity; (3) the cell reshapes with the concomitant adhesion processes; and (4) the cell adheres in an immature form of the focal adhesions (Fig. 9.13) [71, 72]. Nascent adhesions are generally small and short-lived, often lining the edge of the lamellipodium. A small fraction of nascent adhesions mature into focal complexes, and then into more stable, larger structures termed focal adhesions [73]. Focal adhesions connect the inner workings of the cell to the ECM to control cell adhesion, migration and mechanosensing and are composed of multiple distinct actin filament layers, with each layer having specific roles in coupling adhesions to the cytoskeleton, or in controlling adhesion dynamics (Fig. 9.14) [74]. At the external surface, integrin receptors are activated through binding to ECM ligands, which leads to the recruitment of a large, heterogenous multiprotein

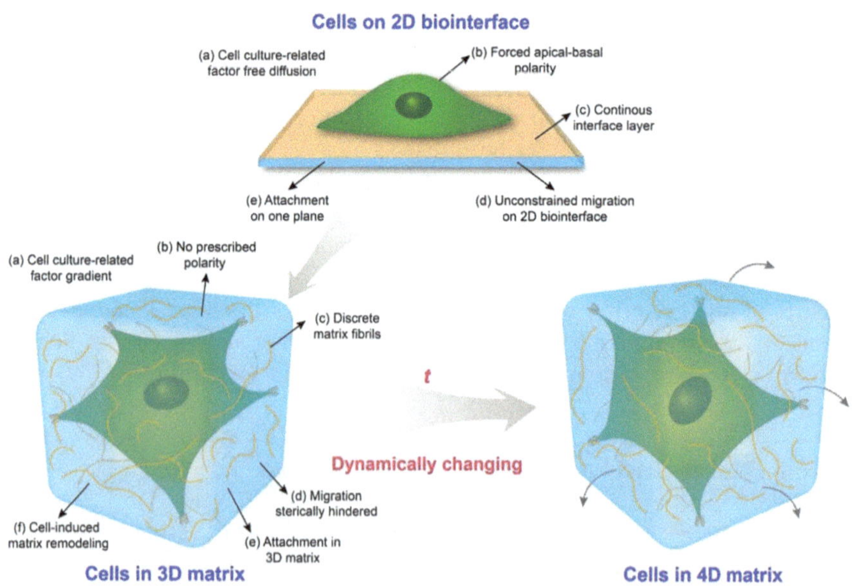

Fig. 9.12 Schematic diagram of the interactions between cells and biointerface/matrix. (*Adapted from* [66] *with permission from American Chemical Society (*https://pubs.acs.org/doi/10.1021/acs.chemrev.0c00752*))*

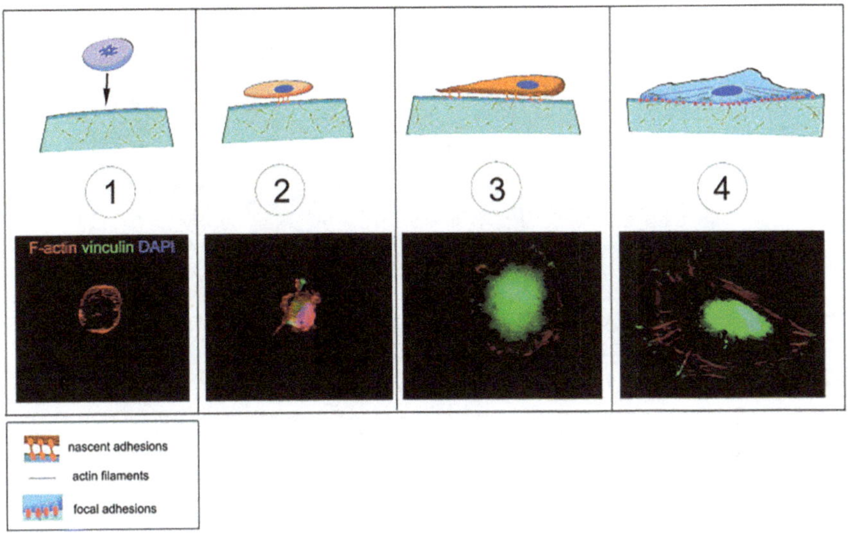

Fig. 9.13 Schematic of cell adhesion. Vinculin: a membrane-cytoskeletal protein in focal adhesion plaques. (*Adapted from* [72] *with permission from Taylor & Francis*)

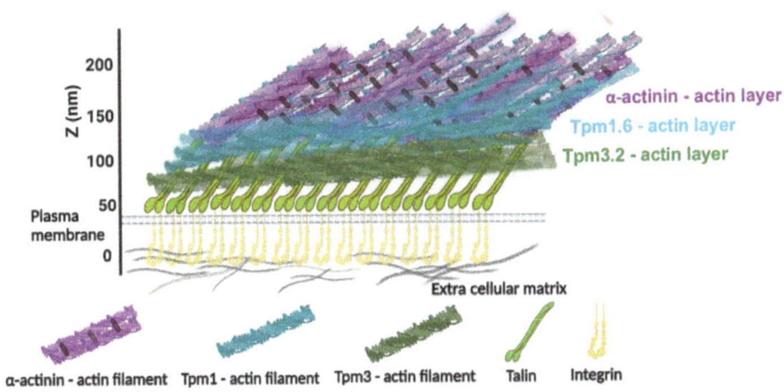

Fig. 9.14 Schematic model of the molecular 3D architecture of a focal adhesion displaying the localizations of three 'actin filament layers' in the nanoscale strata. (*Adapted from* [74]. *The original image was created with* BioRender.com)

signaling platform—the integrin adhesion complex (Fig. 9.15) [73]. Integrins and associated proteins mediate mechanotransduction (i.e., the process of converting mechanical stimuli into a biochemical response), causing the activation of several downstream signaling pathways that regulate cell proliferation, apoptosis, wound healing, and development [75, 76]. There is a network of known interactions and cellular components constituting the focal adhesion complex in mammalian cells—The Adhesome: A Focal Adhesion Network (https://adhesome.org/). Cell migration is essential for physiological processes as diverse as development, immune defense,

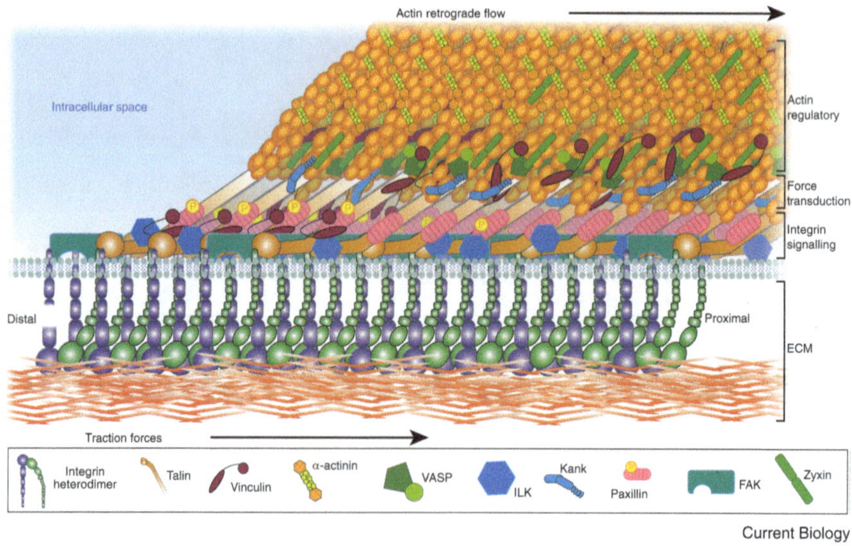

Fig. 9.15 Schematic depicting the z-stratified organization of components within integrin adhesion complexes (IACs), defined by super-resolution microscopy. VASP: vasodilator-stimulated phosphoprotein, ILK: integrin-linked kinase, FAK: focal adhesion kinase, ECM: extracellular matrix. (*Adapted from* [73] *with permission form Elsevier*)

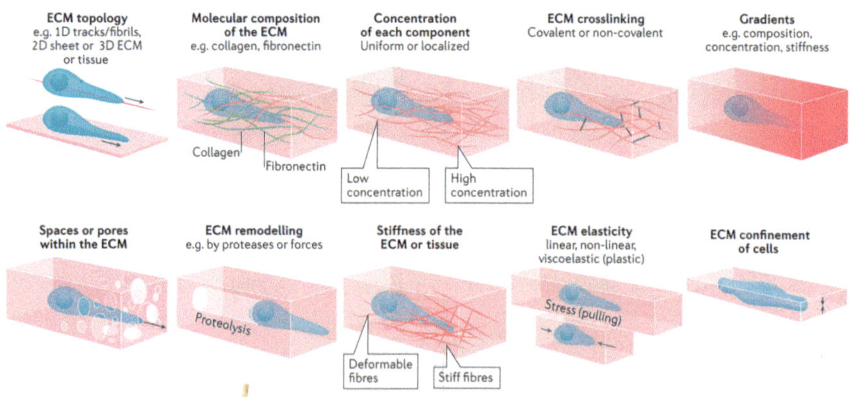

Fig. 9.16 Extracellular matrix (ECM) features that can regulate or modulate cell migration. (*Adapted from* [78] *with permission from Springer Nature*)

wound healing, and regeneration, as well as a hallmark of cancer malignancy. The ability of a cell to migrate depends on the interplay between intracellular states and extracellular environments. ECM features that can regulate or modulate cell migration include ECM topology, molecular composition, concentrations of each ECM component, ECM crosslinking (covalent vs. non-covalent), gradients (such as stiffness and ligand concentration), presence and size of spaces or pores with the 3D environment, ECM remodeling, ECM stiffness, ECM elasticity, and levels of confinement of cells mediated by the ECM (Fig. 9.16) [77, 78]. Cell migration is very different between 2D substrates and more physiological 3D matrices. In more complex environments, cells utilize multiple modes of migration, including mesenchymal, lobopodial, and amoeboid (Fig. 9.17) [78–81].

Biocompatibility is a very important aspect of bioscaffolds. Biocompatibility is defined as the ability of a material to perform its desired functions with respect to a

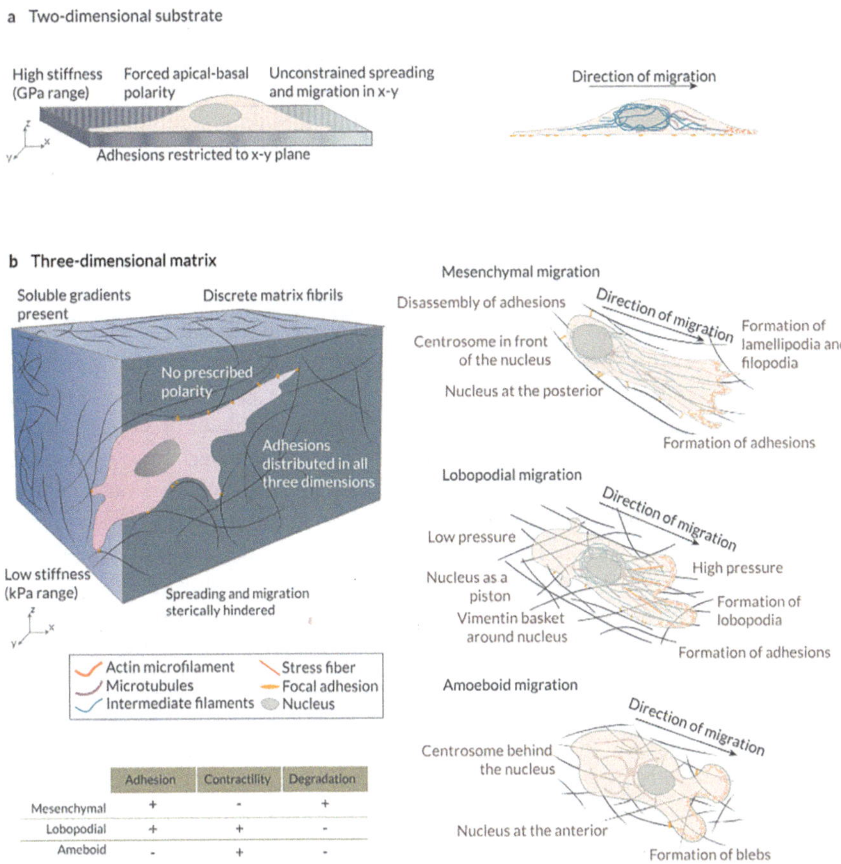

	Adhesion	Contractility	Degradation
Mesenchymal	+	-	+
Lobopodial	+	+	-
Amoeboid	-	+	-

Fig. 9.17 Cells in 2D vs. 3D Cells in (**a**) 2D and (**b**) 3D microenvironments interact differently with their surroundings. Three modes of 3D migration have been identified so far: mesenchymal, amoeboid, and lobopodial migration. In mesenchymal migration, cells attach very strongly to the extracellular matrix through mature stress fiber-linked focal adhesions. These cells also exhibit high matrix-degrading activity. The centrosome is located in front of the nucleus, and the cytoskeletal networks are polarized in the direction of migration. In contrast, amoeboid migration involves very few adhesions and low protease activity. Cells migrate through the formation of contraction-based blebs or use actin-driven protrusions to glide on the substrate. The centrosome is usually positioned behind the nucleus during amoeboid migration. Lastly, during lobopodial migration, tightly adherent cells use actomyosin contractility, hydrostatic pressure, and nuclear pistoning to form bleb-like blunt protrusions called lobopodia. These cells exhibit very low protease activity. (*Adapted from* [79]. *The original image was adapted from* [80, 81])

Table 9.1 International standards for biocompatibility testing of biomaterials[a]

Biological response	Standards[b] ISO	ASTM
Cytotoxicity	10993-5	F813-07; F895-84; F1027-06
Sensitization	10993-10	F720-81; F2147-01; F2148-07
Irritation	10993-10	F719-81; F749-98
Acute systemic toxicity	10993-11	F750-87
Subacute toxicity	10993-11	–
Genotoxicity	10993-3	E1262-88
Immunoresponsiveness	10993-20	F1906-98
Hemocompatibility	10993-4	F756-08
Chronic toxicity	10993-11	–
Carcinogenicity	10993-3	F1439-03
	10993-9; 10993-13	F1983-14
Degradation	10993-14; 10993-15	F1408-97; F763-04; F1904-98
Implantation	10993-6	F981-04; F1983-99

[a] Adapted from [85] with permission from Spandidos Publications
[b] ISO, International Standards Organizations; ATSM, American Society for Testing and Materials

medical therapy, to induce an appropriate host response in a specific application and to interact with living systems without having any risk of injury, toxicity, or rejection by the immune system and undesirable or inappropriate local or systemic effects [18]. The definition has been evolving with time. The author strongly agrees that bioscaffold biocompatibility is a characteristic of a material-biological host system rather than a property of a material, and the term is relative. For example, monolithic polytetrafluoroethylene (PTFE) products tested and used clinically have been found to pass all preclinical biological safety tests and generally considered biocompatible. However, PTFE is not compatible when cell adhesion is required as it is not preferred for cell attachment/adhesion [82–84]. To test biocompatibility, various standards from different organizations, such as the International Standards Organization (ISO) and the American Society for Testing and Materials (ASTM), should be consulted (Table 9.1) [84, 85]. For a comprehensive assessment, the intended application and potential exposure scenarios for selecting relevant assays, cell types, and testing parameters should be considered. Characterization of the composition and properties of should also be conducted [86].

Biodegradability is the degradation of materials after interactions with biological elements [85]. Biodegradability is an essential property for bioscaffolds design. The rate of degradation of the biomaterial should correspond to the rate of the healing process or the rate of formation of new tissue. Moreover, the degradation products should be non-toxic and easily resorbed or eliminated by the body [85]. Regarding polymeric material degradation, it is a multifaceted process leading to the reduction in the physical properties of a polymer due to the breakdown of its molecular and chemical structure over time. This process is dictated by both the polymer properties and the external surrounding environment. Mostly, the degradation *in vivo* is driven by chemical hydrolysis, oxidative, and/or enzymatic processes targeting the polymer backbones [87, 88]. The factors affecting the biodegradability include chemical composition (such as chemical bonds and functional groups), crosslinking, molar mass, morphology (degree of crystallinity and surface-to-volume ratio)

[84, 87]. There are two main types of degradation: surface (degradation occurs on the surface) and bulk degradation (degradation occurs throughout the entire system) (Fig. 9.18) [88]. Regarding bioceramics, the degradation depends on various factors: crystalline features (such as degree of crystallinity and crystal size), physical structure and geometry, presence of additives (e.g., the presence of Zn can inhibit osteoclastic resorption), chemistry, surface properties (such as reactivity) [58].

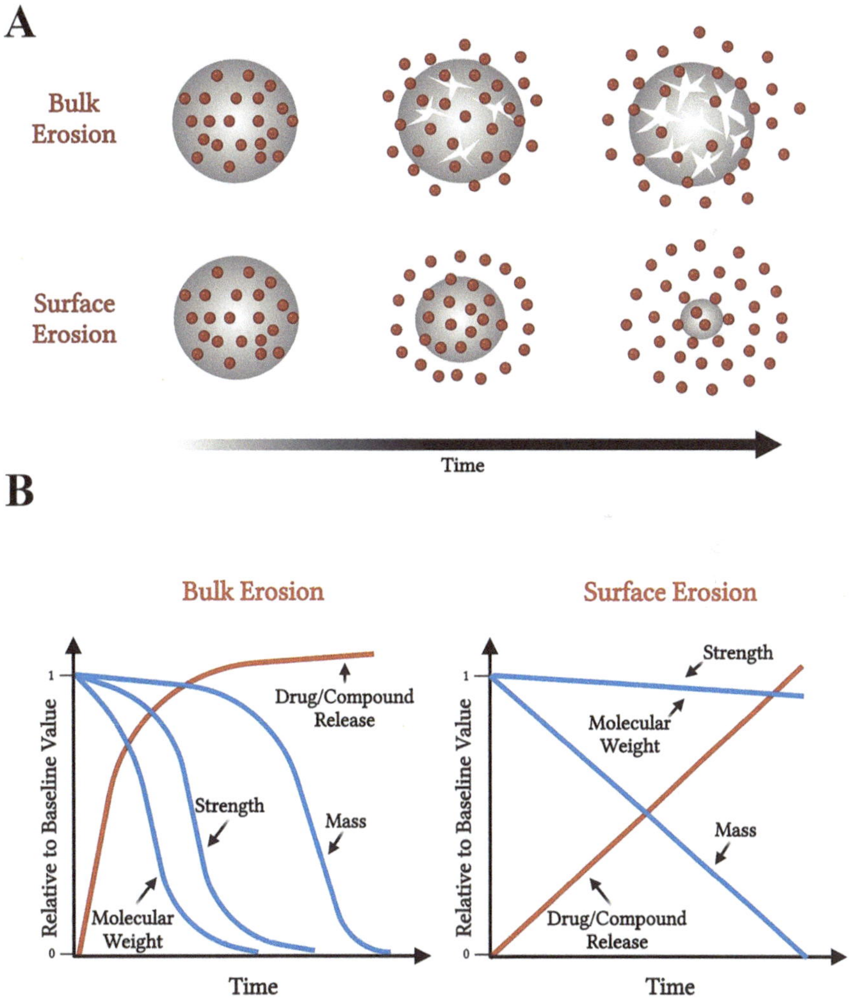

Fig. 9.18 Bulk versus surface erosion dynamics. (**a**) Depiction of polymer structure (grey circle) that contains small molecules (red circles) within the polymeric structure. (Top) Demonstration of bulk erosion and more rapid (burst) release of small molecules due to the rate of solvent absorption being greater than polymer degradation, relatively. (Bottom) Demonstration of surface erosion and a more gradual controlled release of small molecules due to the rate of degradation being greater than solvent absorption, relatively. (**b**) Graphical representation of polymeric dressing properties and drug/small molecular release kinetics over time *via* bulk (left) and surface (right) erosion. (*Adapted from* [88])

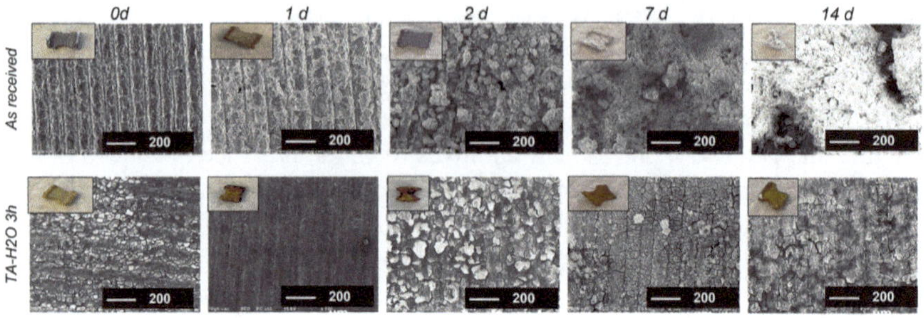

Fig. 9.19 Visual appearance and surface topography of the samples using scanning electron microscopy (SEM). (*Adapted from* [89])

Biodegradable metals are defined as metals which are expected to corrode gradually *in vivo*, with an appropriate host response elicited by released corrosion products, which may pass through, be metabolized or assimilated by cells and/or tissue, and then dissolve completely upon fulfilling the mission to assist with tissue healing with no implant residues [58]. For example, Mg and its alloys are receiving increasing interest due to their biodegradability, mechanical properties closer to one of the human bones, and desired biological effects of Mg ions (Mg^{2+}). Silvia Spriano et al. investigated the degradation control of AZ91 Mg alloy porous structures through tannic acid coatings. The results indicated that the optimized coating significantly reduced the degradation rate (Fig. 9.19) [89].

Biodegradation of bioceramics represents both the physical process of disintegration and fragmentation and resorption *via* cellular mechanism (Fig. 9.20). It is believed that cell mediated CaP resorption occurs as a consequence of particle formation resulting from scaffold disintegration. Following implantation, monocytes/macrophages are among the first cells to colonize the biomaterial surface. Subsequently, multinucleated osteoclasts participate in the resorption of the calcified cement matrix, a process that progresses gradually over time [90]. Osteoclast-mediated bioresorption can be used as an efficient approach to incorporating the dissolution of biomaterials into the bone-remodeling process. For example, the effects of electrical polarization on the surface characteristics and interactions with human osteoclasts of HA and carbonate-substituted apatite (CA) have been investigated. The results indicated that the surface characteristics of the CA promoted the osteoclast resorption (Fig. 9.21) [91].

Regarding surface properties, factors such as chemistry, mechanics (e.g., stiffness), and topography must all be considered. Surface chemistry refers to the chemical characteristics of a material surface and any modifications applied to it; together these features modulate cellular responses [69]. Relevant aspects include hydrophobicity, surface charge, and presence of functional groups (e.g., carboxyl and amino groups). From a mechanical perspective, the surface and/or bulk mechanical properties of the scaffold should mimic the native tissue. Stiffness (or rigidity) represents a material's resistance to deformation under external force. The stiffness of human tissue can vary from 0.1 kPa to 2 GPa. Various polymers have been

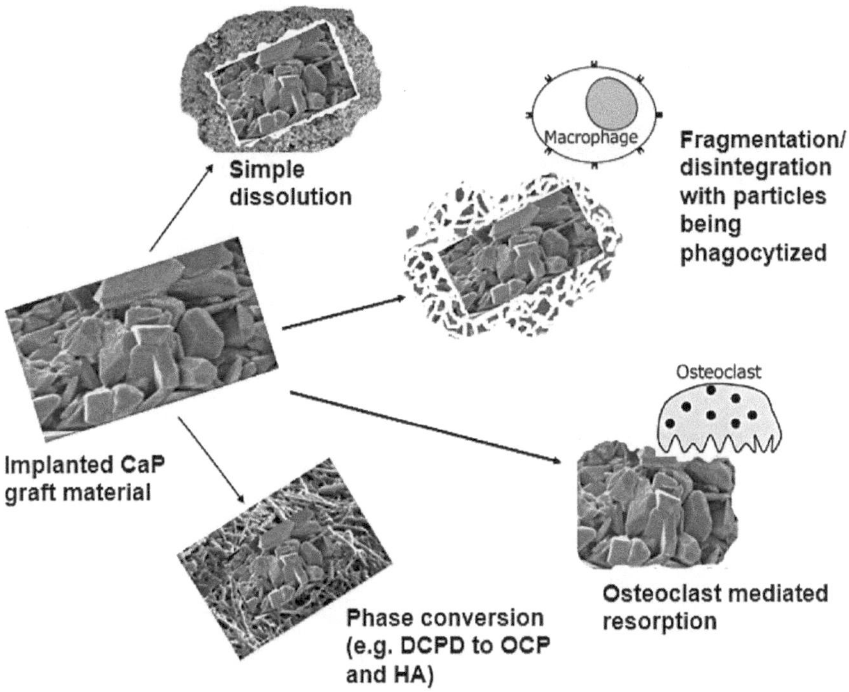

Fig. 9.20 The fate of CaP biomaterials after implantation. CaP: calcium phosphate, DCPD: dicalcium phosphate dihydrate, OCP: octacalcium phosphate, HA: hydroxyapatite. (*Adapted from* [90])

engineered—through adjustments in composition, concentration, and fabrication methods—to better match the intrinsic mechanics of specific tissue environment (Fig. 9.22) [92]. Substrate stiffness has been shown to significantly influence cell behavior, including cell spreading, proliferation, and differentiation (Fig. 9.23) [93–97]. Topology describes the macro-geometry and micro-/nano-scale surface features of the scaffolds [92]. These physical cues play a key role in directing cell behavior. For instance, the width of the microgroove on PLGA substrates has been demonstrated to strongly affect the myogenic differentiation of C2C12 cells (Fig. 9.24) [98].

Regarding scaffold architecture, scaffolds should possess an interconnected pore structure and high porosity to enable cell infiltration and adequate diffusion of nutrients to cells within the construct and the extracellular matrix they produce. In addition, an interconnected porous structure is essential for the efficient removal of metabolic waste products from the scaffold [1]. A pore is defined as a void space within a scaffold, and porosity refers to the proportion of total scaffold volume occupied by pores. Although high porosity is generally desirable, it can conversely decrease mechanical strength [99]. In tissue engineering, several classifications of pore sizes have been proposed. Macropores can be defined as pores larger than 50 μm, while micropores are defined either as pores <50 μm or <10 μm. Another

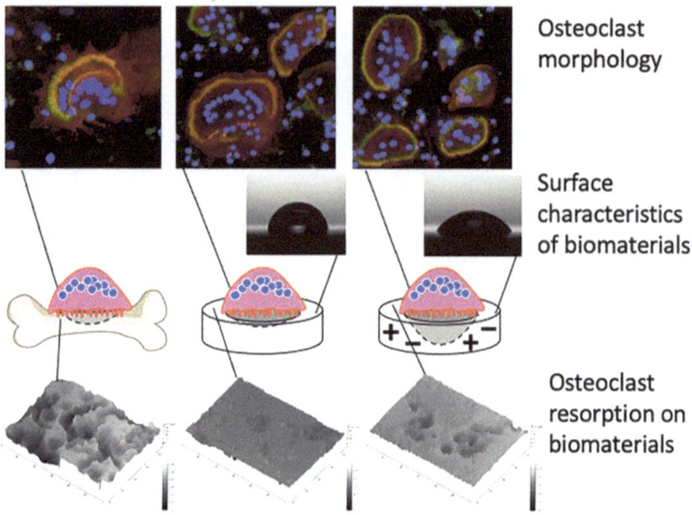

Fig. 9.21 The surface characteristics of the carbonate-substituted apatite promoted osteoclast resorption. Top: Morphologies of osteoclasts adhering to bone slices (left) and to biomaterial samples with (right) and without polarization treatments (middle), as visualized by fluorescent staining of nuclei (blue), vinculin (green), and actin filaments (red). Actin-based sealing rings are a marker of active bone resorption. Bottom: Images of the resorption pits formed on bone slices (left) and on biomaterial samples with (right) and without polarization treatments (middle). (*Adapted from* [91])

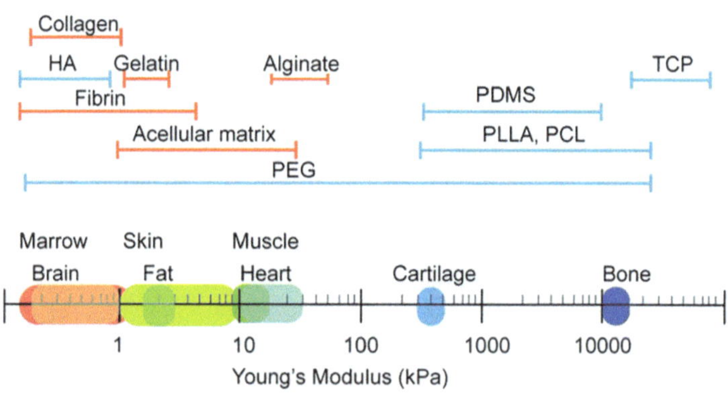

Fig. 9.22 Young's modulus of different tissues and materials for scaffold fabrication. Brain, fat, and skin represent the soft tissue in bodies and share a similar stiffness range with the majority of natural polymers. Cartilage and bone are hard tissues and require substrates with stiffness over 100 kPa. (*Adapted from* [92] *with permission from RSC Publishing*)

classification describes pores as macro (>100 μm), micro (100 μm–100 nm), and nano (<100 nm) [100]. Optimal pore size depends on the target tissue. For example, in bone tissue engineering, scaffolds with smaller pores (50–100 μm) enhance cell attachment, while larger pores (200–400 μm) improve nutrient diffusion and

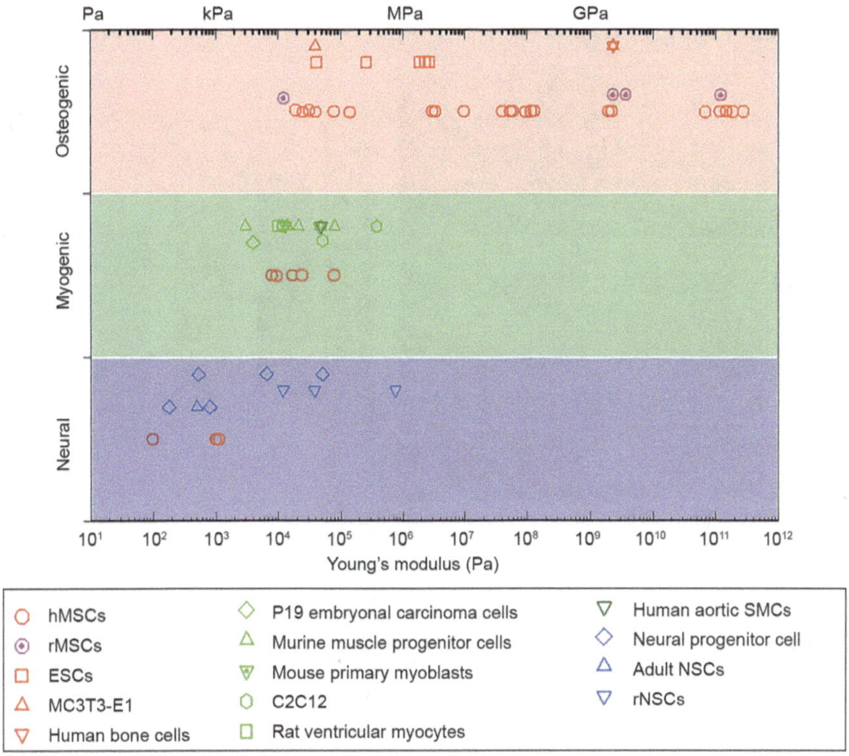

Fig. 9.23 Relationship between stem cell differentiation and substrate stiffness; each symbol represents one cell type. (*Adapted from* [94])

angiogenesis [101]. Pore interconnectivity, the network of channels connecting individual pores (reflecting the extent to which pores are linked throughout a scaffold), governs scaffold permeability by regulating fluid flow throughout the structure. Open pores are connected pores forming continuous pathways accessible to gases, liquids, and suspended particulates, while closed pores, isolated pores with no connection to the scaffold surface, do not contribute to the cell microenvironment, but primarily influence mechanical behavior [100]. High pore interconnectivity promotes cell infiltration, nutrient flow, and integration with host tissue (Fig. 9.25) [102]. Scaffold architectures can be mathematically defined using computational design tools [103]. The unit cell serves as the fundamental building block of a porous architecture and is used in additive manufacturing for scaffold fabrication. The design can be classified as non-parametric (structural and geometric design, such as diamond and the body-centered cubic) or parametric (cellular structures are generated according to specific algorithms; such as the Triply Periodic Minimal surface) (Fig. 9.26) [104].

There are various approaches for fabricating bioscaffolds. For porous bioscaffolds specifically, established techniques include gas foaming, porogen-leaching,

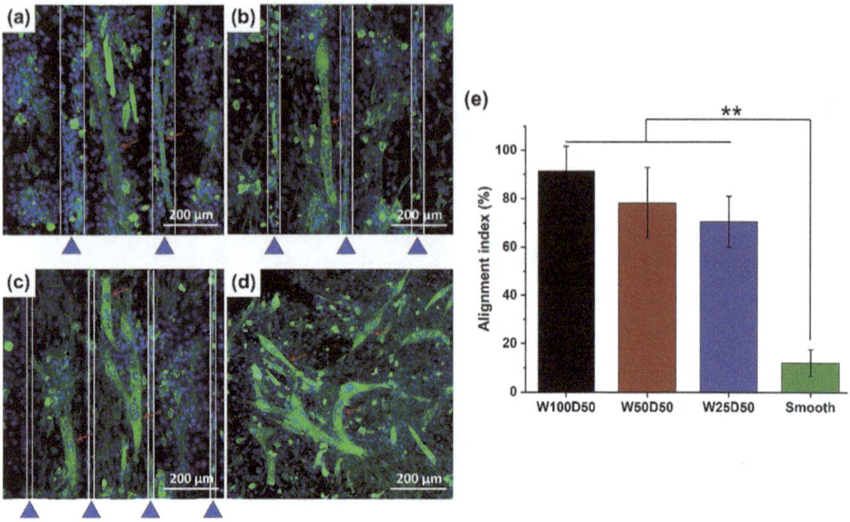

Fig. 9.24 Morphological observation of myotubes on microgrooved PLGA substrates with different groove widths. (**a**) W100D50, (**b**) W50D50, (**c**) W25D50, and (**d**) smooth substrate. The depth of the microgrooves was 50 μm for all samples, and their locations are indicated in the images by blue triangles. The locations of the myotubes in the images are indicated with red arrows. (**e**) Percentage of aligned myotubes out of the total number of myotubes (alignment index). PLGA: poly(lactic-*co*-glycolic acid), W: width, D: depth. (*Adapted from* [98] *with permission from American Chemical Society (*https://pubs.acs.org/doi/10.1021/acsomega.1c02347*)*)

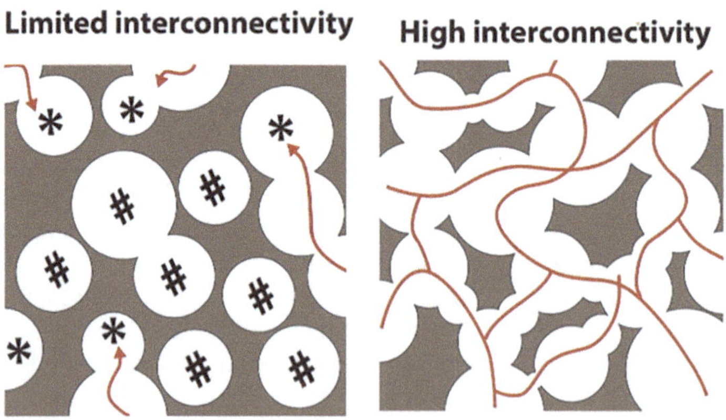

Fig. 9.25 Low interconnectivity limits cell infiltration due to blind (*) and inaccessible (#) pores, while higher interconnectivity provides enhanced permeability and cell penetration. (*Adapted from* [102])

emulsification, freeze-dying, templating, additive manufacturing, and electrospinning (Table 9.2) [100]. Traditional scaffold fabrication methods, such porogen-leaching and freeze-drying, are illustrated in Fig. 9.27 [105]. Additive manufacturing (or 3D printing) is a process in which parts are created layer-by-layer from 3D

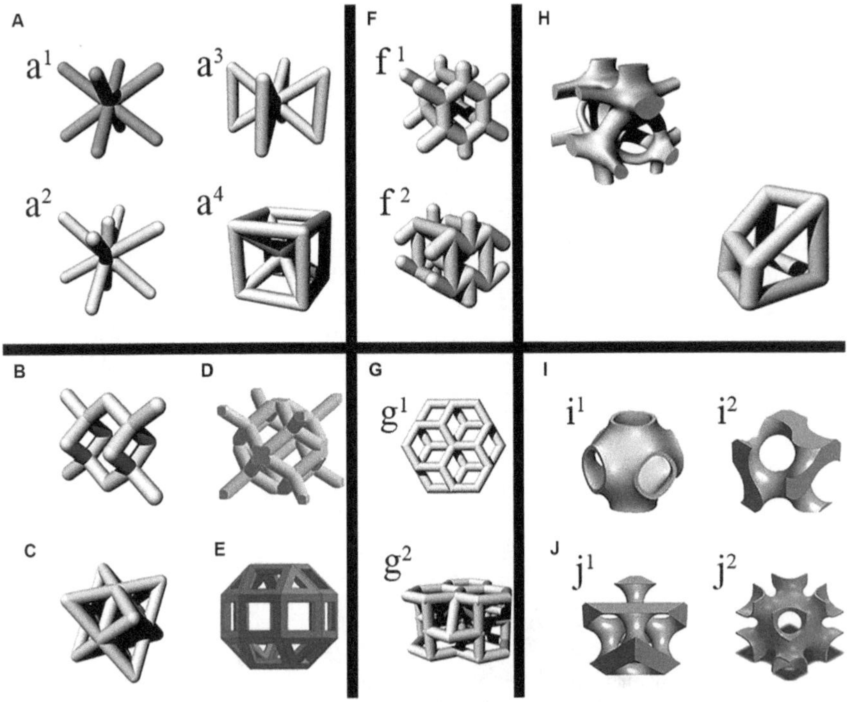

Fig. 9.26 Units of various non-parametric and parametric designs. (**a**) The body-centered cubic (BCC) unit and its modified units. (**a**[1]) BCC unit. (**a**[2]) A type of modified BCC unit by adding vertical stiffeners through the center of the BCC/OC unit. (**a**[3]) The BCCZ unit. (**a**[4]) The pillar BCC. (**b**) The Diamond unit. (**c**) The Octet unit. (**d**) The rhombic dodecahedron unit. (**e**) The rhombic cube octahedron unit. (**f**) The Honeycomb units. (**f**[1]) The common Honeycomb unit. (**f**[2]) The modified Honeycomb unit. (**g**) A new type Honeycomb unit. (**g**[1]) Positional relationship of adjacent layers of new type Honeycomb unit. (**g**[2]) Structural characteristics of the layered slice and rod-connected structure of new-type Honeycomb unit. (**h**) The Voronoi units. (**i, j**) The classification of TPMS. (**i**) The surface dominated by stretching. (**i**[1]) Primitive surface. (**i**[2]) I-WP surface. (**j**) The surface dominated by bending. (**j**[1]) Diamond surface. (**j**[2]) Gyroid surface. (*Adapted from* [104])

computer model data (based on digital slicing). Additive manufacturing techniques can be broadly categorized into four groups: material extrusion, vat photopolymerization, binder jetting, and powder bed fusion (Fig. 9.28) [84, 106]. Additive manufacturing enables fabrication of complex structures with high degree of design freedom, high efficiency, and, critically, supports personalized medicine approaches [107, 108]. Electrospinning is one of the most widely used techniques for generating fibrous scaffolds. Nanofibers possess a high surface-to-volume ratio and can be organized into architectures that closely mimic the extracellular matrices, promoting cell adhesion, proliferation, migration, and differentiation [109]. Electrospinning is an electrohydrodynamic process that employs viscous solutions or melts to produce micro- and nano-scale fibers. As shown in Fig. 9.29, a typical electrospinning system consists of three major components: a high voltage power supply, a syringe pump quipped with a spinneret (mostly a metallic blunt needle), and a grounded

Table 9.2 Manufacturing techniques for the generation of pore within substrate biomaterials[a]

Technique	Using	Advantages	Limitations
Porogen leaching	Particulates, fibers, meshes, salt (NaCl), sugar, paraffin, gelatin, ice crystals	The amount and size of the porogen can modify the pore size and porosity; a minimal amount of polymer is required	Poor reproducibility; residual porogens; poor pore interconnectivity
Gas forming	Gas generation (i.e., CO_2), surface active agents (i.e., tween)	No need for leaching step; reduce overall fabrication time	Low initial strength; poor reproducibility; risk of embolism; difficult control of pore connectivity and pore sizes
Emulsion	Oil-water mixtures	Allow modification of pore size and percentage	May require a cleaning step; risk of embolism
Freeze-drying	Water freezing + sublimation	Porosity and pore size can be modified by controlling the porosity parameters, polymer/water ratio, and freezing temperature; no need for a further rinsing step	Energy and time-consuming; the parameters need to be controlled to increase scaffold homogeneity
Templates	Positive replica (polymeric foams), negative replica (indirect rapid prototyping)	Porosity is determined by the porosity of the template	Very low strength; residuals from the firing step; crystal changes from the firing step; limited pore connectivity; problems with residual solvents
3D printing	Direct rapid prototyping	Allows fabrication of a more complex 3D structure with a controlled internal structure	Micro-architecture limited by the particle's size; applicable to a limited number of polymers due to high temperatures; ceramic scaffold requires sintering to improve on the brittle toughness
Selective laser sintering	Layer-by-layer fusion of a powder bed by laser beam	Allows fabrication of complex scaffolds geometries; fast, cost-effective, and does not require the use of any organic solvents	High energy consumed; chance of degradation and decomposition of materials due to heat generation
Electrospinning	A potential difference between a polymeric solution and a collector	Simple and cost-effective; generate high surface area and high porosity; possibility to control the diameter of each electrospun fiber	Involvement of toxic organic solvents during fabrication

Table 9.2 (continued)

Technique	Using	Advantages	Limitations
Encapsulation	Cell entrapment during scaffold gelation	Ability to provide immunoshielding for cells; possibility of fabricating injectable forms	The gelation process and encapsulating materials should be biocompatible; difficulty in control of diffusion coefficient to/from cells; lack of internal porous structure

[a] Adapted from [100] with permission from Springer Nature

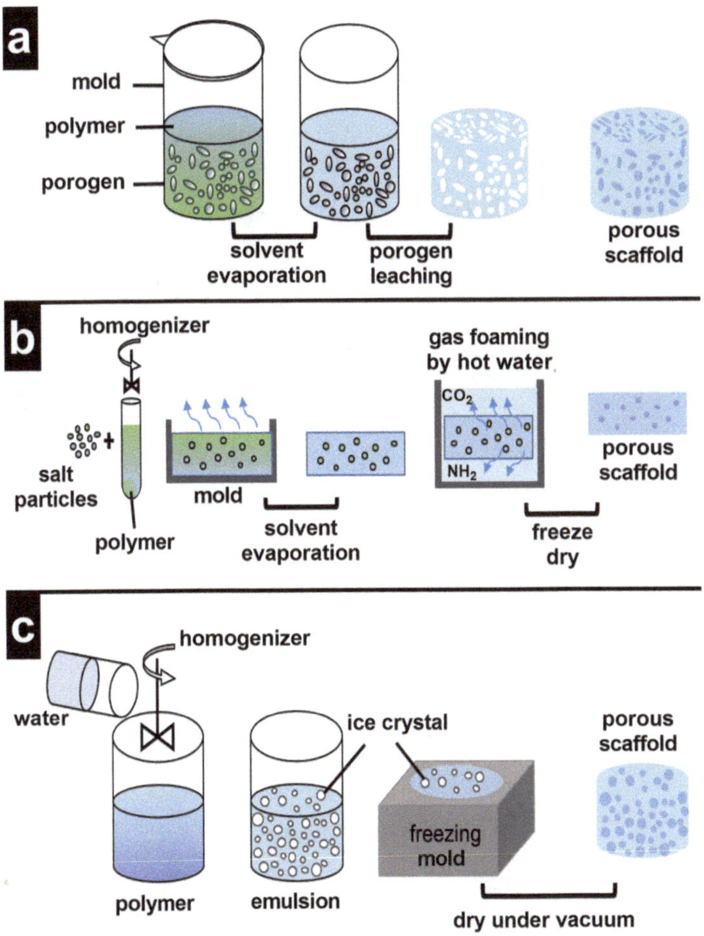

Fig. 9.27 Various porous scaffold fabrication techniques. (**a**) Porogen leaching, (**b**) Gas foaming, (**c**) Freeze-drying. (*Adapted from* [105])

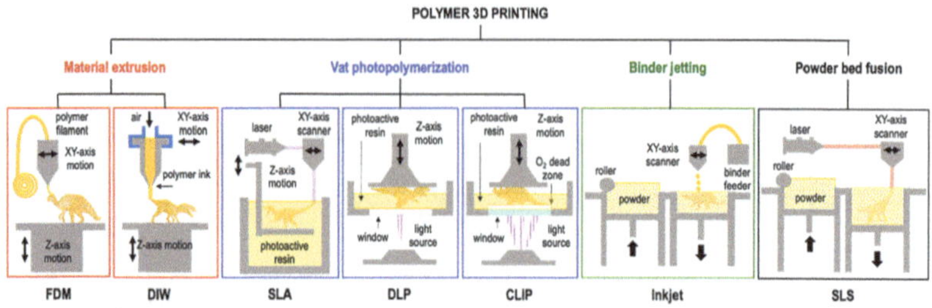

	FDM	DIW	SLA	DLP	CLIP	Inkjet	SLS
Advantages	Cheapest technology, strong parts, variety of nontoxic materials	Relatively cheap, variety of materials, mild print conditions	High resolution, complex geometries, smooth finish	High resolution, complex geometries, smooth finish, fast build speeds	Fast build speeds, high resolution, complex geometries, smooth finish, no "peel" step	Inexpensive, mild print conditions, no support structures	Fast build speeds, relatively low cost, no support structures
Disadvantages	Low resolution, poor surface finish, relatively slow build speeds, need for support structures, poor reproducibility	Relatively slow build speeds, poor mechanical strength, low resolution, limited geometric complexity, need for support structures, poor reproducibility	Post-curing steps, need for support structures, limited range of materials, products prone to shrinkage, slow build speed	Post-curing steps, need for support structures, limited range of materials, products prone to shrinkage, low upper limit for viscosity of ink	Post-curing steps, parts often require support structures, limited range of materials, products prone to shrinkage	Limited reproducibility, poor surface finish, post-processing steps, low upper limit for viscosity of ink, long drying times, poor mechanical strength	Resolution dependent on particle size, powdery surface finish, limited range of materials
Common polymers	PLA, PCL, Polycarbonates, PLGA	PLA, naturally occurring polymers	PPF, Polyacrylates	PPF, Polyacrylates	PPF, Polyacrylates	PCL, PLA/PEG	Polycarbonates, nylons, poly(styrene)

Fig. 9.28 Overview of the common 3D printing (additive manufacturing) techniques used for polymer processing and their advantages/disadvantages. FDM: fused deposition modeling, DIW: direct ink write, SLA: stereolithography, DLP: digital light processing, CLIP: continuous liquid interface production, SLS: selective laser sintering, PLA: polylactide, PCL: polycaprolactone, PLGA: poly(lactic-*co*-glycolic acid), PPF: poly(propylene fumarate), PEG: poly(ethylene glycol). (*Adapted with permission from* [106]. *Copyright (2021) American Chemical Society*)

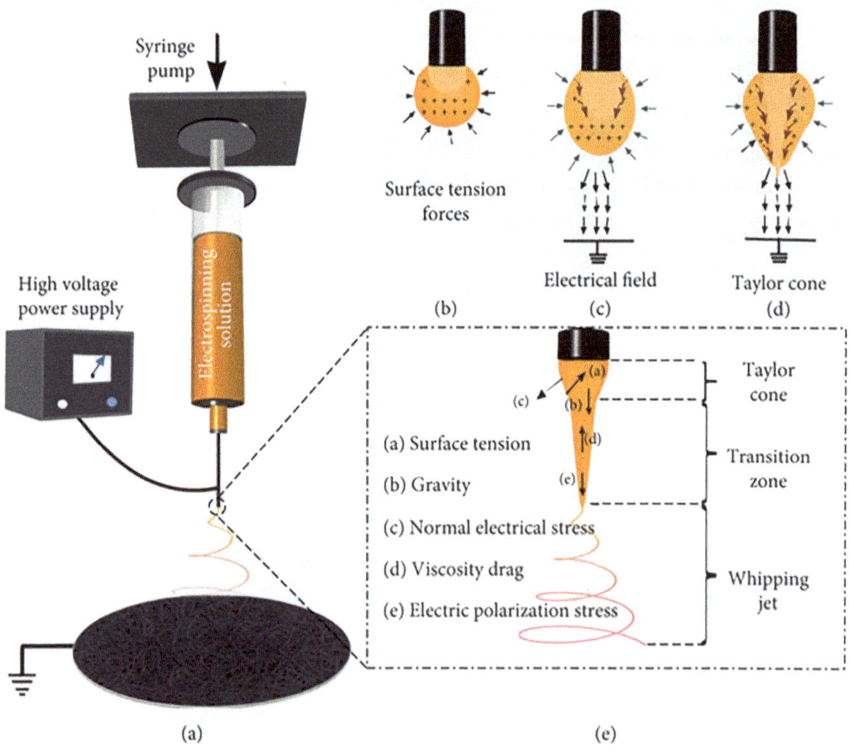

Fig. 9.29 Cartoon illustration of a typical electrospinning setup with important parameters involved in the electrospinning process: (a) a general setup, (b–d) the formation of Taylor cone from the droplet, and (e) factors affecting the formation of the jet. (*Adapted from* [111])

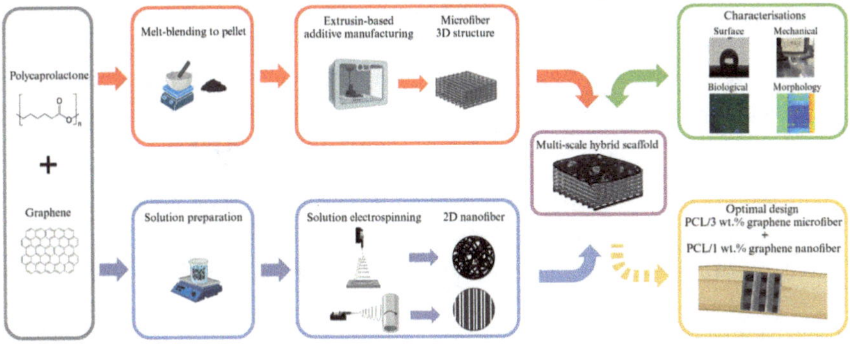

Fig. 9.30 Schematic illustration of the hybrid scaffold manufacturing and characterization. PCL: polycaprolactone. (*Adapted from* [112])

collector. Upon high-voltage application, a pendant droplet forms at the spinneret tip. When electrostatic repulsion exceeds surface tension, the droplet elongates into a conical shape, known as a Taylor cone. At a critical voltage value, a charged jet is ejected from the cone apex. The jet undergoes stretching and bending instabilities, leading to rapid solvent evaporation and stream thinning. As the jet is stretched into finer diameters, it solidifies rapidly, resulting in the deposition of solid fibers on the grounded collector [110, 111]. A recent trend in scaffold fabrication is the integration of multiple manufacturing techniques to leverage their complementary advantages and mitigate individual limitations [99]. For example, hybrid scaffolds combining extrusion-based additive manufacturing and solution electrospinning have been developed. In one such system, polycaprolactone (PCL) served as the base material, while graphene nanosheets were incorporated as functional fillers to enhance mechanical, electrical, surface, and biological properties (Fig. 9.30) [112].

In summary, tissue-engineering scaffolds must provide adequate initial mechanical strength and stiffness, maintain structural integrity during tissue growth and remodeling, and present surface properties optimized for cell–material interactions. Additionally, for clinical translation, scaffold fabrication should follow cGMP principles, ensuring consistent quality while accounting for practical considerations such as production speed and cost [99].

Review and Practice Questions

1. What is a bioscaffold, and what are its major functions in tissue engineering?
2. What are the three major structural components of the extracellular matrix (ECM), and what are their functions?
3. What is the primary function of proteoglycans, and what is their typical structure?
4. What are the benefits of using ECM as a bioscaffold?
5. What are biomaterials, and what are the three major classes of biomaterials?
6. Conduct a literature search and write a short essay on a specific tissue engineering application of a polymeric biomaterial not discussed in this chapter.
7. What are the major tissue engineering application areas of biometals?

8. Why is biodegradation of biometals preferred for certain tissue engineering applications?
9. What are ceramics and glasses, and what are their major tissue engineering applications?
10. What parameters should be considered when designing a polymer array?
11. What are the key considerations when designing bioscaffolds?
12. What is the process of cell adhesion, and what is focal adhesion?
13. What is mechanotransduction, and why is it significant?
14. What are the influencing factors and modes of cell migration?
15. What is biocompatibility, and why is it considered a property of a **system** rather than a material alone?
16. What are some of the tests and standards for evaluating biocompatibility?
17. What is biodegradability, and what are some of the key factors affecting biodegradability?
18. Compare surface erosion and bulk erosion.
19. What surface properties should be considered when designing bioscaffolds? What are stiffness and topography?
20. What are porosity and pore interconnectivity in porous bioscaffolds?
21. Compare different bioscaffold fabrication techniques.
22. Describe the electrospinning process.

References

1. O'Brien, F.J., *Biomaterials & scaffolds for tissue engineering.* Materials Today, 2011. **14**(3): 88–95.
2. Valdoz, J.C., et al., *The ECM: To Scaffold, or Not to Scaffold, That Is the Question.* International Journal of Molecular Sciences, 2021. **22**(23):12690.
3. Karamanos, N.K., et al., *A guide to the composition and functions of the extracellular matrix.* The FEBS Journal, 2021. **288**(24):6850–6912.
4. Sicari, B.M., et al., *Chapter 5 - Extracellular matrix as a bioscaffold for tissue engineering*, in *Tissue Engineering (Third Edition)*, J. De Boer, et al., Editors. 2023, Academic Press.
5. Naba, A., *Mechanisms of assembly and remodelling of the extracellular matrix.* Nature Reviews Molecular Cell Biology, 2024. **25**(11):865–885.
6. Mouw, J.K., G. Ou, and V.M. Weaver, *Extracellular matrix assembly: a multiscale deconstruction.* Nature Reviews. Molecular Cell Biology, 2014. **15**(12):771–85.
7. Wang, Y., Z. Wang, and Y. Dong, *Collagen-Based Biomaterials for Tissue Engineering.* ACS Biomaterials Science & Engineering, 2023. **9**(3):1132–1150.
8. Aumailley, M., *The laminin family.* Cell Adhesion & Migration, 2013. **7**(1):48–55.
9. Nonnast, E., E. Mira, and S. Mañes, *The role of laminins in cancer pathobiology: a comprehensive review.* Journal of Translational Medicine, 2025. **23**(1):83.
10. Dalton, C.J. and C.A. Lemmon, *Fibronectin: Molecular Structure, Fibrillar Structure and Mechanochemical Signaling.* Cells, 2021. **10**(9):2443.
11. Schaefer, L. and R.M. Schaefer, *Proteoglycans: from structural compounds to signaling molecules.* Cell and Tissue Research, 2010. **339**(1):237–246.
12. Alcaide-Ruggiero, L., R. Cugat, and J.M. Domínguez, *Proteoglycans in Articular Cartilage and Their Contribution to Chondral Injury and Repair Mechanisms.* International Journal of Molecular Sciences, 2023. **24**(13):10824.

13. Bu, C. and L. Jin, *NMR Characterization of the Interactions Between Glycosaminoglycans and Proteins.* Frontiers in Molecular Biosciences, 2021.**8**:646808.
14. Vilaça-Faria, H., et al., *Extracellular matrix-derived materials for tissue engineering and regenerative medicine: A journey from isolation to characterization and application.* Bioactive Materials, 2024. **34**:494–519.
15. Vallet-Regí, M., *Evolution of Biomaterials.* Frontiers in Materials, 2022. **9**:864016.
16. *Resource Guide*. Available from: https://biomaterials.org/about-society/resource-guide#C.
17. *About the journal.* Available from: https://www.sciencedirect.com/journal/biomaterials.
18. Ghasemi-Mobarakeh, L., et al., *Key terminology in biomaterials and biocompatibility.* Current Opinion in Biomedical Engineering, 2019. **10**:45–50.
19. *I - Introduction*, in *Definitions of Biomaterials for the Twenty-First Century*, D. Williams and X. Zhang, Editors. 2019, Elsevier.
20. Birla, R., *Biomaterials for Tissue Engineering*, in *Introduction to Tissue Engineering*. 2014, Wiley-IEEE.
21. Biobaku-Mutingwende, B., *Introduction to Biomaterials*, in *Applications of 3D printing in Biomedical Engineering*, N.R. Sharma, et al., Editors. 2021, Springer.
22. Hu, T., et al., *Advances of naturally derived biomedical polymers in tissue engineering.* Frontiers in Chemistry, 2024. **12**:1469183.
23. Wu, Y., et al., *E2-Loaded Microcapsules and Bone Marrow–Derived Mesenchymal Stem Cells with Injectable Scaffolds for Endometrial Regeneration Application.* Tissue Engineering Part A, 2023. **30**(3–4):115–130.
24. Socci, M.C., et al., *Polymeric Materials, Advances and Applications in Tissue Engineering: A Review.* Bioengineering, 2023. **10**(2):218.
25. Arif, Z.U., et al., *Recent advances in 3D-printed polylactide and polycaprolactone-based biomaterials for tissue engineering applications.* International Journal of Biological Macromolecules, 2022. **218**:930–968.
26. Eldeeb, A.E., S. Salah, and N.A. Elkasabgy, *Biomaterials for Tissue Engineering Applications and Current Updates in the Field: A Comprehensive Review.* AAPS PharmSciTech, 2022. **23**(7):267.
27. Doberenz, F., et al., *Thermoresponsive polymers and their biomedical application in tissue engineering – a review.* Journal of Materials Chemistry B, 2020. **8**(4):607–628.
28. Brouns, J.E.P. and P.Y.W. Dankers, *Introduction of Enzyme-Responsivity in Biomaterials to Achieve Dynamic Reciprocity in Cell–Material Interactions.* Biomacromolecules, 2021. **22**(1):4–23.
29. Dee, K.C., D.A. Puleo, and R. Bizios, *Biomaterials*, in *An Introduction To Tissue-Biomaterial Interactions*. 2002, Wiley.
30. Kumar, S., et al., *Advancements in alginate-derived biomaterials for tissue engineering and regenerative medicine: Emerging innovations and translational outlooks.* Journal of Drug Delivery Science and Technology, 2025. **114**:107569.
31. Farshidfar, N., S. Iravani, and R.S. Varma, *Alginate-Based Biomaterials in Tissue Engineering and Regenerative Medicine.* Marine Drugs, 2023. **21**(3):189.
32. Martău, G.A., M. Mihai, and D.C. Vodnar, *The Use of Chitosan, Alginate, and Pectin in the Biomedical and Food Sector—Biocompatibility, Bioadhesiveness, and Biodegradability.* Polymers, 2019. **11**(11):1837.
33. Zhang, W., *Chapter 6 Engineered carriers in drug delivery*, in *Fundamentals and Design of Drug Delivery Systems: A Textbook with Lab Activities*. 2026, Academic Press.
34. David, N., *24 - Pectin in tissue engineering*, in *Natural Biopolymers in Drug Delivery and Tissue Engineering*, R. Jayakumar and V.P. Murali, Editors. 2023, Woodhead Publishing.
35. Abualsoud, B.M., et al., *Pectin-based composites: a promising approach for tissue engineering and wound healing.* International Journal of Polymeric Materials and Polymeric Biomaterials, 2025. **75**(1):69–96.
36. Wu, Y., et al., *E2-Loaded Microcapsules and Bone Marrow–Derived Mesenchymal Stem Cells with Injectable Scaffolds for Endometrial Regeneration Application.* Tissue Engineering Part A, 2024. **30**(3–4):115–130.

37. de Sousa Victor, R., et al., *A Review on Chitosan's Uses as Biomaterial: Tissue Engineering, Drug Delivery Systems and Cancer Treatment.* Materials, 2020. **13**(21):4995.
38. Haider, A., et al., *Chitosan as a tool for tissue engineering and rehabilitation: Recent developments and future perspectives – A review.* International Journal of Biological Macromolecules, 2024. **278**:134172.
39. Satchanska, G., S. Davidova, and P.D. Petrov, *Natural and Synthetic Polymers for Biomedical and Environmental Applications.* Polymers, 2024. **16**(8):1159.
40. Liu, S., et al., *Current applications of poly(lactic acid) composites in tissue engineering and drug delivery.* Composites Part B: Engineering, 2020. **199**:108238.
41. Dukle, A. and M.R. Sankar, *3D-printed polylactic acid scaffolds for bone tissue engineering: Bioactivity enhancing strategies based on composite filaments and coatings.* Materials Today Communications, 2024. **40**:109776.
42. Ghaedamini, S., et al., *PCL/Agarose 3D-printed scaffold for tissue engineering applications: fabrication, characterization, and cellular activities.* Research in Pharmaceutical Sciences, 2023. **18**(5):566–579.
43. Chowdhury, S.K., V. Nagarjuna, and B. Bhaskar, *Metallic Biomaterials in Tissue Engineering: Retrospect and Prospects*, in *Biomaterials in Tissue Engineering and Regenerative Medicine: From Basic Concepts to State of the Art Approaches*, B. Bhaskar, et al., Editors. 2021, Springer.
44. Karunakar, K.K., et al., *Comprehensive Review of Biomedical Metals and Strategies for Advancing Regenerative Medicine.* Biomedical Materials & Devices, 2026. **4**:1483–1515.
45. Al-allaq, A.A., et al., *Review of In Vivo Investigations on Metal Implants for Bone Tissue Engineering.* Regenerative Engineering and Translational Medicine, 2025. **11**(1):132–156.
46. Glenske, K., et al., *Applications of Metals for Bone Regeneration.* International Journal of Molecular Sciences, 2018. **19**(3):826.
47. Chen, W., W.-q. Zhu, and J. Qiu, *Impact of exogenous metal ions on peri-implant bone metabolism: a review.* RSC Advances, 2021. **11**(22):3152–13163.
48. Luo, X., et al., *The Roles of Exosomes upon Metallic Ions Stimulation in Bone Regeneration.* Journal of Functional Biomaterials, 2022. **13**(3):126.
49. Lin, T., et al., *Manufacturing of porous magnesium scaffolds for bone tissue engineering by 3D gel-printing.* Materials & Design, 2021. **209**:109948.
50. Ye, J., et al., *3D printed porous magnesium metal scaffolds with bioactive coating for bone defect repair: enhancing angiogenesis and osteogenesis.* Journal of Nanobiotechnology, 2025. **23**(1):160.
51. Boccaccini, A.R., *3 - Ceramics*, in *Biomaterials, Artificial Organs and Tissue Engineering*, L.L. Hench and J.R. Jones, Editors. 2005, Woodhead Publishing.
52. Hasanuzzaman, M., et al., *Properties of Glass Materials*, in *Encyclopedia of Materials: Technical Ceramics and Glasses*, M. Pomeroy, Editor. 2016, Elsevier.
53. Benrashed, M.A., et al., *7 - Conventional to advanced endodontics: Use of bioactive materials*, in *Biomaterials in Endodontics*, Z. Khurshid, M.S. Zafar, and S. Najeeb, Editors. 2022, Woodhead Publishing.
54. Huang, J., *Chapter 20 - Design and Development of Ceramics and Glasses*, in *Biology and Engineering of Stem Cell Niches*, A. Vishwakarma and J.M. Karp, Editors. 2017, Academic Press.
55. Arango-Ospina, M. and A.R. Boccaccini, *Chapter 4 - Bioactive glasses and ceramics for tissue engineering*, in *Tissue Engineering Using Ceramics and Polymers (Third Edition)*, A.R. Boccaccini, P.X. Ma, and L. Liverani, Editors. 2022, Woodhead Publishing.
56. Perrotti, V., et al., *1 - Introduction to bone response to dental implant materials*, in *Bone Response to Dental Implant Materials*, A. Piattelli, Editor. 2017, Woodhead Publishing.
57. Cinici, B., et al., *Fabrication Strategies for Bioceramic Scaffolds in Bone Tissue Engineering with Generative Design Applications.* Biomimetics, 2024. **9**(7):409.
58. Hynes, C.G., et al., *Chapter 7 - Degradation of biomaterials*, in *Tissue Engineering (Third Edition)*, J. De Boer, et al., Editors. 2023, Academic Press.

59. Cannio, M., et al., *Bioactive Glass Applications: A Literature Review of Human Clinical Trials.* Materials, 2021. **14**(18):5440.
60. de Carvalho, A.B.G., et al., *Personalized bioceramic grafts for craniomaxillofacial bone regeneration.* International Journal of Oral Science, 2024. **16**(1):62.
61. Dejob, L., et al., *Electrospinning of in situ synthesized silica-based and calcium phosphate bioceramics for applications in bone tissue engineering: A review.* Acta Biomaterialia, 2021. **123**:123–153.
62. Zafar, M.J., D. Zhu, and Z. Zhang, *3D Printing of Bioceramics for Bone Tissue Engineering.* Materials, 2019. **12**(20):3361.
63. Zhao, Y., et al., *Recent Progress of 3D Printing Bioceramic Scaffolds for Bone Regeneration.* Tissue Engineering Part B: Reviews, 2025. Epub ahead of print. https://doi.org/10.1177/19373341251372914.
64. Liu, Y., et al., *Optimal structural characteristics of osteoinductivity in bioceramics derived from a novel high-throughput screening plus machine learning approach.* Biomaterials, 2025. **321**:123348.
65. Cranford, S.W., et al., *Materiomics: An -omics Approach to Biomaterials Research.* Advanced Materials, 2013. **25**(6):802–824.
66. Yang, L., et al., *High-Throughput Methods in the Discovery and Study of Biomaterials and Materiobiology.* Chemical Reviews, 2021. **121**(8):4561–4677.
67. Hook, A.L., et al., *Chapter 9 - Biomaterials discovery: experimental and computational approaches*, in *Tissue Engineering (Third Edition)*, J. De Boer, et al., Editors. 2023, Academic Press.
68. Burroughs, L., et al., *Discovery of synergistic material-topography combinations to achieve immunomodulatory osteoinductive biomaterials using a novel in vitro screening method: The ChemoTopoChip.* Biomaterials, 2021. **271**:120740.
69. Donnelly, H., et al., *Chapter 8 - Cell–material interactions*, in *Tissue Engineering (Third Edition)*, J. De Boer, et al., Editors. 2023, Academic Press.
70. Neffe, A.T., *Cell–Material Interactions 2022.* International Journal of Molecular Sciences, 2023. **24**(7):6057.
71. Buitrago, J.O., B.M. Bosch, and R.A. Pérez, *Cell-Materials Interaction*, in *Biomaterials and Tissue Engineering*, O. Gunduz, et al., Editors. 2023, Springer.
72. Van Tam, J.K., et al., *Mesenchymal stem cell adhesion but not plasticity is affected by high substrate stiffness.* Science and Technology of Advanced Materials, 2012. **13**(6):064205.
73. Chastney, M.R., J.R.W. Conway, and J. Ivaska, *Integrin adhesion complexes.* Current Biology, 2021. **31**(10):R536–R542.
74. Kumari, R., et al., *Focal adhesions contain three specialized actin nanoscale layers.* Nature Communications, 2024. **15**(1):2547.
75. Katoh, K., *Integrin and Its Associated Proteins as a Mediator for Mechano-Signal Transduction.* Biomolecules, 2025. **15**(2):166.
76. Alenghat, F.J. and D.E. Ingber, *Mechanotransduction: All Signals Point to Cytoskeleton, Matrix, and Integrins.* Science's STKE, 2002. **2002**(119):pe6.
77. Xu, K.L., R.L. Mauck, and J.A. Burdick, *Hydrogel Design to Understand and Guide 3D Cell Migration.* Regenerative Engineering and Translational Medicine, 2025. **11**:802–813
78. Yamada, K.M. and M. Sixt, *Mechanisms of 3D cell migration.* Nature Reviews. Molecular Cell Biology, 2019. **20**(12):738–752.
79. Merino-Casallo, F., et al., *Unravelling cell migration: defining movement from the cell surface.* Cell Adhesion & Migration, 2022. **16**(1):25–64.
80. Duval, K., et al., *Modeling Physiological Events in 2D vs. 3D Cell Culture.* Physiology, 2017. **32**(4):266–277.
81. Seetharaman, S. and S. Etienne-Manneville, *Cytoskeletal Crosstalk in Cell Migration.* Trends in Cell Biology, 2020. **30**(9):720–735.
82. Williams, D.F., *There is no such thing as a biocompatible material.* Biomaterials, 2014. **35**(38):10009–10014.

83. Williams, D.F., *On the mechanisms of biocompatibility.* Biomaterials, 2008. **29**(20):2941–2953.
84. Zhang, W., *Polymer and Biopolymer Processing and Characterization*, in *Biopolymer Engineering*. 2026, Springer.
85. Huzum, B., et al., *Biocompatibility assessment of biomaterials used in orthopedic devices: An overview (Review).* Experimental and Therapeutic Medicine, 2021. **22**(5):1315.
86. Rosa, V., et al., *Guidance on the assessment of biocompatibility of biomaterials: Fundamentals and testing considerations.* Dental Materials, 2024. **40**(11):1773–1785.
87. Umemori, K. and D. Little, *Impact of polymer degradation on cellular behavior in tissue engineering.* Bioprinting, 2025. **50**:e00429.
88. Hodge, J.G., et al., *Evaluating polymeric biomaterials to improve next generation wound dressing design.* Biomaterials Research, 2022. **26**(1):50.
89. Spriano, S., et al., *Tannic Acid Coatings to Control the Degradation of AZ91 Mg Alloy Porous Structures.* Metals, 2023. **13**(2):200.
90. Sheikh, Z., et al., *Mechanisms of in Vivo Degradation and Resorption of Calcium Phosphate Based Biomaterials.* Materials, 2015. **8**(11):7913–7925.
91. Bergara-Muguruza, L., et al., *Surface Electric Fields Increase Human Osteoclast Resorption through Improved Wettability on Carbonate-Incorporated Apatite.* ACS Applied Materials & Interfaces, 2021. **13**(49):58270–58278.
92. Su, T., et al., *Adipogenesis or osteogenesis: destiny decision made by mechanical properties of biomaterials.* RSC Advances, 2022. **12**(38):24501–24510.
93. Baker, B.M., et al., *Cell-mediated fibre recruitment drives extracellular matrix mechanosensing in engineered fibrillar microenvironments.* Nature Materials, 2015. **14**(12): 1262–1268.
94. Yang, Y., et al., *Biophysical Regulation of Cell Behavior—Cross Talk between Substrate Stiffness and Nanotopography.* Engineering, 2017. **3**(1):36–54.
95. Llewellyn, J., et al., *Substrate stiffness alters layer architecture and biophysics of human induced pluripotent stem cells to modulate their differentiation potential.* iScience, 2024. **27**(8):110557.
96. Even-Ram, S., V. Artym, and K.M. Yamada, *Matrix Control of Stem Cell Fate.* Cell, 2006. **126**(4):645–647.
97. Miyoshi, H., et al., *Guideline for design of substrate stiffness for mesenchymal stem cell culture based on heterogeneity of YAP and RUNX2 responses.* Biophysics and Physicobiology, 2023. **20**(2):e200018.
98. Gao, H., et al., *Regulation of Myogenic Differentiation by Topologically Microgrooved Surfaces for Skeletal Muscle Tissue Engineering.* ACS Omega, 2021. **6**(32):20931–20940.
99. Hutmacher, D.W., B. Tandon, and P.D. Dalton, *Chapter 11 - Scaffold design and fabrication*, in *Tissue Engineering (Third Edition)*, J. De Boer, et al., Editors. 2023, Academic Press.
100. Ebrahimi, M., *Porosity parameters in biomaterial science: Definition, impact, and challenges in tissue engineering.* Frontiers of Materials Science, 2021. **15**(3):352–373.
101. Mukasheva, F., et al., *Optimizing scaffold pore size for tissue engineering: insights across various tissue types.* Frontiers in Bioengineering and Biotechnology, 2024. **12**: 1444986.
102. Aldemir Dikici, B. and F. Claeyssens, *Basic Principles of Emulsion Templating and Its Use as an Emerging Manufacturing Method of Tissue Engineering Scaffolds.* Frontiers in Bioengineering and Biotechnology, 2020. **8**:875.
103. Melchels, F.P.W., et al., *Mathematically defined tissue engineering scaffold architectures prepared by stereolithography.* Biomaterials, 2010. **31**(27):6909–6916.
104. Chen, H., et al., *Porous Scaffold Design for Additive Manufacturing in Orthopedics: A Review.* Frontiers in Bioengineering and Biotechnology, 2020. **8**:609.
105. Abbasi, N., et al., *Porous scaffolds for bone regeneration.* Journal of Science: Advanced Materials and Devices, 2020. **5**(1):1–9.
106. Kirillova, A., et al., *Fabrication of Biomedical Scaffolds Using Biodegradable Polymers.* Chemical Reviews, 2021. **121**(18):11238–11304.
107. Vesvoranan, O., A. Anup, and K.R. Hixon, *Current Concepts and Methods in Tissue Interface Scaffold Fabrication.* Biomimetics, 2022. **7**(4):151.

108. Wu, Y., et al., *A Critical Review of Additive Manufacturing Techniques and Associated Biomaterials Used in Bone Tissue Engineering*. Polymers, 2022. **14**(10):2117.
109. Zulkifli, M.Z.A., et al., *Overview of Electrospinning for Tissue Engineering Applications*. Polymers, 2023. **15**(11):2418.
110. Zhang, W., *Nanofabrication I: Electrospinning of Nanofibers*, in *Nanotechnology for Bioengineers*. 2020, Springer.
111. Topuz, F. and T. Uyar, *Electrospinning of Cyclodextrin Nanofibers: The Effect of Process Parameters*. Journal of Nanomaterials, 2020. **2020**(1):7529306.
112. Ucan, M.T., et al., *Design, Fabrication, and Evaluation of Hybrid Polycaprolactone/Graphene Scaffold Based on Additive Manufacturing and Electrospinning*. Macromolecular Materials and Engineering. **310**:e00236.

Transport Phenomena in Tissue Engineering

10

Learning Objectives

- Discuss how transport phenomena are relevant to different aspects of tissue engineering
- Understand the governing equations related to flow in a circular tube
- Understand the three modes of heat transfer and determine the thermal conductivity of an engineered tissue
- Estimate the diffusivity of a spherical particle and determine the diffusion coefficient of an engineered tissue

10.1 Introduction to Transport Phenomena in Tissue Engineering

The field of transport phenomena encompasses three closely related topics: fluid dynamics (momentum transport), heat transfer (transport of energy), and mass transfer (transport of mass). These processes generally occur simultaneously [1]. Figure 10.1 summarizes different aspects of transport phenomena in tissue engineering. For example, the tissue maturation process within a bioreactor, cell seeding, nutrient delivery, waste removal, and fluid-induced physical stimuli (e.g., shear stress and hydrostatic pressure) involve fluid flow. Heat transfer is necessary to maintain the temperature at 37 °C, while oxygen and nutrient delivery, as well as controlled release of bioactive molecules, involve mass transfer. Bioprinting using a thermo-responsive bioink is another good example—bioink dispensing involves both fluid dynamics and mass transfer, while heat transfer drives scaffold formation such as through gelation. It is not practical to cover all the aspects of transport phenomena relevant to tissue engineering within a single chapter. Readers are encouraged to consult books focus on transport phenomena in biological systems.

W. Zhang, *Cell Culture and Tissue Engineering*,
https://doi.org/10.1007/978-3-032-32096-4_10

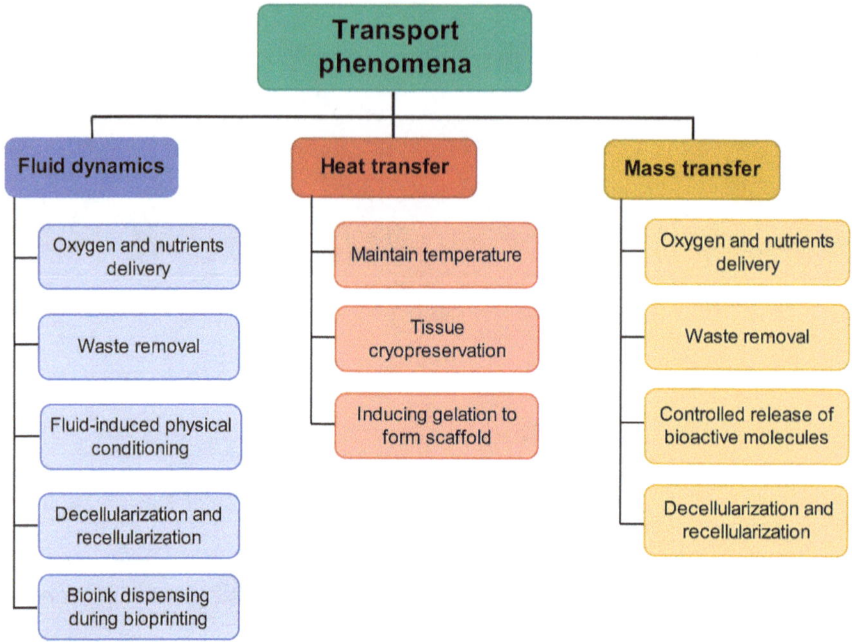

Fig. 10.1 Transport phenomena in tissue engineering

10.2 Flow in a Cylindrical Tube

Laminar Newtonian flow in a cylindrical tube can be described using the Hagen-Poiseuille equation:

$$Q = \frac{\pi R^4 (\Delta P)}{8 \mu L}$$

where Q is the volumetric flow rate, R is the radius, ΔP (= P_L—P_0) is the pressure drop, μ is the viscosity, and L is the length. To obtain a more generalized equation that can be applied to turbulent flow and accounts for the effects of changes of elevation and/or tube cross-section, and energy loss due to friction, the Bernoulli equation is used:

$$\frac{P_1}{\rho} + gZ_1 + \frac{\alpha_1 V_1^2}{2} + W_{device} = \frac{P_2}{\rho} + gZ_2 + \frac{\alpha_2 V_2^2}{2} + h_{friction}$$

where P is pressure, ρ is fluid density, Z is elevation (relative to a reference plane), V is average fluid velocity, α is kinetic energy correction term ($\alpha = 2$ for laminar flow (Reynolds number (Re) < 2100) and $\alpha = 1$ for turbulent flow), W_{device} is work per unit mass of fluid, and $h_{friction}$ represents friction effect. The expression for $h_{friction}$ is:

$$h_{friction} = \left(4\sum_i f_i \frac{L_i V_i^2}{2D_i} + \sum_j \frac{V_j^2}{2} K_{fittingj} \right)$$

where f_i is the friction factor of the fluid in tube segment i (length L_i and diameter D_i), and $K_{fitting}$ accounts for energy loss in the fluid due to tube contractions, expansions, or fittings (e.g., valves). Fitting losses are generally not considered for laminar flow. The friction factor can be determined using the following equations based on Re:

$$f = \frac{16}{\mathrm{Re}} (\mathrm{Re} < 2100)$$

$$f = \frac{0.0791}{\mathrm{Re}^{1/4}} \left(2.1 \times 10^3 < \mathrm{Re} < 10^5\right)$$

$$\frac{1}{\sqrt{f}} = 4.0 \log_{10} \mathrm{Re}\sqrt{f} - 0.4\left(2.3 \times 10^3 < \mathrm{Re} < 4 \times 10^6\right)$$

and $\frac{1}{\sqrt{f}} = 3.6 \log_{10} \left[\frac{6.9}{\mathrm{Re}} + \left(\frac{k/D}{3.7} \right)^{10/9} \right] \left(4 \times 10^4 < \mathrm{Re} < 10^8 \, and \, 0 < \frac{k}{D} < 0.05 \right)$, for rough tubes with roughness k. Additionally, mass balance must be satisfied: $\rho VA = \text{constant}$, with V is average velocity and A is cross-section area [2, 3].

10.3 Heat Transfer in Tissues and Scaffolds

Heat transfer can occur through conduction (the movement of thermal energy through a medium from more energetic particles to the less energetic ones, i.e., through physical contact), convection (the transfer of heat through a medium as a result of the net motion of material in the medium, such as fluid flow over a surface), and radiation (the spontaneous emission of electromagnetic waves by all matter) [4]. Equations for determining the rate of heat flow (q) are summarized in Table 10.1 [4, 5].

Table 10.1 Equations for determining rate of heat flow

Mode of heat transfer	Equation(s)	Notes
Conduction	$q_x = -Ak\frac{dT}{dx}$ $q_r = -Ak\frac{dT}{dr}$ (radial heat transfer)	A: area normal to the x-direction k: thermal conductivity T: temperature
Convection	$q = Ah\Delta T$	h: heat transfer coefficient A: area normal to the direction of heat flow ΔT: temperature difference between fluid and surface
Radiation	$q = A\sigma T^4$	A: energy-emitting area σ: Stefan-Boltzmann constant = 5.670×10^{-8} W/(m^2 × K^4)

10.4 Diffusive Mass Transfer in Tissues and Scaffolds

Diffusive mass transfer refers to the movement of one mass component from a region of higher concentration to a region of lower concentration. For molecular diffusion with a concentration gradient as the driving force, Fick's first law applies:

$$j_{A,x} = -D_{AB}\frac{dc_A}{dx},$$

where $j_{A,x}$ is the diffusive flux (mass transport per unit area) in the x direction, D_{AB} is the mass diffusivity (or diffusion coefficient) of component A in B, and dc_A/dx is the concentration gradient [6].

The diffusivity (D) can be estimated using the Stokes-Einstein equation:

$$D = \frac{RT}{6\pi\mu\alpha N_A} = \frac{k_B T}{6\pi\mu\alpha},$$

where R is the ideal gas constant (8.314 J/(mol × K)), T is temperature in K, α is solute radius, μ is viscosity, N_A is Avogadro's number (6.023×10^{23} mol^{-1}), and k_B is Boltzmann's constant (equals R/N_A). If the solute's molar mass (M) is known, and it can be approximated as a sphere with a density close to 1000 kg/m^3, its radius can be estimated using the equation:

$$\alpha = \left(\frac{3M}{4\pi\rho N_A}\right)^{1/3}.$$

Using this radius, diffusivity can then be computed [7]. Fick's second law describes non steady-state diffusion, describing how concentration changes with time: $\frac{\partial c}{\partial t} = D\frac{\partial^2 c}{\partial x^2}$. In three dimensions, the equation is $\frac{\partial c}{\partial t} = D\left(\frac{\partial^2 c}{\partial x^2} + \frac{\partial^2 c}{\partial y^2} + \frac{\partial^2 c}{\partial z^2}\right)$ [8].

Porous scaffolds are commonly used in tissue engineering. Molecules diffuse inside pores through pore diffusion (transport in pore volume) and surface diffusion (molecules adsorbed migrate along pore surface). For diffusion in porous materials, the effective diffusivity (D_e) can be calculated using:

$$D_e = D_b \frac{\varepsilon}{\tau} F(\lambda),$$

where D_b is bulk diffusivity (diffusivity in bulk fluids), ε is porosity (pore volume/total volume), τ is tortuosity of porous materials (reflecting sinuosity and the interconnection between the pores [9]), and F(λ), ranges from 0 to 1, is the restrictive factor to account for the hindered effect within pores, where λ is the ratio of solute to pore diameter. The commonly used factor expression, proposed by E. M. Renkin, is: $F(\lambda) = (1 - \lambda)^2(1 - 2.104\lambda + 2.09\lambda^3 - 0.95\lambda^5)$ [10]. The Bruggeman relation is commonly used to describe the tortuosity-porosity relationship in porous materials: $\tau^2 = \frac{1}{\varepsilon^{0.5}}$ [11].

Review and Practice Questions

1. To measure the thermal conductivity of a skin graft (~3 mm thick) that you just fabricated, you set up a system where one side of the graft is at 23 °C and the other at 37 °C. Heat flows through the thickness of graft with a flux of 20 W/m^2. What is its thermal conductivity?
2. Using the same skin graft from Q1, a system is set up to determine the diffusion coefficient of O_2, where the oxygen concentration on one side of the graft is 0.3 kg/m^3 and the other side is 2.5 kg/m^3. What is its diffusion coefficient, if the flux is 3.0×10^{-6} $kg/(m^2 \times s)$?
3. Estimate the diffusivity of a spherical protein with a molar mass of 45,000 g/mol in a cell culture medium at 37 °C. The viscosity of the culture medium is 0.93 mPa × s.

References

1. Bird, R.B., et al., *Chapter 0 The Subject of Transport Phenomena*, in *Introductory Transport Phenomena*. 2015, Wiley.
2. Fournier, R.L., *Chapter 4 The Physical and Flow Properties of Blood and Other Fluids*, in *Basic Transport Phenomena in Biomedical Engineering*. 2011, CRC Press.
3. Bird, R.B., et al., *Chapter 6 Interphase Transport in Isothermal Systems*, in *Introductory Transport Phenomena*. 2015, Wiley.
4. Datta, A.K., *Chapter 2 Modes of Heat Transfer*, in *Biological and Bioenvironmental Heat and Mass Transfer*. 2002, CRC Press.
5. Freeman, J.W. and D. Banerjee, *Chapter 2 Heat Transfer and Diffusion*, in *Building Tissues: An Engineer's Guide to Regenerative Medicine*. 2018, CRC Press.
6. Datta, A.K., *Chapter 10 Modes of MassTransfer*, in *Biological and Bioenvironmental Heat and Mass Transfer*. 2002, CRC Press.
7. Fournier, R.L., *Chapter 5 Solute Transport in Biological Systems*, in *Basic Transport Phenomena in Biomedical Engineering*. 2011, CRC Press.

8. Zhang, W., *Diffusion-Controlled Drug Delivery Systems*, in *Fundamentals and Design of Drug Delivery Systems: A Textbook with Lab Activities*. 2026, Academic Press.
9. da Silva, M.T.Q.S., et al., *Tortuosity: A brief review.* Materials Today: Proceedings, 2022. **58**:1344–1349.
10. Sun, W., *A model for hindered diffusion in pores.* Chemical Engineering Research and Design, 2024. **209**:299–310.
11. Holzer, L., et al., *Tortuosity-Porosity Relationships: Review of Empirical Data from Literature*, in *Tortuosity and Microstructure Effects in Porous Media: Classical Theories, Empirical Data and Modern Methods*. 2023, Springer.

Bioreactors for Tissue Engineering 11

Learning Objectives

- Discuss various types of bioreactors, with particular emphasis on perfusion-based systems
- Examine the major applications of bioreactors in tissue engineering
- Design bioreactors for specific tissue engineering applications, including strategies for scale-up and scale-out

11.1 Introduction to Bioreactors

Defining bioreactors can be challenging because their contents and intended applications vary widely. In general, a bioreactor can be described as a self-contained entity capable of supporting the growth and/or sustaining the biological functions of living organisms [1]. From a tissue engineering standpoint, bioreactors are devices used to enable the fabrication of 3D artificial tissues and support the growth, maturation, and development of artificial tissues during controlled *in vitro* cultures. This definition emphasizes the role of bioreactors as enabling and supporting technologies in tissue engineering [2].

There are various types of bioreactors, including static (e.g., T-flask), mixing, and perfusion bioreactors (Fig. 11.1) [3, 4]. In an ideal well-mixed bioreactor, the fluid is homogenous throughout the system. Mixing can be achieved through stirring, rotating and rocking [5]. Perfusion bioreactors, by contrast, are designed to perfuse culture medium through the artificial tissue/organs. In addition to enhancing the mass transport, perfusion can also provide mechanical stimulation to cells and tissues [6]. The perfusion-based bioreactors exhibit considerable versatility in configuration, such as closed loop systems (in which medium is recirculated) and single pass systems (which supply fresh medium while removing waste) (Fig. 11.2) [7, 8]. Perfusion may be indirect, where medium flows around and only partially through

W. Zhang, *Cell Culture and Tissue Engineering*,
https://doi.org/10.1007/978-3-032-32096-4_11

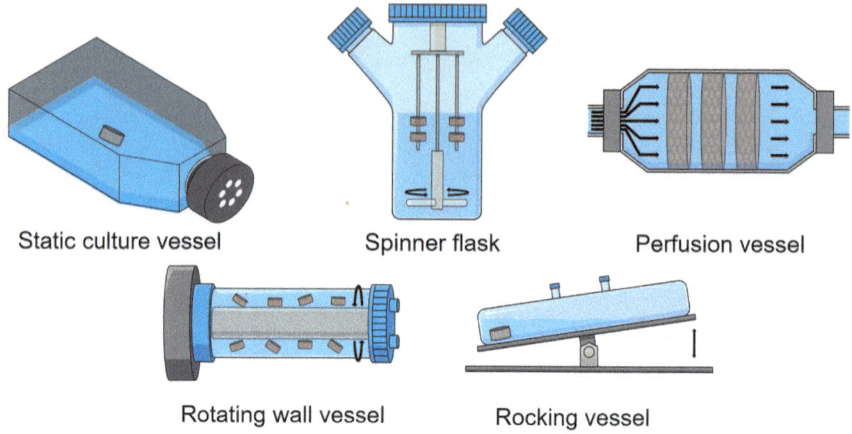

Fig. 11.1 Different types of bioreactors

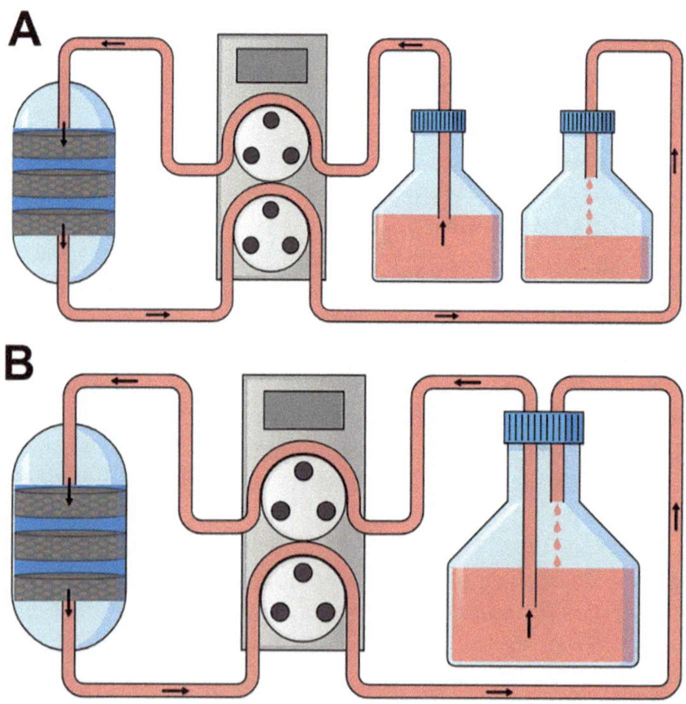

Fig. 11.2 Schematic comparison of perfusion-based system configurations: single-pass (**a**) versus closed-loop (**b**) setups

the construct, or direct, in which medium flow is forced, allowing full passage through the tissue [9]. Additionally, pulsatile perfusion bioreactors have also been developed to replicate the cyclic mechanical forces and cardiovascular conditions that vascular cells experience during vasculogenesis [10].

11.2 Bioreactors in Tissue Engineering Applications

Bioreactors play a critical role in the production of functional 3D tissues by maintaining efficient and uniform cell distribution within the scaffold during seeding, facilitating mass transfer, and regulating microenvironmental parameters (both physical and biochemical) [3, 4, 11, 12]. Moreover, bioreactors can be used to standardize manufacturing processes to meet regulatory and commercialization requirements through the integration of computational modeling, sensors and automation [11].

In the context of cell seeding, Sarah K. Saunders et al. developed a perfusion-based seed-and-culture bioreactor to compare three seeding strategies: bioreactor-mediated perfusion (BMP), static injection counterpart (SIC), and static seeding. The results indicated that BMP seeding produced the highest cell viability and most uniform distribution across samples. Although SIC seeding yielded a similar total cell number compared to BMP seeding, cell distribution remained inconsistent (Fig. 11.3) [13]. One notable advantage of perfusion-based seeding is its compatibility with integrated systems, allowing scaffolds to be both seeded and perfused within a single bioreactor unit (Fig. 11.4) [14, 15].

Mass transfer within a bioreactor serves three major functions: (1) delivery of nutrients, (2) removal of waste products, and (3) facilitation of cell communication *via* soluble factors. Dynamic *in vitro* culture using perfusion bioreactors is generally preferred for improving mass transport, as perfusion enables direct medium flow through tissue constructs, rather than relying solely on diffusion, which is limited by tissue size/thickness and concentration gradients [4]. However, applying perfusion to volumetric and anatomically complex scaffolds remains challenging. David P. Forrestal et al. addressed this limitation by designing custom bioreactors that combine 3D-printed scaffolds with flow-focusing inserts in patient specific shapes, ensuring optimal medium delivery regardless of scaffold geometry (Fig. 11.5) [16].

Bioreactors can also be used to regulate microenvironmental parameters, such as pH, temperature, as well as controlled physiological conditioning. Various types of stimuli, such as mechanical (e.g., shear, stretch and compress) and electrical/magnetic cues, have been incorporated to enhance tissue maturation [3, 14, 17]. It is essential to mimic *in vivo*-like conditions when engineering tissues or organs. For example, both mechanical and electrical stimulation are required in cardiac tissue engineering, while air-liquid interface stimulation is critical for lung and skin models [4, 18]. Commonly applied mechanical forces include compression, hydrostatic pressure, shear stress, and tensile stress [19]. A recent study introduced a modular bioreactor that synchronizes mechanical and electrical stimulation for cardiac tissue engineering. This system applied a cyclic mechanical load replicating left ventricle volume inflation, along with an electrical stimulus with an action potential profile to mirror the cellular microenvironment and electrophysiological conditions in the heart. Both stimuli were synchronized in accordance with cardiac physiology and controlled through continuous feedback (Fig. 11.6). This bioreactor enhanced mesenchymal stem/

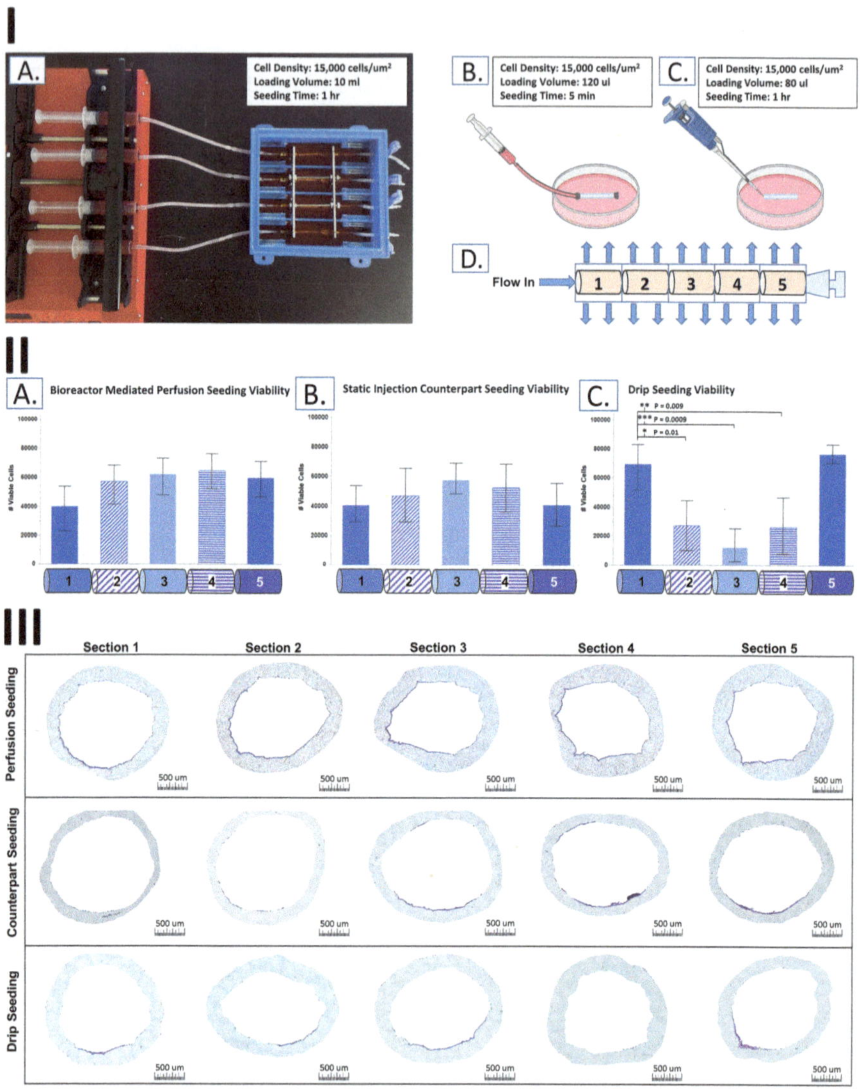

Fig. 11.3 Comparing different seeding approaches—bioreactor-mediated perfusion, static injection counterpart, and static seeding. I: Seeding method diagrams for (**a**) the bioreactor-mediated perfusion seeding method, (**b**) static injection counterpart seeding method and (**c**) drip seeding. (**d**) Operating principle of bioreactor mediated perfusion seeding method. Numbers indicate sample organization for analysis. II: Section average viable cell numbers, based on alamarBlue assay, for (**a**) perfusion bioreactor seeding, (**b**) static counterpart seeding, and (**c**) drip seeding. III: Representative hematoxylin and eosin-stained cross sections of each seeding method 24 h after seeding in longitudinal order. (*Adapted from* [13])

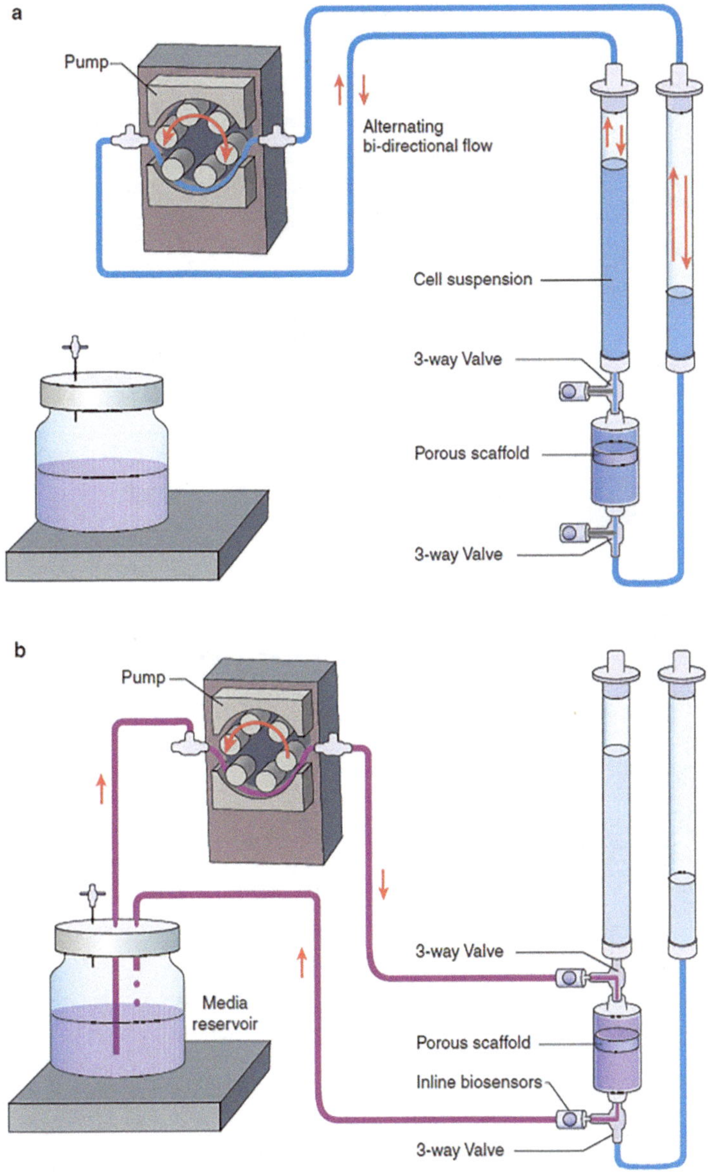

Fig. 11.4 Bioreactor unit for perfusion seeding and prolonged cell culture. (**a**) 3D scaffolds are first seeded with cells using a fluid flow path designed to distribute the cells uniformly throughout the porous scaffold. (**b**) Following the cell-seeding phase, the flow is diverted to an alternative path for prolonged culture. (*Adapted from* [14] *with permission from Springer Nature*)

stromal cell proliferation, promoted dense tissue-like structure formation, and uniquely induced expression of early cardiac progenitor markers only under stimulated conditions [20].

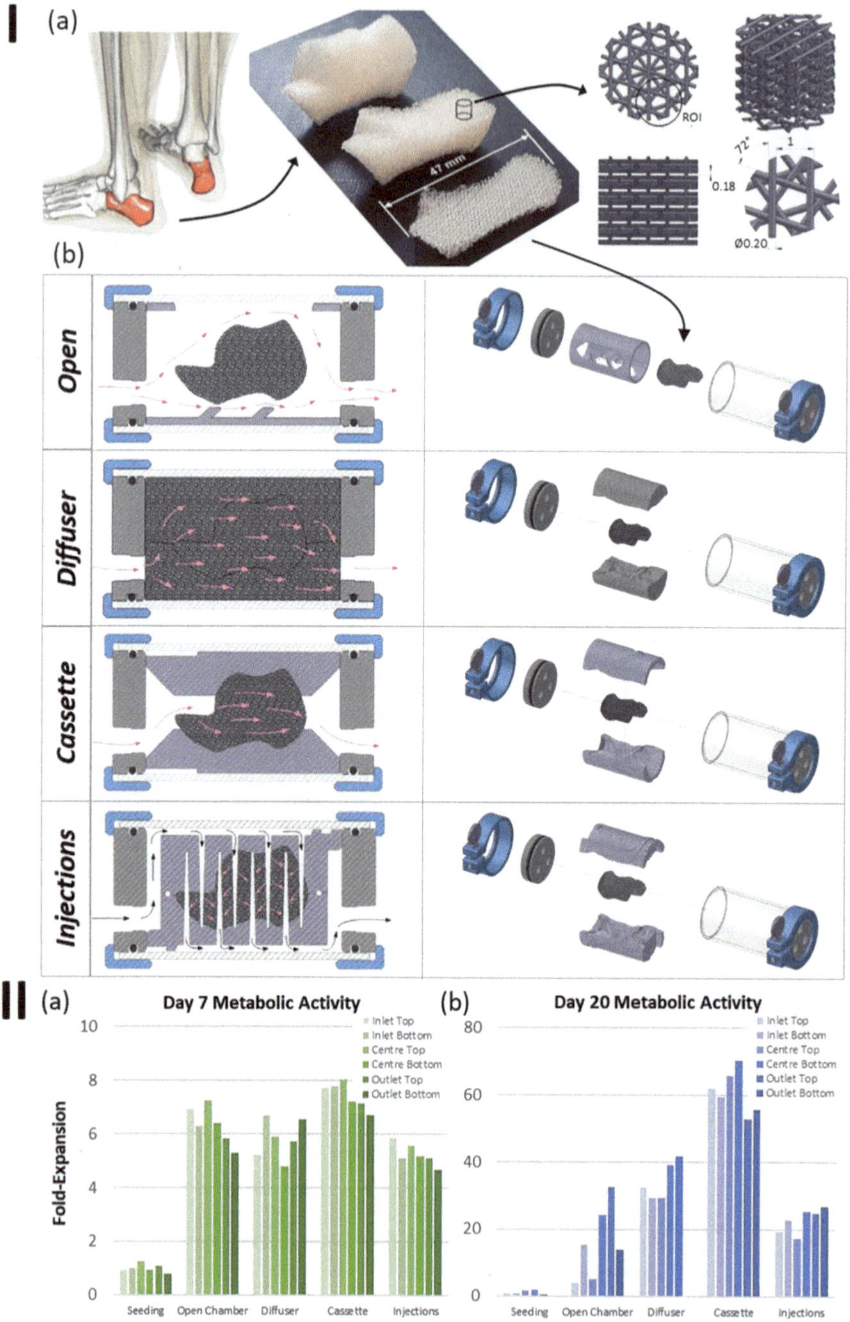
I
(a)
ROI
47 mm
72°
1
0.18
Ø0.20
(b)
Open
Diffuser
Cassette
Injections
II
(a)
Day 7 Metabolic Activity
Fold-Expansion
0
2
4
6
8
10
Inlet Top
Inlet Bottom
Centre Top
Centre Bottom
Outlet Top
Outlet Bottom
Seeding
Open Chamber
Diffuser
Cassette
Injections
(b)
Day 20 Metabolic Activity
0
20
40
60
80
Inlet Top
Inlet Bottom
Centre Top
Centre Bottom
Outlet Top
Outlet Bottom
Seeding
Open Chamber
Diffuser
Cassette
Injections

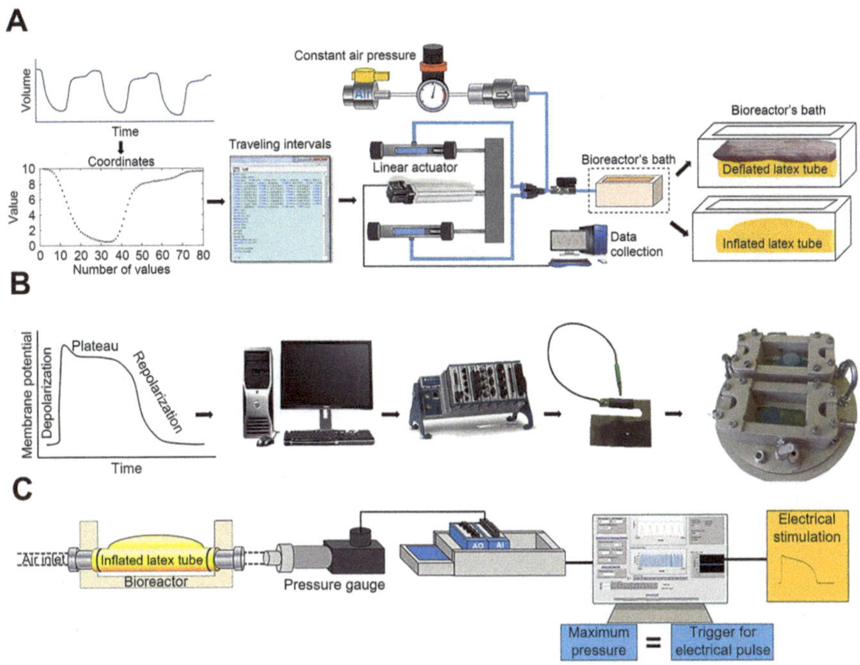

Fig. 11.6 (**a**) Design of the system for mechanical stimulation. The physiological changes in the left ventricle volume over time were used to create a set of coordinates in Matlab®. Values were input into the designated software to create a point-by-point path that was sent to a servo motor connected to a linear actuator, activating four pistons that inflated and deflated a latex tube beneath the scaffold. Constant air pressure compensated for any air loss due to instrument deterioration. (**b**) Design of the system for electrical stimulation. The action potential wave-shaped pulse was created using LabVIEW and transferred through a digital-to-analog converter to electrically stimulate the matrix with platinum foil electrodes placed within the bioreactor. (**c**) Scheme of the synchronization system. After the air inflates the latex tube, a pressure gauge connected to the other end of the tube transformed the pressure in bars into analog voltages ranging from 0 to 10 V. Subsequently, a data acquisition device converted the analog voltages into digital data, which were imported into LabVIEW software and filtered. In this software, the maximum values were then used to trigger a time suspension, followed by an electrical signal shaped as an action potential that was applied to the tissue inside the bioreactor. (*Adapted from* [20])

Fig. 11.5 I: The universal design of 3D-printed scaffolds and chamber inserts to control the transport of perfused cell culture medium. (**a**) Using an example human calcaneus (heel bone) and optimized 3D-printed structures, tissue scaffolds were designed to fit within a universal bioreactor chamber using flow-controlling inserts, which could be easily assembled to the scaffold, completely enclosing it and conforming to its external shape. (**b**) Four different flow-focusing inserts were designed and compared against each other and static controls during computational fluid dynamics (CFD) simulations and 7-day, 20-day, and 24-day bioreactor cell culture. On the left, a cross-sectional view of bioreactor designs illustrates the flow of medium using pink arrows. On the right, an exploded computer-aided design (CAD) rendering illustrates how the scaffold, inserts, and bioreactor shell were assembled. II: Mass transport limitations in volumetric, long-term tissue growth can be overcome through flow-focusing bioreactors. (**a**) Cellular fold expansion within scaffold biopsies after 7 days of bioreactor culture, normalized to metabolic activity of day 0 scaffolds seeded with 7 million cells, as assessed by alamarBlue assay. (**b**) Cellular fold expansion within scaffold biopsies after 20 days of bioreactor culture, normalized to metabolic activity of day 0 scaffolds seeded with 1.6 million cells. (*Adapted from* [16])

11.3 Design of Bioreactors for Tissue Engineering Applications

To design bioreactors, multiple aspects should be considered, including identifying culture variables, defining stimuli and stimulation protocols, and determining sensor technologies for monitoring the conditions and processes [2]. As shown in Fig. 11.7, general bioreactor design begins with examining the *in vivo* conditions of the specific tissue/organ engineered. Next, the needs for a bioreactor should be identified, including whether cell seeding is necessary. If the cells are already seeded within the scaffold, the bioreactor's major function is to promote cell proliferation, differentiation and tissue maturation. Culture variables and stimuli, along with corresponding stimulation protocols, should be determined. Typical cultural variables include T, pH, and oxygen level. If a specific stimulus (stimuli) is required, the method and protocol for its incorporation must also be established. For example, in muscle tissue engineering, it is essential to start with mild stretching since early-stage constructs may not withstand rigorous mechanical loading [2]. All of these factors should be taken into account when constructing, selecting, or modifying a bioreactor for a specific tissue engineering application.

When a bioreactor is in operation, physical parameters such as temperature are easy to monitor and control. However, monitoring cell growth and tissue maturation is far more challenging. Sampling is not always straight forward (e.g., due to contamination risk) and destructive monitoring approaches are undesirable.

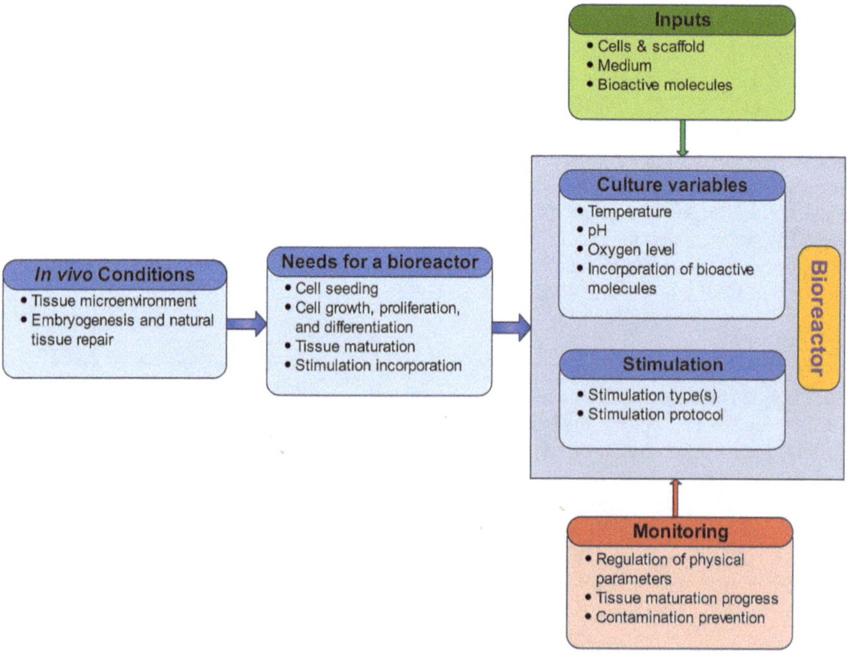

Fig. 11.7 Schematic illustration of the bioreactor design considerations and process

Therefore, non-destructive and real-time monitoring is preferred. Advancements in sensor technology are crucial, as they provide high potential for real-time, non-invasive monitoring and feedback control [21, 22]. For instance, ultrasound imaging is a rapid, label-free, and quantitative vibrational spectroscopy method for obtaining information on a tissue matrix, the structure and function of vasculature, and the viscoelastic properties, often in combination with optical imaging and elastography tools. Electrochemical impedance spectroscopy measures electrical conductivity to determine scaffold and tissue properties [21]. For monitoring cell viability and proliferation, in addition to direct imaging, oxygen and nutrient consumption and metabolite production can be measured to infer cellular activity. When monitoring cell differentiation and function, surface plasmon resonance-based biosensors have been used to evaluate stem cell differentiation and detect specific cell markers [23]. Jimin Lee et al. developed a smart bioreactor system that enables multi-spatial sensing for real-time, wireless culture monitoring, allowing simultaneous assessments of pH, dissolved oxygen, glucose, and temperature [24]. More recently, Adit Mehta et al. integrated both stimulation and sensing components into a uniaxial bioreactor. In this design, a sensor loop and media exchange system were used to monitor culture conditions and metabolite levels over time, while a camera system positioned above the tissue construct enabled macroscopic visualization of tissue maturation (Fig. 11.8) [25].

Scale-up is commonly employed when transitioning from lab research to industry production to enable the biofabrication of large-size tissues as well as high product yields. Moreover, scale-up can reduce the operator-dependent variability and thus ensure consistent product quality [4]. Take a stirred-tank bioreactor as an example, key process conditions affected by scale-up include aeration (O_2 and CO_2), mixing, and hydrodynamic shear stress. Hence, many operating parameters must be considered, and these can be categorized as scale-dependent or scale-independent (Table 11.1). For example, scale-independent variables such as temperature, pH, and dissolved oxygen should be kept constant. Scale-dependent variables include stir speed, mixing time, power input, and superficial gas velocity [26]. During scale-up, it is critical to consider both geometric and dynamic similarities. Geometric similarity requires identical impeller geometry and location. Dynamic similarity, evaluated through dimensional analysis, ensures that dimensionless groups (such as Reynolds number, Froude number, and Power number; Table 11.2) in the governing equations and boundary conditions remain the same [26, 27]. Regarding the oxygen transfer coefficient, it serves as a measure of a systems' oxygen transfer capabilities (k_La). For a given bioreactor configuration, it is mainly correlated with the volumetric power input (P/V) and the superficial gas velocity (ϑ_S), according to the relationship:

$$k_L a = K\left(\frac{P}{V}\right)^{\alpha} \vartheta_S{}^{\beta},$$

where K is a constant and exponents α and β are approximately 0.4 and 0.5, respectively [26].

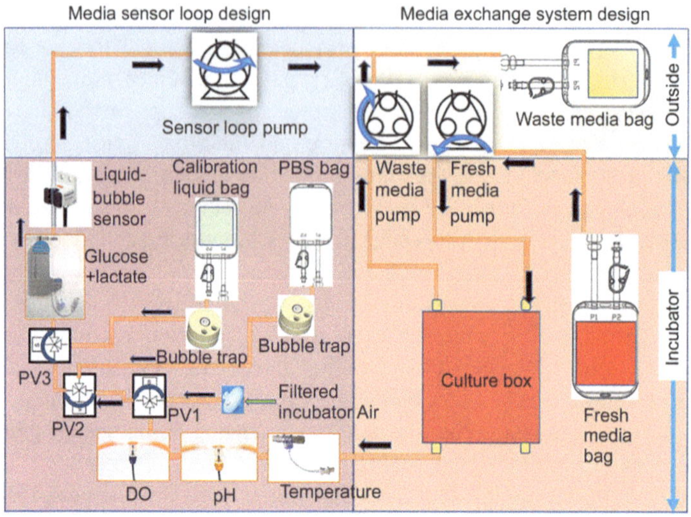

Fig. 11.8 Design of sensor loop and media exchange system. The sensor loop is designed to automate fluid movement for media sensing, where sensors are exposed to media from the culture box, PBS buffer (for intermittent washing), or glucose-and-lactate sensor calibration solution by controlling the sensor loop pump and three three-way solenoid pinch valves (PV1–PV3). Different liquids were isolated without mixing using PV1 at the beginning of the sensor loop to pull in incubator air, separating the fluids going to the sensor. These fluids were received by the glucose-and-lactate sensor, followed immediately by an ultrasonic liquid-bubble sensor before reaching the peristaltic pump. The media exchange system contains two pumps that can be manually turned on or off for media filling and emptying purposes. PBS: phosphate-buffered saline, DO: dissolved oxygen. (*Adapted from* [25] *with permission from Springer Nature*)

Table 11.1 Scale-dependent and scale-independent variables

Scale-dependent	Scale-independent
Impeller tip speed	Temperature
Oxygen mass transfer coefficient	pH
Gas residence time	Dissolved oxygen
Mixing time	Medium composition
Superficial gas velocity	Seeding density

Note: information adapted from [26]

Table 11.2 Important dimensionless groups relevant to bioreactor scale up

Dimensionless group	Expression	Variable
Reynolds number (Re)	$\mathrm{Re} = \frac{N_i D_i^2 \rho}{\mu}$	N_i: stirrer speed D_i: impeller diameter ρ: density μ: viscosity g: gravitational acceleration P: power
Froude number (Fr)	$Fr = \frac{D_i N_i^2}{g}$	
Power number (N_p)	$N_P = \frac{P}{\rho N_i^3 D_i^5}$	

For tissue engineering, scale-out is more relevant in the context of personalized medicine and autologous therapies. During scale-out, multiple bioreactors operate in parallel [4, 28]. Customization of tissue-engineered grafts requires the optimizing of cultivation conditions based on image-based geometries and using of patient-derived cells to create autologous grafts that evade immune rejection, while also accounting for other patient-specific factors (Fig. 11.9) [11].

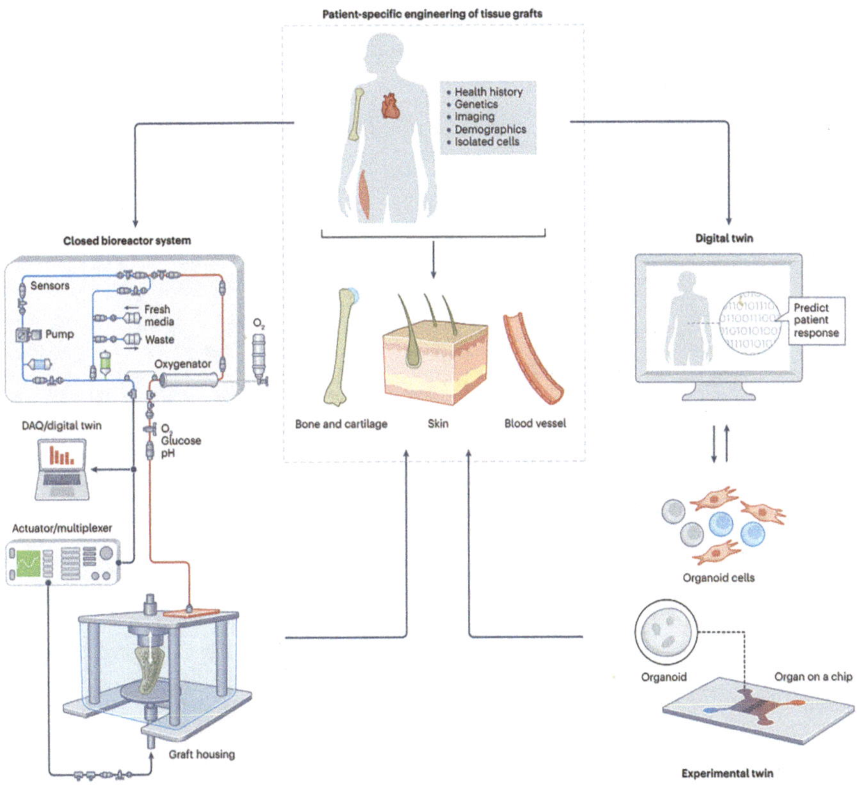

Fig. 11.9 Bioreactors for patient-specific tissue engineered grafts. Patient specificity can be integrated into tissue-engineered grafts through imaging (anatomic geometry) and the use of autologous cells. In addition, patient demographics, health history, and genetics should be taken into account when engineering patient-specific grafts or the digital or experimental twin (an avatar that integrates patient data through multiscale computer modelling and simulation, and uses these digital tools to optimize treatment approaches). The digital twin provides input to and receives input from microfluidic bioreactor systems (the experimental twin), in which patient-derived organoids can be cultured on integrated organ-on-a-chip systems to model individual organs or multiple organs. Closed bioreactor systems incorporate a housing chamber (which may also serve as a media reservoir) for controlled perfusion and biophysical cues. The closed loop includes a pump, ports for medium exchange, in-line biosensors and oxygenation. The tubing, which is in contact with the tissue product and the culture media, is made of disposable materials. The system is fully closed and separated from other components by an aseptic boundary. Ideally, a data acquisition (DAQ) system facilitates *in silico* modelling and real-time monitoring of graft status. Automation can be achieved using an actuator and controller to modify input parameters based on sensor readouts. (*Adapted from* [11] *with permission from Springer Nature*)

Several commercial bioreactors have been developed for cell culture and tissue engineering (Table 11.3). These systems are designed for diverse purposes and constructed from various materials. In particular, single-use bioreactors have gained popularity, and application-specific systems for particular tissue engineering needs have also been commercialized.

Table 11.3 Different commercial bioreactors for cell culture and tissue engineering

Company	Product(s)	Application(s)
CELLEC BIOTEK	U-CUP perfusion bioreactor	3D cell and tissue cultures; generation of 3D cell-scaffold constructs
Corning	Ascent® Fixed Bed Bioreactor (FBR) System	Cell culture
Culture Biosciences	Stratyx™ 250 bioreactor	Cell culture
Curi Bio	Stingray™	Electrical stimulation of 3D tissue cultures
Cytiva	ReadyToProcess WAVE™ 25 Rocker; Xcellerex™ XDR 50–2000 L single-use bioreactor; iCELLis™ fixed-bed bioreactor; Xcellerex™ X-platform 50–2000 L cell culture bioreactor	Cell culture
EBERS Medical Technology	TC-3 tension/compression bioreactor	Application of load on the cell culture substrate to reproduce *in vivo* conditions
Eppendorf	BioBLU® c Single-Use Bioreactors; DASbox® Mini Bioreactor; DASGIP® Bioblock/SciVario® twin Spinner Vessels; DASGIP® Bioblock/SciVario® twin Stirrer Vessels; SciVario® twin Benchtop Bioreactors for Cell Culture; BioFlo® 120 Bioreactors; BioFlo® 320 Bioreactors	Cell culture
Flexcell	Flexcell® Tension Systems; Flexcell's® Compression System	Mechanical stimulation of cells
Getinge	Applikon BioBench; AppliFlex ST GMP; Applikon AppliFlex ST; Applikon Bio	Cell culture
PBS Biotech	Vertical-Wheel bioreactors	Cell culture
Regemat 3D	BMAP® KNEE bioreactor	Maturation of knee cartilage
Sartorius	Ambr® 15 Cell Culture; Ambr® 250 High Throughput Perfusion; Biostat® RM TX & Flexsafe® RM TX Bags; Univessel® SU	Cell culture
SKE Research E	CC100 InFlow bioreactor	Perfusion bioreactor for controlled, bidirectional and interstitial perfusion
ST3	ST3's bioreactor systems	Tissue-engineered medical product manufacturing
SYNTHECON	Custom bioreactors	3D cell culture and tissue engineering
TA Instruments	3DCulturePro bioreactors; BioDynamic bioreactors	Tissue culture; mechanical stimulation
The Cultivated B	AUXO V® bioreactors	Cell culture
Thermo Fisher Scientific	DynaDrive single-use bioreactor; HyPerforma single-use bioreactor; single-use rocker bioreactor	Cell culture

Review and Practice Questions

1. What is a bioreactor, and what are the different types of bioreactors?
2. Why are perfusion-based bioreactors preferred in tissue engineering applications?
3. What are the major functions of bioreactors in tissue engineering?
4. Conduct a literature search and write a short essay on a novel perfusion-based bioreactor developed for tissue engineering applications not discussed in this chapter.
5. Conduct a literature search and write a short essay on a novel bioreactor that provides stimuli (e.g., mechanical, electrical) developed for tissue engineering (not discussed in this chapter).
6. What are design considerations and process steps for a bioreactor intended for tissue engineering?
7. Compare scale-up with scale-out in the context of bioreactor design and application.
8. What are some of the scale-dependent and scale-independent variables, respectively, in bioreactor design?
9. Why is scale-out particularly important in tissue engineering?
10. Conduct a literature search and write a short essay on a commercial bioreactor used for tissue engineering applications.

References

1. Valdevit, A., *Bioreactors: System Design and Application for Regenerative Engineering*, in *Encyclopedia of Biomedical Engineering*, R. Narayan, Editor. 2019, Elsevier.
2. Birla, R., *Bioreactors for Tissue Engineering*, in *Introduction to Tissue Engineering*. 2014, Wiley.
3. Salehi-Nik, N., et al., *Engineering Parameters in Bioreactor's Design: A Critical Aspect in Tissue Engineering*. BioMed Research International, 2013. **2013**(1):762132.
4. Egger, D., et al., *Chapter 13 - Bioreactors: enabling technologies for research and manufacturing*, in *Tissue Engineering (Third Edition)*, J. De Boer, et al., Editors. 2023, Academic Press.
5. Hu, W.-S., *Chapter 8 - Cell Culture Bioreactors*, in *Cell Culture Bioprocess Engineering*. 2020, CRC Press.
6. Dahlin, R.L., et al., *Design of a high-throughput flow perfusion bioreactor system for tissue engineering*. Tissue Eng Part C Methods, 2012. **18**(10):817–20.
7. Castro, N., et al., *Physically Active Bioreactors for Tissue Engineering Applications*. Advanced Biosystems, 2020. **4**(10):2000125.
8. Moheb Afzali, A., M.A. Kheradmand, and S.M. Naghib, *Bioreactor design-assisted bioprinting of stimuli-responsive materials for tissue engineering and drug delivery applications*. Bioprinting, 2024. **37**:e00325.
9. Rauh, J., et al., *Bioreactor Systems for Bone Tissue Engineering*. Tissue Engineering Part B: Reviews, 2011. **17**(4):263–280.
10. Ahata, B., et al., *Bioreactors for Tissue Engineering*, in *Biomaterials and Tissue Engineering*, O. Gunduz, et al., Editors. 2023, Springer International Publishing.
11. *Bioreactors for engineering patient-specific tissue grafts*. Nature Reviews Bioengineering, 2023. **1**(5):361–377.
12. Zhao, J., et al., *Bioreactors for tissue engineering: An update*. Biochemical Engineering Journal, 2016. **109**:268–281.

13. Saunders, S.K., et al., *Evaluation of perfusion-driven cell seeding of small diameter engineered tissue vascular grafts with a custom-designed seed-and-culture bioreactor.* PLOS One, 2022. **17**(6):e0269499.
14. Maxson, S., D. Orr, and K.J.L. Burg, *Bioreactors for Tissue Engineering*, in *Tissue Engineering: From Lab to Clinic*, N. Pallua and C.V. Suscheck, Editors. 2011, Springer.
15. Wendt, D., M. Jakob, and I. Martin, *Bioreactor-based engineering of osteochondral grafts: from model systems to tissue manufacturing.* Journal of Bioscience and Bioengineering, 2005. **100**(5):489–494.
16. Forrestal, D.P., et al., *Personalized Volumetric Tissue Generation by Enhancing Multiscale Mass Transport through 3D Printed Scaffolds in Perfused Bioreactors.* Advanced Healthcare Materials, 2022. **11**(24):2200454.
17. Van Houten, S.K., M.T.K. Bramson, and D.T. Corr, *A Bioreactor for Controlled Electrical and Mechanical Stimulation of Developing Scaffold-Free Constructs.* Journal of Biomechanical Engineering, 2022. **144**(9):094501.
18. Bravo-Olín, J., et al., *Analysis of the role of perfusion, mechanical, and electrical stimulation in bioreactors for cardiac tissue engineering.* Bioprocess and Biosystems Engineering, 2024. **47**(6):767–839.
19. Ravalli, S., et al., *New Insights on Mechanical Stimulation of Mesenchymal Stem Cells for Cartilage Regeneration.* Applied Sciences, 2020. **10**(8):2927.
20. Markish, M.G., et al., *A unique bioreactor that offers synchronized physiological-like electrical and mechanical stimuli for cardiac tissue engineering.* Bio-Design and Manufacturing, 2025. **8**(4):581–594.
21. McCorry, M.C., et al., *Sensor technologies for quality control in engineered tissue manufacturing.* Biofabrication, 2023. **15**(1):012001.
22. Zhou, Y., et al., *Chapter 18 - Biosensors in tissue engineering*, in *Tissue Engineering*, C.P. Sharma, et al., Editors. 2022, Academic Press.
23. Li, Y.-C.E. and I.-C. Lee, *The Current Trends of Biosensors in Tissue Engineering.* Biosensors, 2020. **10**(8):88.
24. Lee, J., et al., *Large-scale smart bioreactor with fully integrated wireless multivariate sensors and electronics for long-term in situ monitoring of stem cell culture.* Science Advances, 2024. **10**(7):eadk6714.
25. Mehta, A., et al., *Development of a multifunctional uniaxial bioreactor with real-time monitoring of culture conditions and tissue health.* Bio-Design and Manufacturing, 2025. **8**(2):310–330.
26. Lemire, L., et al., *Practical Considerations for the Scale-Up of Chinese Hamster Ovary (CHO) Cell Cultures*, in *Cell Culture Engineering and Technology: In appreciation to Professor Mohamed Al-Rubeai*, R. Pörtner, Editor. 2021, Springer.
27. Bird, R.B., W.E. Stewart, and E.N. Lightfoot, *Chapter 3. The Equations of Change for Isothermal Systems*, in *Transport Phenomena*. 2006, Wiley.
28. Bellani, C.F., et al., *Scale-Up Technologies for the Manufacture of Adherent Cells.* Frontiers in Nutrition, 2020. **7**: 575146.

12 Controlled Release of Bioactive Molecules in Tissue Engineering

Learning Objectives

- Discuss the importance of controlled release of bioactive molecules in tissue engineering
- Analyze how heparin and fibronectin can be applied in controlled release of growth factors
- Design controlled release strategies for specific tissue engineering applications

12.1 Introduction to Controlled Release

Controlled drug release, which aims to optimize therapeutic delivery to maximum efficacy (avoiding subtherapeutic concentrations) and minimize side effects (preventing toxic levels that may damage health cells and tissues), refers to the release of drugs at a desired rate, duration, and location [1]. Controlled release can be sustained (at a programmed rate), delayed (release of a discrete portion or portions of the drugs after administration rather than immediately), or extended (at least a two-fold reduction in dosing frequency compared to a conventional dosage form) (Fig. 12.1) [2, 3].

With regard to release mechanisms, diffusion-controlled, erodible, osmotic-controlled or combination systems are most common (Fig. 12.2). For tissue engineering applications, particularly *in vivo*, erodible systems are especially attractive, since they eliminate the need for follow-up procedures. Furthermore, targetability and responsiveness of the delivery systems are critical for achieving maximum therapeutic effects while mitigating potential toxicity [3, 4].

W. Zhang, *Cell Culture and Tissue Engineering*,
https://doi.org/10.1007/978-3-032-32096-4_12

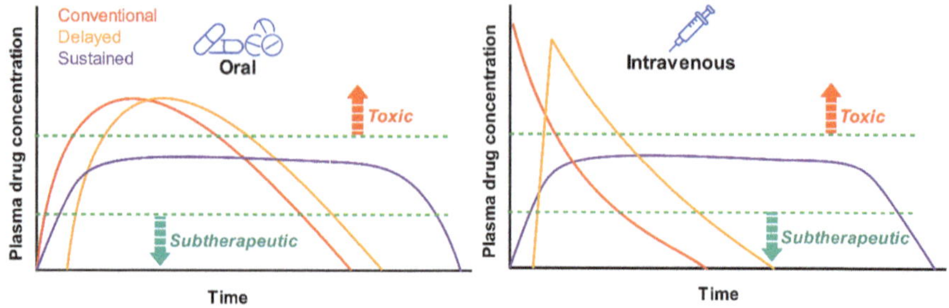

Fig. 12.1 Controlled release through oral (left) and intravenous injection (right) drug administration routes

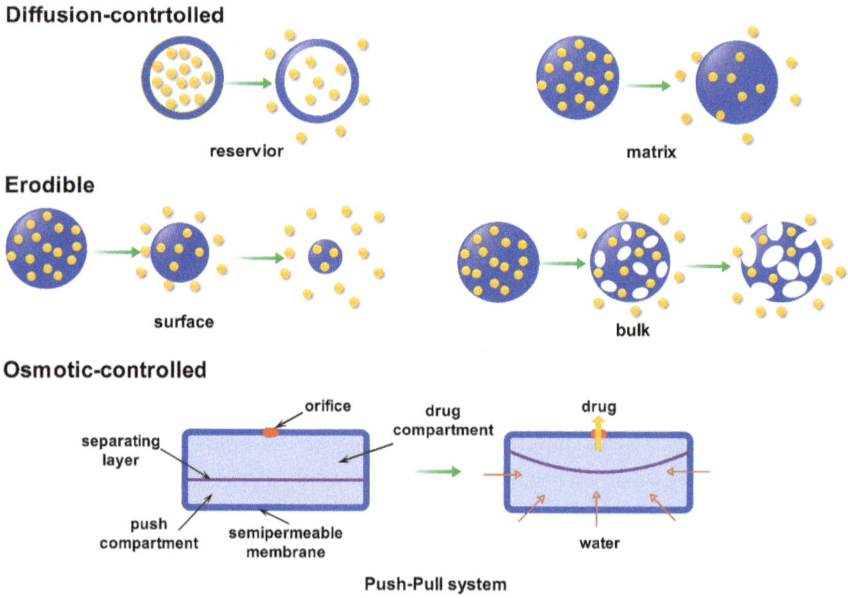

Fig. 12.2 Representative controlled delivery systems. (*Adapted from* [3] *with permission from Springer Nature*)

12.2 Controlled Release of Bioactive Molecules in Tissue Engineering

In tissue engineering, especially for *in vivo* applications, it is crucial to deliver bioactive molecules locally rather than systemically [5]. Systemic administration may cause toxicity and/or unintended effects on healthy cells and tissues. For instance, overdose of bone morphogenetic protein 2 (BMP-2) can even induce carcinogenesis. Bioactive molecules, such as growth factors, generally have short biological half-lives, as they are easily degraded or inactivated within the physiological microenvironment [4, 6]. Moreover, repeated systemic dosing is not cost-effective, given the high expense of many bioactive molecules [5].

Bioactive molecules such as growth factors, cytokines, peptides, and small molecules are commonly utilized in tissue engineering [5, 7]. A variety of strategies have been exploited for their controlled release (Fig. 12.3). These include physical entrapment within a gel network, encapsulation within a hydrophobic degradable polymer-based matrix or microparticles, and covalent or non-covalent immobilization to a matrix. For non-covalent binding, bioactive molecule-matrix affinity can be specific (e.g., growth factors interactions with fibronectin-containing matrices) or non-specific (e.g., electrostatic interactions) [5].

Cytokines and growth factors are fundamental to numerous physiological processes, including embryonic development, tissue homeostasis, wound healing, and immune system regulations. Cytokines constitute broad and heterogeneous groups of small proteins and polypeptides that mediate intercellular communication, particularly in the immune and hematopoietic systems. Growth factors, a specialized subset of cytokines, are distinguished by their specific ability to promote cellular growth, proliferation, differentiation, and survival [8]. Several growth factors-based products have already been approved by the FDA (Table 12.1). Among them, becaplermin (Regranex®) was the first [9, 10].

Various materials, including hydrogels, microencapsulation systems, coatings, membranes, and porous scaffolds, have been used for growth factor delivery. Examples include sodium alginate/platelet-rich plasma, gelatin, hyaluronic acid, fibroin, fibrinogen, poly(lactic-*co*-glycolic acid), chitosan, heparin, poly(lactic acid), and octacalcium phosphate [4]. Physiologically, interactions between growth factors and the extracellular matrix (ECM) enable localized and spatially regulated signaling. However, the lack of ECM binding in clinically used forms of growth factors limits their efficacy and translation [11]. Consequently, biomimetic strategies inspired by ECM-growth factor interactions have gained momentum. The ECM contains several molecules that sequester and regulate growth factor activity,

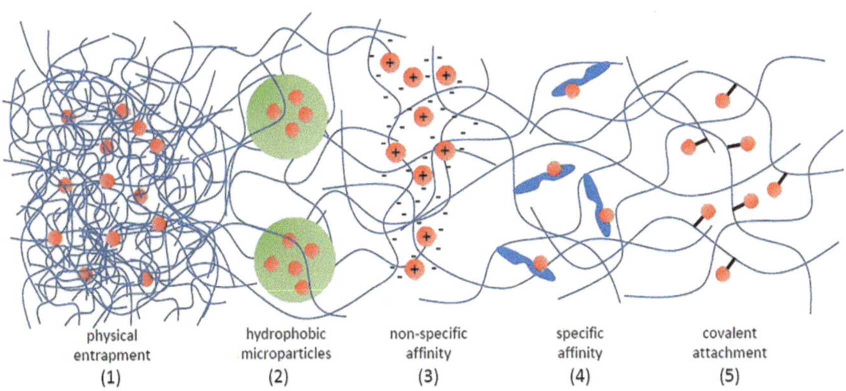

Fig. 12.3 Strategies for controlled release of bioactive molecules in tissue engineering. (*Adapted from* [5] *with permission from Elsevier*)

Table 12.1 FDA-approved growth factor-based products

Product and trade name	Growth factor	Matrix	Targeted disease(s)
REGRANEX® (becaplermin)	Recombinant human platelet-derived growth factor (rhPDGF-BB)	Sodium carboxymethylcellulose gel	Lower extremity diabetic neuropathic ulcers
INFUSE® Bone Graft	Recombinant human bone morphogenetic protein-2 (rhBMP-2)	Absorbable collagen sponge	Anterior lumbar interbody fusion; tibial nonunion; alternative to autogenous bone graft for sinus augmentations, and for localized alveolar ridge augmentations for defects associated with extraction sockets
GEM 21S™	Recombinant human platelet-derived growth factor (rhPDGF-BB)	Synthetic β-tricalcium phosphate	Bone and periodontal regeneration
Augment® Bone Graft	Recombinant human platelet-derived growth factor (rhPDGF-BB)	Synthetic β-tricalcium phosphate	Arthrodesis (i.e., fusion procedures) of the ankle and/or hindfoot
INCRELEX™ (mecasermin [rDNA origin] injection)	Recombinant human insulin-like growth factor-1 (rhIGF-1)	NA	Growth failure associated with IGF-1 deficiency
Kepivance (Palifermin)	Modified keratinocyte growth factor (KGF)	NA	Severe mucositis
OXERVATE™ (cenegermin-bkbj)	Recombinant human nerve growth factor (rhNGF)	NA	Neurotrophic keratitis

Note: This table is subject to change

particularly heparin and fibronectin [12]. Notably, collagen, although the most abundant ECM protein, has little affinity for growth factors due to its lack of appropriate binding sites [11].

Heparin, a highly sulfated glycosaminoglycan, owes its strong affinity for numerous growth factors to its high negative charge density [12, 13]. Structurally, heparin is a linear, unbranched polysaccharide composed of disaccharide units of uronic acid 1,4-linked to D-glucosamine (Fig. 12.4) [14]. Heparin can be incorporated into scaffolds by simple mixing or covalent crosslinking with scaffolding materials, such as collagen, silk fibroin, gelatin, hyaluronic acid, and decellularized matrix [12]. For

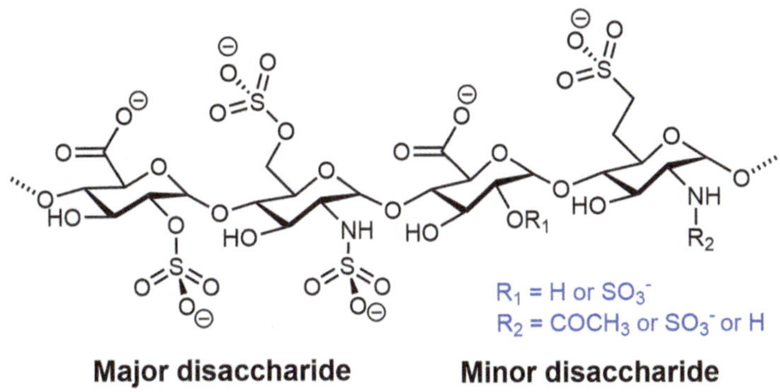

Fig. 12.4 Chemical structure of heparin. (*Generated based on* [14])

example, heparin-conjugated collagen nanofibers have been used for the sustained release of platelet derived growth factor (PDGF-BB), transforming growth factor-β1 (TGF-β1), and connective tissue growth factor (CTGF) [15]. Heparin-conjugated collagen scaffold has been used for immobilization of basic fibroblast growth factor (bFGF), vascular endothelial growth factor (VEGF), and hepatocyte growth factor (HGF) [16]. A heparin-functionalized bioink (GH/HepMA), composed of hyaluronic acid, gelatin methacrylate (GelMA), and heparin methacrylate (HepMA), has been reported for the sustained delivery of VEGF in 3D bioprinting of prevascularized dermal constructs. In this design, the top layer contained human umbilical vein endothelial cells (HUVECs) embedded GH/HepMA with VEGF, while the bottom layer contained human dermal fibroblasts (HDFs) in GH/HepMA. VEGF enhanced HUVEC proliferation, migration, and capillary-like network formation (Fig. 12.5) [17]. Some limitations of using heparin include a lack of specificity due to molecular heterogeneity and potential risks in certain biomedical applications arising from its anticoagulant activity. To address these issues, heparin mimetics have been developed [12].

Fibronectin contains several constitutive and conformational-dependent binding motifs for various growth factors [12, 18]. For example, fibronectin-incorporated 3D hydrogel has been employed to present VEGF and bone morphogenetic protein-2 (BMP-2) in a physiologically relevant manner [19]. Moreover, positively charged amphiphilic amino acid copolymers, composed of cationic and hydrophobic residues, have been designed to mimic fibronectin III12–14. Optimized copolymer-immobilized 3D scaffolds enabled the sustained release of BMP-2 and enhanced *in vivo* bone formation (Fig. 12.6) [20].

Growth factors can also be engineered to achieve controlled release. For example, non-glycosylated BMP-2, which is less soluble than the native protein, showed enhanced retention in fibrin matrices and improved bone healing *in vivo* [21]. Priscilla S. Briquez et al. screened 25 biomolecules for binding to six key ECM proteins and identified a domain in placenta growth factor-2, PlGF-$2_{124\text{-}144}$, that binds exceptionally strongly and promiscuously to ECM proteins. By fusing this domain to various growth factors, VEGF-A, PDGF-BB, BMP-2, engineered

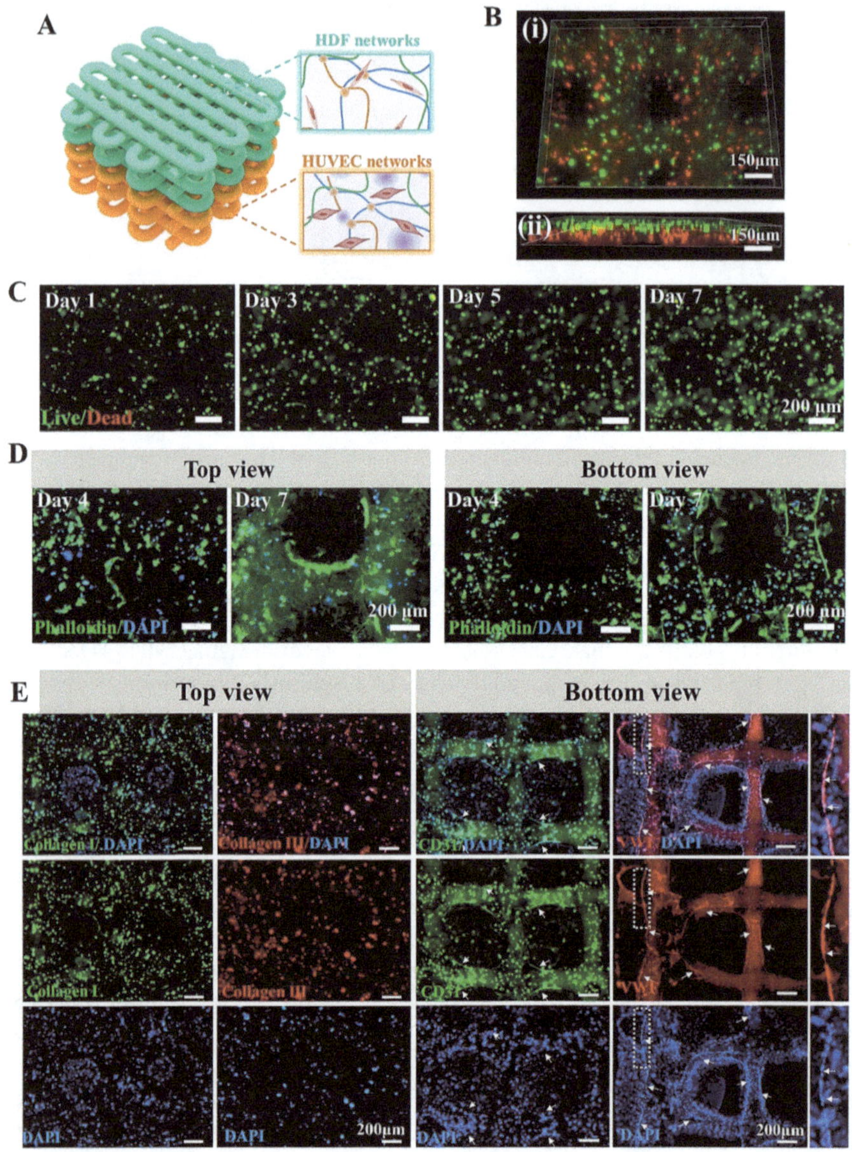

Fig. 12.5 Cytocompatibility, extracellular matrix remodeling, and angiogenesis analysis of 3D-printed prevascularized dermal constructs. (**a**) Structural diagram of a prevascularized dermal construct. (**b**) Dermal constructs printed with HDFs (green) and HUVECs (red), co-cultured for 5 days. (**c**) Live/dead staining of HDFs and HUVECs. (**d**) Cytoskeleton staining after 4 and 7 days of *in vitro* culture, showing the cell morphology of HDFs and HUVECs in the constructs, respectively. (**e**) Immunofluorescent staining of collagen I/III, CD31, and vWF. Collagen deposition observed from top view (green: collagen I; red: collagen III; blue: nuclei). Angiogenesis observed from bottom view (green: CD31; red: vWF; blue: nuclei). HDFs: human dermal fibroblasts, HUVECs: human umbilical vein endothelial cells, CD: cluster of differentiation, vWF: von Willebrand factor. (*Adapted from* [17] *with permission from Elsevier*)

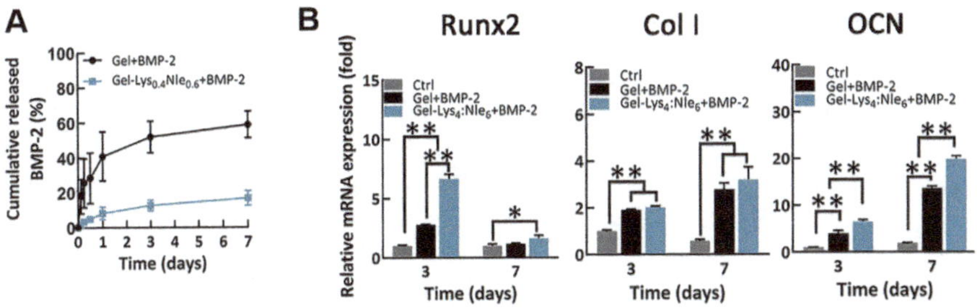

Fig. 12.6 Delivery of BMP-2 within amino acid copolymers-immobilized 3D scaffolds induces the osteogenic differentiation *in vitro* and *in vivo*. (**a**) Release behavior of BMP-2 from different scaffolds. (**b**) Expression of osteogenic marker (corresponding to runt-related transcription factor 2 (Runx2), type I collagen (Col I), and osteocalcin (OCN)) within C2C12 cells cultured within various scaffolds for 3 days and 7 days, with cells cultured on tissue culture plates serving as the control. BMP-2: bone morphogenetic protein-2. (*Adapted from* [20])

variants with super-affinity for the ECM were generated, yielding greatly enhanced repair in rodent models of chronic wounds and bone defects (Fig. 12.7) [11].

Simultaneous or sequential delivery of multiple growth factors has also been investigated. For example, the initiation of angiogenesis requires VEGF, fibroblast growth factor (FGF), and angiopoietin-2 (Ang-2) to disrupt existing blood vessels and promote proliferation and migration of endothelial cells to form new immature vessels. Subsequently, angiopoietin-1 (Ang-1) and PDGF-BB stabilize the newly formed blood vessels [5, 22]. Thomas P. Richardson et al. demonstrated that dual delivery of VEGF and PDGF-BB with distinct kinetics from a polymeric scaffold promoted rapid formation of a mature vascular network [23]. A layer-by-layer coating system has been designed for codelivery of TGF-β1 and PDGF-BB in tissue engineering applications [24]. Another study showed that simultaneous release of rhBMP-2 and rhVEGF165 from collagen-heparin based polyelectrolyte multilayer films (PEM) on micro-rough titanium surfaces significantly improved peri-implant bone formation and bone-implant contact [25]. Similarly, Di Wu et al. reported a bioconstruct for volumetric muscle loss (VML) therapy, using decellularized ECM scaffolds conjugated with nanocapsules of bFGF and IGF-1 (insulin growth factor 1). This system recapitulated endogenous temporal dynamics: early release of bFGF to stimulate the proliferation of muscle stem cells (MuSCs), followed by delayed release of IGF-1 to drive differentiation. Sequential kinetics were achieved by modulating electrostatic interactions between the polymer shell and encapsulated proteins (Figs. 12.8 and 12.9) [26, 27].

Stimuli-responsive systems also show great promise, as they can provide spatiotemporal control and deliver growth factors at physiologically appropriate doses. These scaffolds respond to physical, chemical, or biological stimuli, triggering volumes and structural changes that regulate release [4]. For instance, a recent study incorporated curcumin and BMP-2 loaded mesoporous silica/titania nanoparticles into polycaprolactone/polyethene glycol scaffolds. The scaffolds exhibited enhanced osteogenesis when release was stimulated with low-intensity pulsed ultrasound

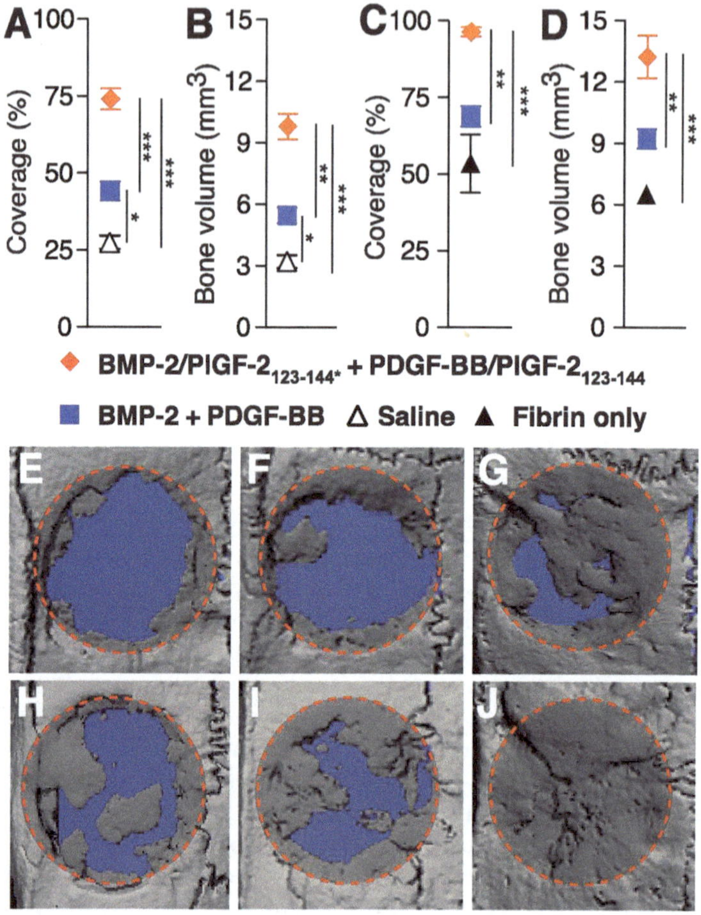

Fig. 12.7 Delivering PDGF-BB/PlGF-$2_{123\text{-}144}$ and BMP-2/PlGF-$2_{123\text{-}144}$ induces greater bone regeneration than wild-type PDGF-BB and BMP-2. Critical-size calvarial defects (6 mm in diameter) were treated with GFs delivered topically to the dura (1 μg of each GF, combined) or in a fibrin matrix (200 ng of each GF, combined). (**a–d**) Four weeks after treatment, bone repair was measured by micro-computed tomography (μCT) as bone volume and coverage of the defect [(**a**, **b**) show topical groups; (**c**, **d**) show fibrin groups]. (**e–j**) Representative calvarial reconstructions. (**e**) Saline vehicle; (**f**) BMP-2 + PDGF-BB; (**g**) BMP-2/PlGF-$2_{123\text{-}144*}$ + PDGF-BB/PlGF-$2_{123\text{-}144}$; (**h**) fibrin only; (**i**) fibrin with BMP-2 + PDGF-BB; (**j**) fibrin with BMP-2/PlGF-$2_{123\text{-}144*}$ + PDGF-BB/PlGF-$2_{123\text{-}144}$. The defect area is shaded with a red dotted outline. PDGF-BB: platelet-derived growth factor-BB, BMP-2: bone morphogenetic protein-2, PlGF-2: placenta growth factor-2, GF: growth factor. (*Adapted from* [11] *with permission from The American Association for the Advancement of Science*)

[28]. Another example is a responsive cascade drug delivery scaffold for peripheral nerve injury repair, in which vitamin B12 was released early to suppress inflammation and nerve growth factor (NGF) was released later to promote nerve regeneration. Release kinetics were controlled by ultrasound-triggered hierarchical opening polymer chains (Fig. 12.10) [29]. Future growth factor delivery strategies will likely

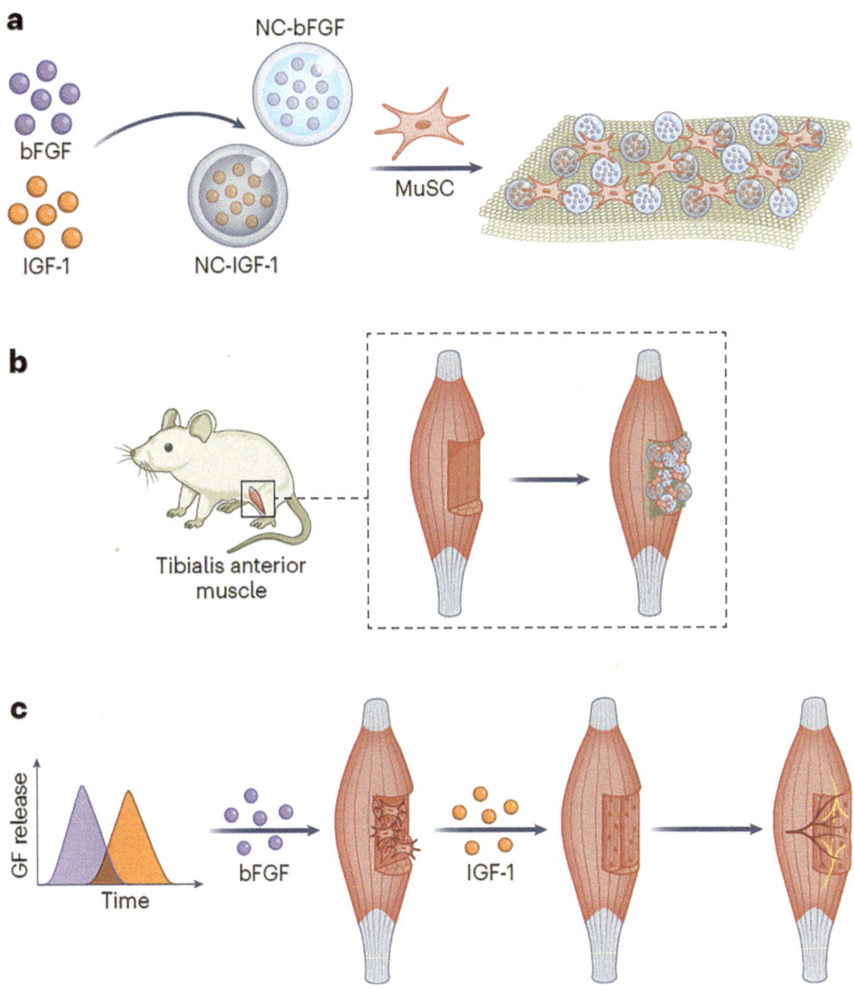

Fig. 12.8 Schematic illustration of bioconstructs designed for staged growth factor (GF) release in the treatment of volumetric muscle loss (VML). (**a**) Formation of nanocapsules (NCs) of bFGF and IGF-1 *via in situ* polymerization and their conjugation with decellularized extracellular matrix (dECM) scaffolds to form NC-dECM bioconstructs for MuSC seeding. (**b**) Schematic illustration of induced VML injury in mouse tibialis anterior muscle for bioconstruct implantation and assessment. (**c**) Sequential release of NC-bFGF (purple, early) and NC-IGF-1 (orange, late) from the bioconstructs facilitated muscle cell activation, proliferation and differentiation, eventually leading to tissue engraftment at 6 weeks post-implantation, characterized by vascular (yellow) and neural (dark red) ingrowth. bFGF: basic fibroblast growth factor, IGF-1: insulin growth factor 1, MuSC: muscle stem cell. (*Adapted from* [26] *with permission from Springer Nature*)

integrate exogenous and endogenous stimuli to achieve precise spatiotemporal control, enabling customized release profiles under specific physiological and pathological conditions. Moreover, intelligent delivery system incorporating biosensors and feedback loops may dynamically adjust release rates and dosages to optimize therapeutic outcomes [4].

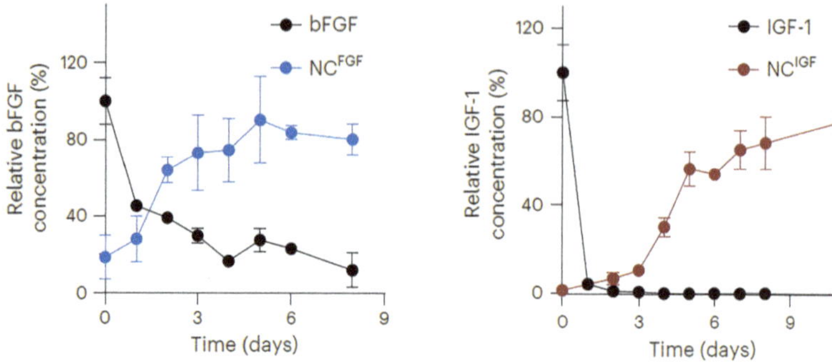

Fig. 12.9 Release profiles of unencapsulated and encapsulated bFGF (left) and IGF-1 (right) during incubation with mouse serum at 37 °C, revealing immediate release of bFGF and a 3-day delayed release of IGF-1. Both encapsulated factors showed sustained release compared with unencapsulated counterparts. bFGF: basic fibroblast growth factor, IGF-1: insulin growth factor 1, dECM: decellularized extracellular matrix, NC: nanocapsule. (*Adapted from* [27] *with permission from Springer Nature*)

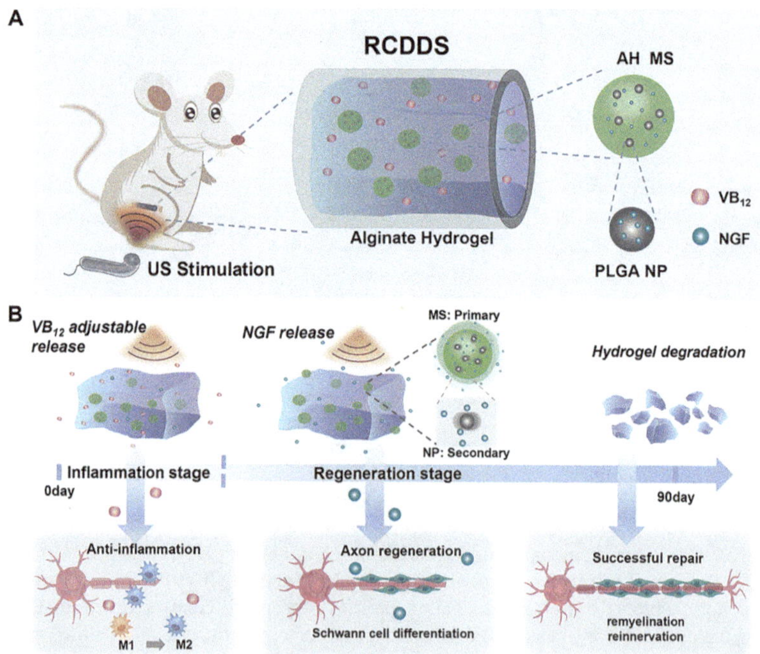

Fig. 12.10 Schematic illustration of the structure and drug release process of the responsive cascade drug delivery scaffold (RCDDS) for peripheral nerve injury repair. (**a**) Structural diagram of the RCDDS implanted in rat. (**b**) Drug release sequence and the corresponding repair stage. Vitamin B_{12} (VB_{12}), loaded in the hydrogel system, can be released during the early stage by ultrasound stimulation to alleviate inflammation, while nerve growth factor (NGF), encapsulated in alginate microspheres and poly(lactic-*co*-glycolic acid nanoparticles) (PLGA) nanoparticles, can be gradually released from the RCDDS to promote axon regeneration.US: ultrasound stimulation, AH: alginate hydrogel, MS: microsphere, NP: nanoparticle. M1: pro-inflammatory phenotype of macrophages, M2: anti-inflammatory phenotype of macrophages. (*Adapted from* [29] *with permission from Royal Society of Chemistry*)

Review and Practice Questions

1. Why is controlled release of bioactive molecules critical in tissue engineering?
2. What are cytokines and growth factors?
3. What are the different controlled release strategies used in tissue engineering?
4. How can heparin be applied in growth factor-controlled release?
5. How can fibronectin be utilized in growth factor-controlled release?
6. How can growth factors be engineered for controlled release? Provide examples.
7. Read the article entitled "Growth Factors Engineered for Super-Affinity to the Extracellular Matrix Enhance Tissue Healing" and answer the following questions:
   (a) According to the authors, what is the physiological reason for the limited clinical translation of growth factors?
   (b) How did the authors discover a domain in placenta growth factor-2 with exceptionally strong and promiscuous binding to ECM proteins?
8. Why is the simultaneously or sequentially release of two or more growth factors important? Provide examples.
9. What are the advantages of using stimuli-responsive systems for controlled release of growth factors?
10. What are the future directions for controlled release of bioactive molecules in tissue engineering?

References

1. Puccetti, M., et al., *Biologics, theranostics, and personalized medicine in drug delivery systems.* Pharmacological Research, 2024. **201**:107086.
2. Zingales, S.K. and J.H. Ferguson, *Chapter 9 - AI in microfabrication technology*, in *A Handbook of Artificial Intelligence in Drug Delivery*, A. Philip, et al., Editors. 2023, Academic Press.
3. Zhang, W., *Biopolymers: Biomedical Applications*, in *Biopolymer Engineering*. 2026, Springer.
4. Chen, T., et al., *Essential elements for spatiotemporal delivery of growth factors within bio-scaffolds: A comprehensive strategy for enhanced tissue regeneration.* Journal of Controlled Release, 2024. **368**:97–114.
5. Rice, J.J., et al., *Chapter 12 - Controlled release strategies in tissue engineering*, in *Tissue Engineering (Third Edition)*, J. De Boer, et al., Editors. 2023, Academic Press.
6. Qu, M., et al., *Stimuli-Responsive Delivery of Growth Factors for Tissue Engineering.* Advanced Healthcare Materials, 2020. **9**(7):e1901714.
7. Wu, G., *Editorial: Bioactive agents for functionalization of biomaterials for precise tissue engineering.* Frontiers in Bioengineering and Biotechnology, 2023. **11**:1141827.
8. Kungwankiattichai, S. and R.T. Maziarz, *The history of cytokines and growth factors development.* Best Practice & Research Clinical Haematology, 2025. **38**(1):101612.
9. Fang, R.C. and R.D. Galiano, *A review of becaplermin gel in the treatment of diabetic neuropathic foot ulcers.* Biologics, 2008. **2**(1):1–12.
10. Yamakawa, S. and K. Hayashida, *Advances in surgical applications of growth factors for wound healing.* Burns & Trauma, 2019. **7**:10.
11. Martino, M.M., et al., *Growth Factors Engineered for Super-Affinity to the Extracellular Matrix Enhance Tissue Healing.* Science, 2014. **343**(6173):885–888.
12. Teixeira, S.P.B., et al., *Biomaterials for Sequestration of Growth Factors and Modulation of Cell Behavior.* Advanced Functional Materials, 2020. **30**(44):1909011.

13. Wang, J.-F., J.-S. Jan, and J.-J. Hu, *Heparin-Based Growth Factor Delivery Platforms: A Review.* Pharmaceutics, 2025. **17**(9):1145.
14. Noti, C. and P.H. Seeberger, *Chemical Approaches to Define the Structure-Activity Relationship of Heparin-like Glycosaminoglycans.* Chemistry & Biology, 2005. **12**(7):731–756.
15. Baek, J., et al., *Collagen fibrous scaffolds for sustained delivery of growth factors for meniscal tissue engineering.* Nanomedicine, 2022. **17**(2):77–93.
16. Ikegami, Y., et al., *Heparin-conjugated collagen as a potent growth factor-localizing and stabilizing scaffold for regenerative medicine.* Regenerative Therapy, 2020. **15**:236–242.
17. Li, M., et al., *A heparin-functionalized bioink with sustained delivery of vascular endothelial growth factor for 3D bioprinting of prevascularized dermal constructs.* International Journal of Biological Macromolecules, 2024. **262**:130075.
18. Zhu, J. and R.A.F. Clark, *Fibronectin at select sites binds multiple growth factors and enhances their activity: expansion of the collaborative ECM-GF paradigm.* Journal of Investigative Dermatology, 2014. **134**(4):895–901.
19. Trujillo, S., et al., *Engineered 3D hydrogels with full-length fibronectin that sequester and present growth factors.* Biomaterials, 2020. **252**:120104.
20. Zhang, W., et al., *Statistic Copolymers Working as Growth Factor-Binding Mimics of Fibronectin.* Advanced Science, 2022. **9**(21):2200775.
21. Schmoekel, H., et al., *Bone healing in the rat and dog with nonglycosylated BMP-2 demonstrating low solubility in fibrin matrices.* Journal of Orthopaedic Research, 2004. **22**(2):376–381.
22. Lee, K., E.A. Silva, and D.J. Mooney, *Growth factor delivery-based tissue engineering: general approaches and a review of recent developments.* Journal of the Royal Society Interface, 2011. **8**(55):153–170.
23. Richardson, T.P., et al., *Polymeric system for dual growth factor delivery.* Nature Biotechnology, 2001. **19**(11):1029–1034.
24. Damanik, F.F.R., et al., *Long-Term Controlled Growth Factor Release Using Layer-by-Layer Assembly for the Development of In Vivo Tissue-Engineered Blood Vessels.* ACS Applied Materials & Interfaces, 2022. **14**(25):28591–28603.
25. Kauffmann, P., et al., *Dual Growth Factor Release From Collagen Based Multilayer Films on Ti-Surfaces Enhances Periimplant Bone Formation and Angiogenic Activity—An Experimental In-Vivo Study.* Clinical Oral Implants Research. 2026. **37**(1):1–17.
26. Fischer, E.O., A. Tsukerman, and S. Levenberg, *Orchestrated growth factor release for muscle restoration.* Nature Materials, 2025. **24**(9):1340–1341.
27. Wu, D., et al., *Bioinstructive scaffolds enhance stem cell engraftment for functional tissue regeneration.* Nature Materials, 2025. **24**(9):1364–1374.
28. Malekmohammadi, S., et al., *Stimuli-Responsive Codelivery System-Embedded Polymeric Nanofibers with Synergistic Effects of Growth Factors and Low-Intensity Pulsed Ultrasound to Enhance Osteogenesis Properties.* ACS Applied Bio Materials, 2024. **7**(7):4293–4306.
29. Shan, Y., et al., *A responsive cascade drug delivery scaffold adapted to the therapeutic time window for peripheral nerve injury repair.* Materials Horizons, 2024. **11**(4):1032–1045.

Part III

Tissue Fabrication Technologies

Scaffold Design and Fabrication: Microfabrication

13

Learning Objectives

- Explain the mechanism of photolithography and provide representative examples
- Interpret the principles and provide examples of soft lithography
- Discuss the advantages and disadvantages of scaffold-free tissue engineering
- Integrate the mechanisms of cell-sheet engineering and evaluate diverse micro-tissue fusion strategies for tissue fabrication

13.1 Photolithography

Photolithography techniques use light (photons) to transfer the geometric pattern of a mask onto a light sensitive surface made of photoresist [1, 2]. Photoresists are categorized as either positive or negative. With positive photoresists, exposure to light causes the material to weaken, for example the material degrades, making the exposed areas more soluble and able to be removed. In contrast, with negative photoresist, exposure causes the material to polymerize, allowing the exposed areas to be retained during the subsequent processing steps (Fig. 13.1) [3, 4]. Interestingly, pure silk fibroin film can serve as a positive photoresist for ArF excimer laser lithography, relying on the phototendering effect, which decreases the crystallinity of the silk fibroin. In this case, water serves as both the solvent and developer (Fig. 13.2) [5]. Notably, silk can be either used as a positive and negative resist for electron-beam lithography [6].

In cell culture and tissue engineering related applications, photolithography can be used to either create 2D scaffolds for growing cells or to encapsulate cells in a 3D

W. Zhang, *Cell Culture and Tissue Engineering*,
https://doi.org/10.1007/978-3-032-32096-4_13

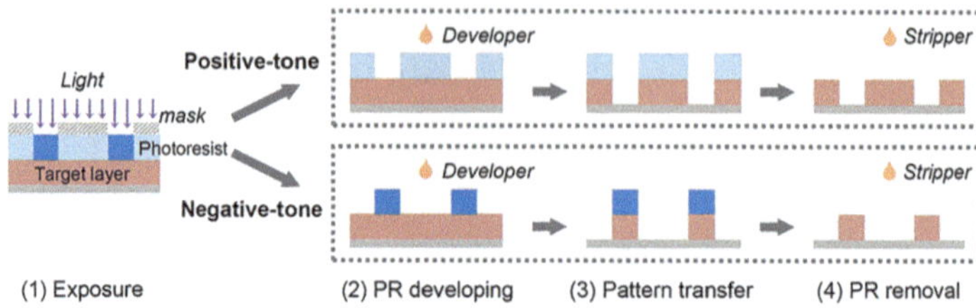

Fig. 13.1 Schematic illustration of the photolithography process using positive and negative photoresists (PRs). (*Adapted from* [4] *with permission from John Wiley and Sons*)

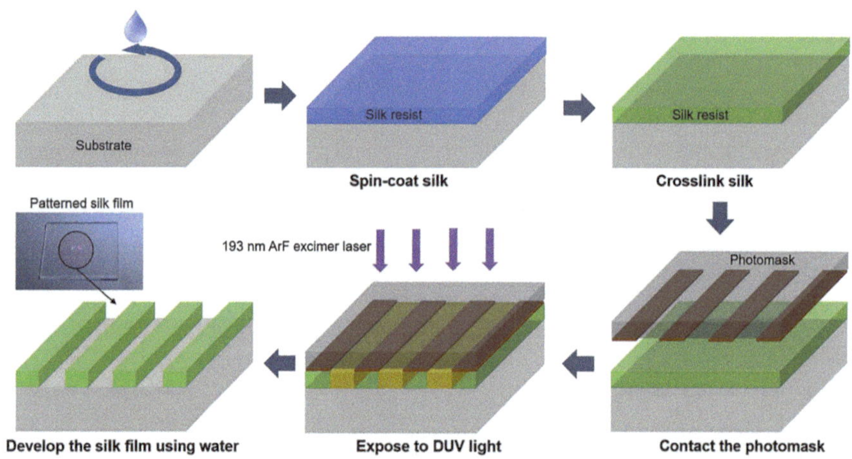

Fig. 13.2 Schematic diagrams showing the ArF excimer laser photolithography process to form high-resolution silk fibroin micropatterns. DUV: deep ultraviolet. (*Adapted from* [5] *with permission from RSC Publishing*)

network [1]. For example, a rapid method for producing microfabricated neural culture systems using a digital micromirror device (DMD) for dynamic mask projection lithography has been reported (Fig. 13.3) [7]. Hanie Kavand et al. introduced a novel approach called direct-cell photolithography, in which cells were guided into the cavities on cell-imprinted substrates. These negative replicas of cell surface topographies were shown to influence cell shape and direct stem cell differentiation [8].

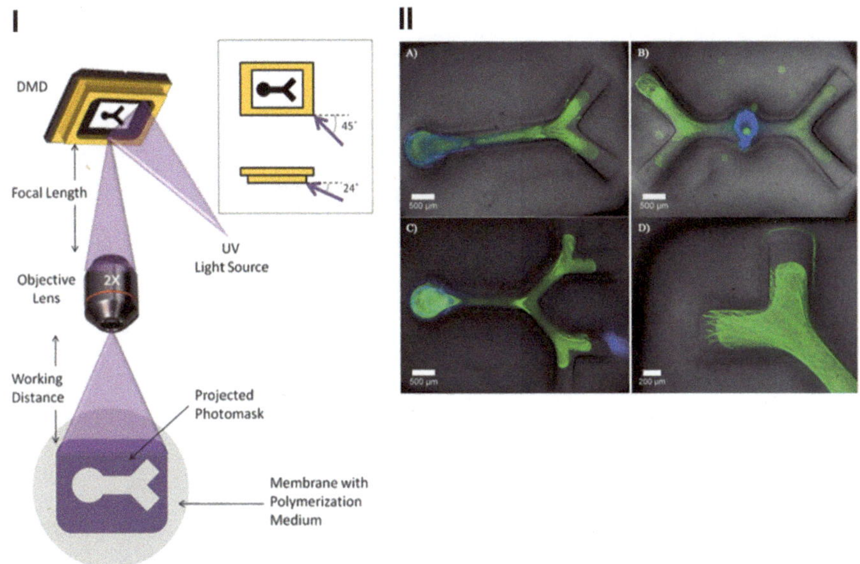

Fig. 13.3 I: Schematic illustration of the light path used for photolithography. Insert: UV light illuminates the DMD at a 45° angle and 24° below the plane of the mirror array. II: Labeled images showing growth and proliferation within dorsal root ganglia (DRG)-containing dual hydrogel constructs. (**a–d**) Polymerized poly(ethylene glycol) (PEG) constructs appear in gray, with neurites labeled with βIII tubulin (green), 4′,6-diamidino-2-phenylindole (DAPI) stained cell nuclei (blue). DRG explants, embedded in PuraMatrix peptide hydrogel, are located in the circular regions of the pattern, with neurites growing towards the bifurcation(s). UV: ultraviolet, DMD: digital micromirror device. (*Adapted from Curley, J. L., Jennings, S. R., Moore, M. J. Fabrication of Micropatterned Hydrogels for Neural Culture Systems using Dynamic Mask Projection Photolithography. J. Vis. Exp. (48), e2636, doi:10.3791/2636 (2011)* [7])

13.2 Soft Lithography

Soft lithography encompasses a group of techniques based on printing, molding and embossing using elastomeric masters. It supports a wide variety of materials, enables precise control of surface chemistries, and is generally compatible with biological applications [9]. The term "soft" refers to the elasticity of the elastomers used for fabricating masters or stamps [10, 11]. Key advantages of soft lithography include relatively low cost, simple setup, and high throughput. However, the fabrication of the stamp typically requires an additional technique, such as photolithography [12] (Fig. 13.4).

Representative soft lithographic techniques include microcontact printing and replica molding. Microcontact printing (μCP) is a process of transferring materials from a patterned elastomeric stamp to a substrate at points of contact. The stamp is first coated by immersion in a solution containing the material of interest, such as extracellular matrix proteins, then brought into conformal contact with the substrate

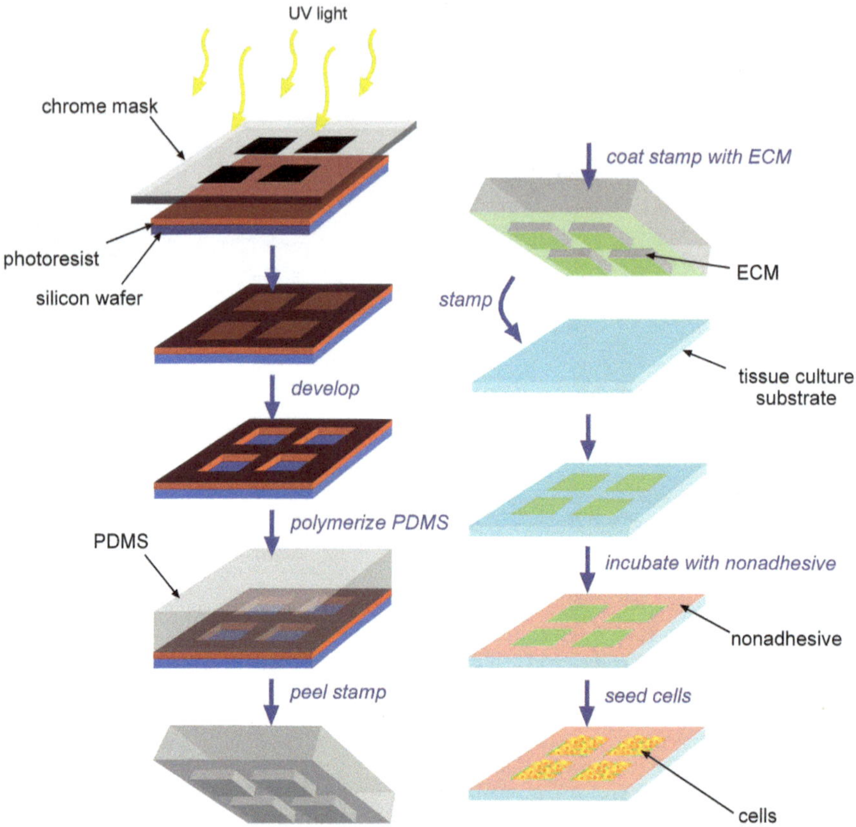

Fig. 13.4 Schematic of microcontact printing (micropatterning) process. Polydimethylsiloxane (PDMS) stamps are fabricated using photolithography (left column). These stamps are then used to transfer extracellular matrix (ECM) proteins onto tissue-culture substrates, creating defined patterns for subsequent cell seeding (right column). UV: ultraviolet. (*Adapted from* [14] *with permission from Elsevier*)

to deposit the material in the desired pattern [13]. To achieve selective cell patterning, unstamped regions can be blocked [14]. Two μCP methods with different protein-substrate adhesion mechanism—stamp-off (direct molecular adsorption) and covalent-bond (covalent linking both proteins and substrates)—have been compared to evaluate their effects on cell behavior. Results showed that insufficient adhesion in the stamp-off method can cause ECM delamination (Fig. 13.5), while protein-substrate adhesion significantly influences cell viability and cytoskeletal arrangements [15].

Replica molding is a method in which surface patterns from a master mold are duplicated on a polymer material. The master mold is generally fabricated using standard photolithography or micromachining techniques [16]. Allison P. McGuigan et al. explored cell encapsulation in gel modules using replica molding, enabling the

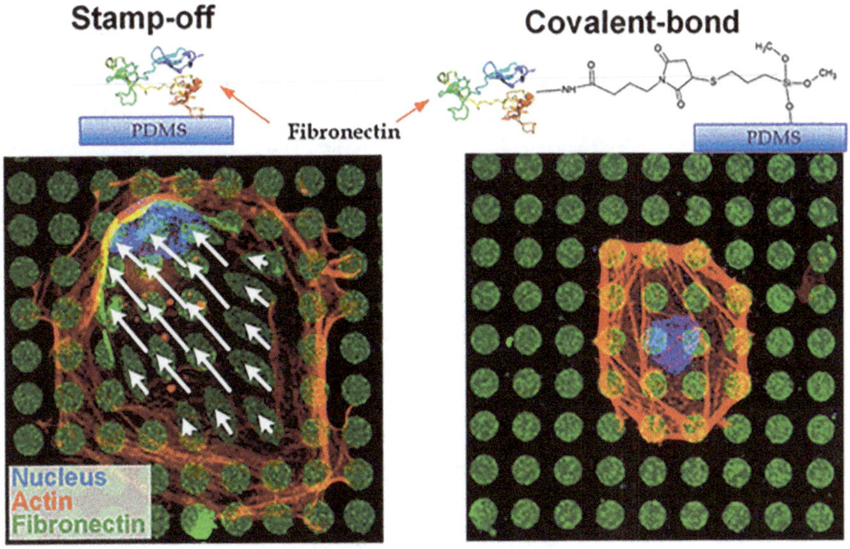

Fig. 13.5 Effects of microcontact printing methods on delamination. Arrows indicate microislands of displaced fibronectin (delamination). PDMS: polydimethylsiloxane. (*Adapted with permission from* [15]. *Copyright© 2018 American Chemical Society*)

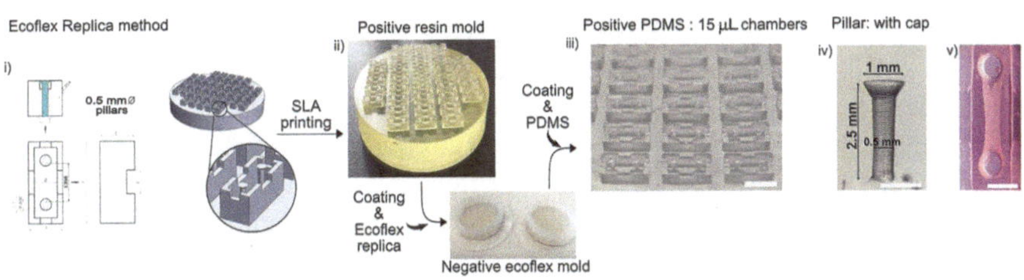

Fig. 13.6 Ecoflex Replica method. (i) Technical drawing of a single chamber and 3D computer-aided design (CAD) model of the positive master mold. (ii) Resin positive master mold fabricated *via* stereolithography (SLA) printing (top), and corresponding negative replica molds cast in Ecoflex (bottom). (iii) Close-up of a PDMS positive replica. (iv) T-shaped pillar with conical cap. (v) Example of engineered skeletal muscles tissue within a PDMS chamber. Scale bars: 5 mm (iii), 1 mm (iv, v). PDMS: polydimethylsiloxane. (*Adapted from* [18] *with permission from John Wiley and Sons*)

generation of modules with high cell-density close to dense native tissues [17]. Replica molding has also been integrated with 3D printing for fabricating skeletal muscle tissue engineering devices. As shown in Fig. 13.6, a positive master mold was fabricated using stereolithography, followed by the creation of a negative replica mold by casting Ecoflex over the master. Polydimethylsiloxane (PDMS) was then poured inside the Ecoflex negative mold, cured, and the resulting PDMS positive replica was peeled off by stretching the Ecoflex negative mold [18].

13.3 Scaffold-Free Tissue Engineering

Scaffold-free tissue engineering refers to strategies that eliminate the need for exogenous 3D materials for cell seeding or adhesion. Instead, the extracellular matrix is produced by the cells themselves [19, 20]. These approaches mimic developmental processes, typically involving cell condensation, cell proliferation, cell differentiation, ECM production, and tissue maturation [19]. Common building blocks include cell sheets, spheroids or tissue strands. Owing to the high initial cell density, tissue formation can proceed more rapidly, as cell proliferation and migration are no longer decisive factors [21]. Additionally, scaffold-free methods can also help eliminate the issues of rejection and tissue failure which are common with scaffolds [22, 23]. However, key limitations include the poor mechanical properties of the building blocks and the re-immobilization time for the initial fusion of building block and deposition of ECM to obtain a cohesive construct [21].

Cell sheet engineering is an emerging scaffold-free approach that leverages the close cell-cell and cell-ECM interactions to autonomously engineer microtissues [24]. It is categorized as a self-organization approach—rather than self-assembly—since external manipulation is required to create the desired structure. Cells are first cultured as a monolayer with high confluence over an extended period. Once sufficient ECM is produced to form a cohesive layer, the intact sheet can be detached from the substrate. The cell sheets are then layered or fused to form continuous tissue constructs [19]. Common methods used for harvesting cell sheets include temperature-responsive culture surfaces, chemically defined cultures, photo-responsive culture surfaces, and magnetic-responsive culture surfaces (Fig. 13.7) [23]. However, scaffold-free approaches are still facing challenges, such as difficulty in generating large constructs and maintaining structural integrity. One study proposed strategies to increase the size of human embryonic stem cells (hESC)-derived cartilage tissue constructs: co-culture on human articular cartilage and culture on polyethylene terephthalate-transwell membrane [25].

To better define the shape and structure of the target construct, templated molding techniques—such as sacrificial material printing and "kenzan" methods—can be used to facilitate tissue fusion. Sacrificial printing methods employe a sacrificial material to help support and/or general specific structure temporarily. After the microtissues have been fused, the sacrificial material is removed, typically by altering temperature. The "kenzan" method, inspired by the Japanese art of flower arrangement in which flower stems are plugged onto an array of vertical needles (i.e., "kenzan"), involves placing microtissues onto needle arrays to support fusion into larger tissue structures [10, 11] (Fig. 13.8).

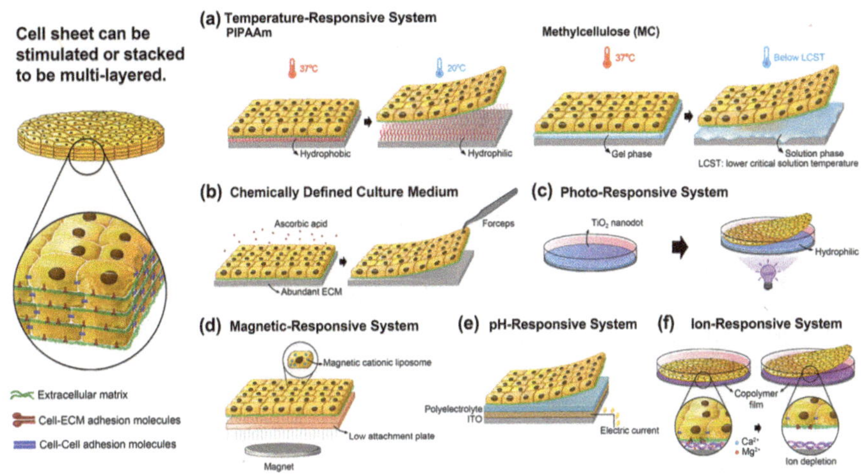

Fig. 13.7 Cell sheet fabrication and harvesting systems. (**a**) Temperature-responsive system: cells cultured on the PIPAAm-coated surface at 37 °C; lowering the temperature to 20 °C makes PIPAAm hydrophilic, enabling cell sheet detachment. Similarly, below the LCST, MC turns hydrophilic, allowing cell release. (**b**) Chemically defined culture medium: ascorbic acid supplementation promotes extracellular matrix production within cell sheets. (**c**) Photo-responsive system: TiO_2 nanodot films enable cell sheet detachment upon exposure to 365-nm ultraviolet light by decreasing the water contact angle. (**d**) Magnetic-responsive system: magnetite cationic liposomes combined with a magnetic force, assisted by low-attachment surfaces to minimize cellular adhesion. (**e**) pH-responsive system: electrochemically induced pH reduction at the interface alters protein microstructure and disrupted cell–ECM interactions for detachment. (**f**) Ion-responsive system: functional copolymers coating, with hydrophobicity determined by the divinylbenzene/4-vinylpyridine ratio, enable detachment through changing the concentration of cations. ECM: extracellular matrix; PIPAAm: N-isopropylacrylamide. LCST: lower critical solution temperature. (*Adapted from* [23])

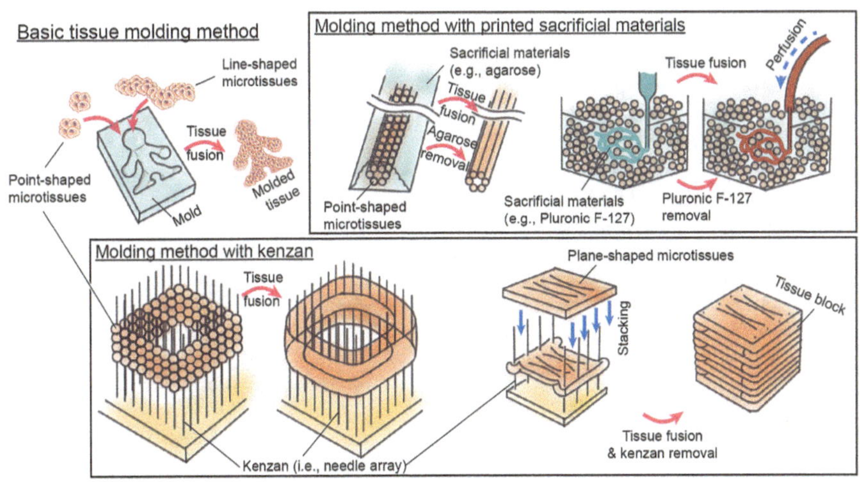

Fig. 13.8 Basic and templated approaches to microtissue molding for tissue fusion. (*Adapted from* [11] *with permission from Elsevier*)

Review and Practice Questions

1. What is photolithography? What are the differences between positive and negative photoresists?
2. Read the article entitled "All-water-based electron-beam lithography using silk as a resist," and answer the following questions:
   (a) What was used as the developer solvent?
   (b) Because of its polymorphic crystalline structure, silk can be used as either a positive or negative resist through interactions with an electron beam. Explain how silk can be used as either a positive or negative resist.
3. What is soft lithography? What are its major advantages and disadvantages?
4. Explain the working mechanism of microcontact printing.
5. Explain the working mechanism of replica molding.
6. What is scaffold-free tissue engineering? What are its advantages and disadvantages?
7. Explain the working mechanism of cell-sheet engineering.
8. Explain the working mechanism of sacrificial material printing for templated molding.
9. Explain the working mechanism of "Kenzan" method for templated molding.

References

1. Bajaj, P., et al., *3D Biofabrication Strategies for Tissue Engineering and Regenerative Medicine.* Annual Review of Biomedical Engineering, 2014. **16**:247–276.
2. Tran, K.T.M. and T.D. Nguyen, *Lithography-based methods to manufacture biomaterials at small scales.* Journal of Science: Advanced Materials and Devices, 2017. **2**(1):1–14.
3. Quero, J.M., F. Perdigones, and C. Aracil, *11 - Microfabrication technologies used for creating smart devices for industrial applications*, in *Smart Sensors and MEMs (Second Edition)*, S. Nihtianov and A. Luque, Editors. 2018, Woodhead Publishing.
4. Park, S.Y., et al., *Patterning Quantum Dots via Photolithography: A Review.* Advanced Materials, 2023. **35**(41):2300546.
5. Park, J., et al., *Eco-friendly photolithography using water-developable pure silk fibroin.* RSC Advances, 2016. **6**(45):39330–39334.
6. Kim, S., et al., *All-water-based electron-beam lithography using silk as a resist.* Nature Nanotechnology, 2014. **9**(4):306–310.
7. Curley, J.L., S.R. Jennings, and M.J. Moore, *Fabrication of Micropatterned Hydrogels for Neural Culture Systems using Dynamic Mask Projection Photolithography.* Journal of Visualized Experiments, 2011(48):e2636.
8. Kavand, H., et al., *Cell-Imprint Surface Modification by Contact Photolithography-Based Approaches: Direct-Cell Photolithography and Optical Soft Lithography Using PDMS Cell Imprints.* ACS Applied Materials & Interfaces, 2019. **11**(11):10559–10566.
9. Qin, D., Y. Xia, and G.M. Whitesides, *Soft lithography for micro- and nanoscale patterning.* Nature Protocols, 2010. **5**(3):491–502.
10. Cha, C., F. Piraino, and A. Khademhosseini, *Chapter 9 - Microfabrication Technology in Tissue Engineering*, in *Tissue Engineering (Second Edition)*, C.A.V. Blitterswijk and J. De Boer, Editors. 2014, Academic Press.

11. Nie, M., R. Truckenmüller, and S. Takeuchi, *Chapter 10 - Microfabrication technology in tissue engineering*, in *Tissue Engineering (Third Edition)*, J. De Boer, et al., Editors. 2023, Academic Press.
12. Sahin, O., M. Ashokkumar, and P.M. Ajayan, *3 - Micro- and nanopatterning of biomaterial surfaces*, in *Fundamental Biomaterials: Metals*, P. Balakrishnan, S. M S, and S. Thomas, Editors. 2018, Woodhead Publishing.
13. Otsuka, H., *Chapter 11 - Micropatterning of cell aggregate in three dimension for in vivo mimicking cell culture*, in *Colloid and Interface Science in Pharmaceutical Research and Development*, H. Ohshima and K. Makino, Editors. 2014, Elsevier.
14. Liu, W.F. and C.S. Chen, *Engineering biomaterials to control cell function.* Materials Today, 2005. **8**(12):28–35.
15. Hu, S., et al., *Protein–Substrate Adhesion in Microcontact Printing Regulates Cell Behavior.* Langmuir, 2018. **34**(4):1750–1759.
16. Lakshminarayanan, S., *Micro/Nano Patterning on Polymers Using Soft Lithography Technique*, in *Micro/Nanolithography - A Heuristic Aspect on the Enduring Technology*, J. Thirumalai, Editor. 2017, IntechOpen.
17. McGuigan, A.P., et al., *Cell Encapsulation in Sub-mm Sized Gel Modules Using Replica Molding.* PLOS One, 2008. **3**(5):e2258.
18. Iuliano, A., et al., *Coupling 3D Printing and Novel Replica Molding for In House Fabrication of Skeletal Muscle Tissue Engineering Devices.* Advanced Materials Technologies, 2020. **5**(9):2000344.
19. DuRaine, G.D., et al., *Emergence of Scaffold-Free Approaches for Tissue Engineering Musculoskeletal Cartilages.* Annals of Biomedical Engineering, 2015. **43**(3):543–554.
20. Birla, R., *Tissue Fabrication Technology*, in *Introduction to Tissue Engineering: Applications and Challenges*. 2014, John Wiley & Sons.
21. Ovsianikov, A., A. Khademhosseini, and V. Mironov, *The Synergy of Scaffold-Based and Scaffold-Free Tissue Engineering Strategies.* Trends in Biotechnology, 2018. **36**(4):348–357.
22. Alblawi, A., et al., *Scaffold-free: A developing technique in field of tissue engineering.* Computer Methods and Programs in Biomedicine, 2020. **185**:105148.
23. Liu, K.-C., et al., *Scaffold-free 3D culture systems for stem cell-based tissue regeneration.* APL Bioengineering, 2024. **8**(4):041501.
24. De Pieri, A., Y. Rochev, and D.I. Zeugolis, *Scaffold-free cell-based tissue engineering therapies: advances, shortfalls and forecast.* npj Regenerative Medicine, 2021. **6**(1):18.
25. Griffith, L.A., et al., *A scaffold-free approach to cartilage tissue generation using human embryonic stem cells.* Scientific Reports, 2021. **11**(1):18921.

Scaffold Design and Fabrication: Organoids and Organs-on-Chips

14

Learning Objectives

- Analyze the potential applications of organoids and organs-on-chips
- Examine and apply the design principles and potential applications of organoids
- Evaluate representative organ-on-a-chip systems and assess their potential applications

14.1 Organoids

Nature Methods named organoids as the Methods of the Year 2017 for their potential to probe human biology in health and disease (Fig. 14.1). Organoids are 3D multicellular *in vitro* tissue constructs that closely mimic their corresponding *in vivo* organs [1]. They exhibit a high degree of similarity to their tissue of origin, replicating and simulating their unique biological characteristics. Organoids can self-renew and self-organize, contain various cell types, perform some specific functions, and form spatial structures that resemble those of native organs [2]. Since the development of small intestinal organoids with crypt-villus architecture from adult intestinal stem cells in 2009, organoid models have been developed for numerous organs: brain, liver, kidney, breast, retina, and organs of the gastrointestinal tract among others [2, 3].

A major challenge in the organoid field is to engineer cellular complexity into organoids in a controlled manner that enables organized assembly and tissue function acquisition. To address this, researchers have drawn inspiration from normal organogenesis to organoid formation (i.e., organoidgenesis) (Fig. 14.2). Narrative engineering, an emerging approach that integrates biological and engineering principles, aims to control organoidgenesis through self-organization. This strategy involves three interrelated design considerations:

W. Zhang, *Cell Culture and Tissue Engineering*,
https://doi.org/10.1007/978-3-032-32096-4_14

Fig. 14.1 Volume 15 Issue 1 cover of *Nature Methods*. (Adapted with permission from Springer Nature)

- Space design: tissue designed as a spatial default for preparedness toward self-organization.
- Biological environmental control: the selection of biology-inspired environmental modulators, such as soluble factors or extracellular matrix.
- Synthetic environmental control: this systems feature modulation of variables such as cell-intrinsic properties, perfusion, and mechanical properties (Fig. 14.3) [4].

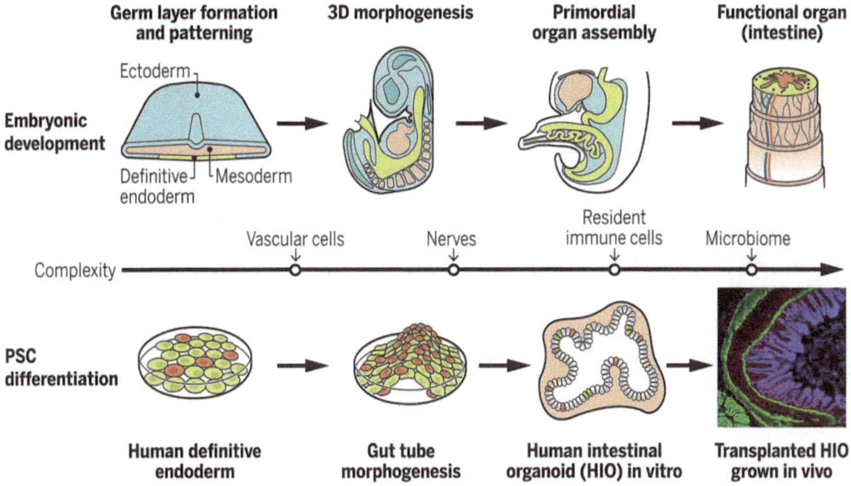

Fig. 14.2 Applying principles of organogenesis into generating cellular complexity during organoidgenesis. Upper panel: some of the main stages that drive assembly of organ primordia. Middle line: additional cell types that get incorporated into developing organs (vascular cells, nerves, immune cells) and how these cell types have been experimentally incorporated into developing organoids (lower panel). After transplantation, organoids can become vascularized by the host and continue to grow in size and undergo tissue morphogenesis. (Adapted from [4] with permission from the American Association for the Advancement of Science)

Organoid technology has a wide range of applications, including modeling genetic disease, providing insights into emerging and future health threats, drug discovery and toxicity assessment, cancer research and treatment, precision medicine, gene repair and regenerative medicine [2, 5–7]. For example, Xiaocui Wang et al. developed a cartilaginous organoid system derived from expanded pluripotent stem cells for drug screening and identified α2-adrenerigic receptor signaling as a therapeutic target for cartilage regeneration. The engineered cells carried $COL2A1^{mCherry}$ and $COL10A1^{eGFP}$ dual reporters, allowing real-time tracking of chondrogenesis and hypertrophy, respectively. Organoids were developed using 3D culture, forming spherical microtissue that grew over time—exhibiting intensive red fluorescence from day 28 and green fluorescence at the edge of the pellet on day 42. The cells exhibited a zonal distribution resembling the distinct layers of native articular cartilage: superficial, intermediate, and deep zones [8] (Fig. 14.4).

A recent advancement in organoid technology involves the development of a glia-enriched cortical organoid model by inducing a gliogenic switch in a subset of progenitors, which facilitates accelerated astrogliogenesis (Fig. 14.5). Upon intracerebral transplantation, these organoids produced a diverse repertoire of cortical neurons and anatomical subclasses of human astrocytes [9]. In another study, Eric Carolina et al. reported the generation of human induced pluripotent stem cell (hiPSC)-derived 3D bile duct within liver organoid by incorporating

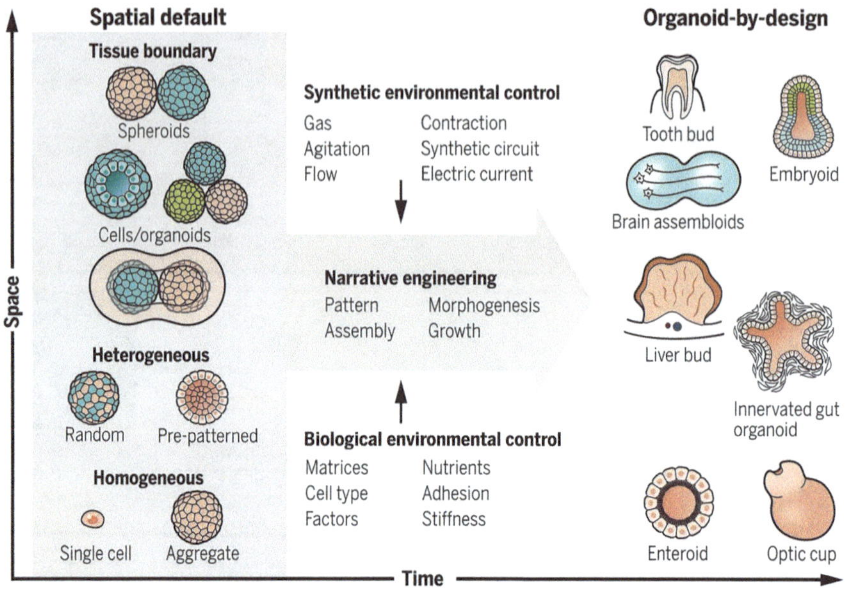

Fig. 14.3 Concept of narrative engineering. Starting from the initial default structures, the timed manipulation of environmental factors will facilitate complex and stereotyped organoid formation from stem cells. The right panel explains some examples of terminal products: Enteroids (gut epithelial organoids) or optic cup organoids develop from single or homogeneous stem cell aggregates, whereas vascularized liver bud or innervated gut organoids self-organize by coculturing heterogeneous progenitors. Recent examples of brain organoids and embryoids were self-assembled from the two distinct, preformed tissue aggregates. These self-organization processes are optimized by temporal modulation of biological and/or synthetic parameters. (Adapted from [4] with permission from the American Association for the Advancement of Science)

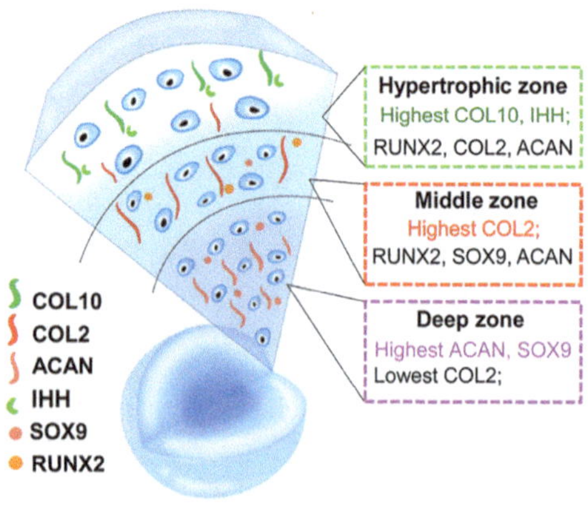

Fig. 14.4 Schematic of the zonal distribution cell types in the organoid. *COL* collagen (*COL2A1* cartilage-specific extracellular matrix, *COL10A1* a hypertrophic marker), *ACAN* aggrecan (a chondrogenic marker), *IHH* Indian hedgehog (a transcriptional factor regulating chondrocyte hypertrophy), *SOX9* sex-determining region Y-box 9 (a critical factor for chondrocyte differentiation), *RUNX2* runt-related transcription factor 2 (a transcriptional factor promoting chondrocyte hypertrophy). (Adapted from [8] with permission from Elsevier)

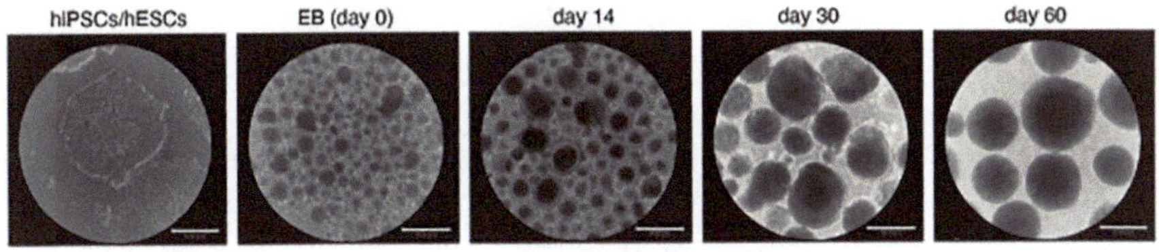

Fig. 14.5 Generation of glia-enriched cortical organoids. *hiPSCs* human-induced pluripotent stem cells, *hESCs* human embryonic stem cells, *EB* embryoid bodies. To generate organoids, embryoid bodies were first produced. (Adapted from [9] with permission from Springer Nature)

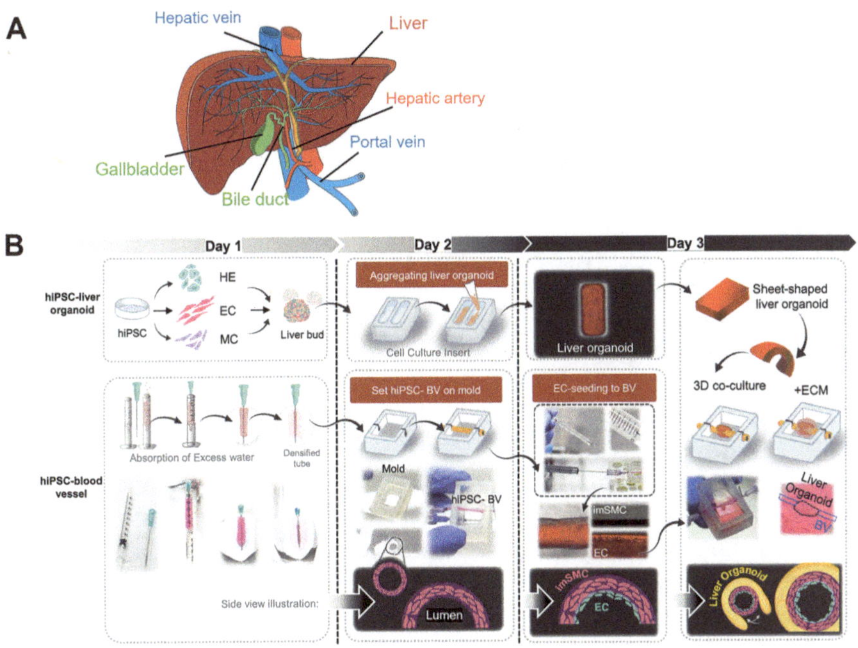

Fig. 14.6 (**a**) Anatomy of human liver. (**b**) Experimental workflow for generating hiPSC-blood vessel incorporated liver organoid for inducing tubular-like bile duct structure. *hiPSC* human induced pluripotent stem cells, *HE* hepatic endoderm cells, *EC* endothelial cells, *MC* mesenchymal cells, *BV* blood vessel, *imMSC* immature smooth muscle cells, *ECM* extracellular matrix. (Adapted from [10] with permission from Springer Nature)

human iPSC-derived blood vessels. As shown in Fig. 14.6, the hiPSC-liver progenitors consisted of iPSC-derived immature smooth muscle and endothelial cells. After 3 weeks, the liver progenitors within hiPSC-blood vessel-incorporated liver organoids differentiated into cholangiocytes (the epithelial component of intrahepatic bile duct which are specialized in secretory function) and acquired characteristic epithelial features [10]. Additionally, it has been shown that treatment with the placenta-derived factors (components of fetal blood) under hypoxia conditions significantly enhances the progenitor expansion in human organoid culture [11]. One of the major challenges of commercialization

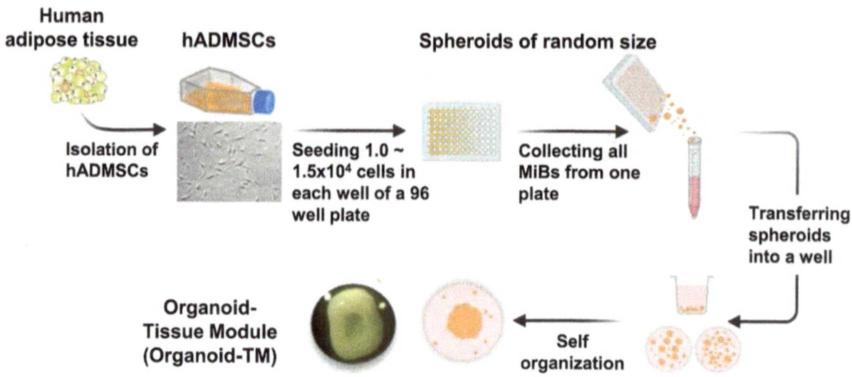

Fig. 14.7 Schematic summary of experiment for Organoid-TM generation. Various sizes of MiBs were produced by seeding ADMSCs into a 96 well plate. Harvested MiBs were then transferred into one well of a 96 well plate. Self-assembly and self-organization of MiBs were induced to form an Organoid-TM. *hADMSCs* human adipose-derived mesenchymal stem cells, *MiBs* microblocks. (Adapted from [12] with permission from Elsevier)

of organoids and other tissue engineered products is scalability. To address this, a recent study demonstrated the generation of organoid-tissue modules through scaffold-free self-assembly of microblocks derived from adipose-derived mesenchymal stem cells (Fig. 14.7) [12].

14.2 Organs-on-Chips

Organs-on-chips are microfabricated cell culture devices designed to replicate key functional units of human organs *in vitro*. Their design follows a reductionist approach—understanding and analyzing the target organ's anatomy to identify basic elements essential for physiological function, designing a cell culture device to replicate the identified features, and then fabricating the designed system [13]. A milestone example is the lung-on-a-chip, which reconstituted organ-level lung functions. As shown in Fig. 14.8, the microfluidic system featured two microchannels separated by a thin, porous, and flexible poly(dimethyl siloxane) (PDMS) membrane coated with fibronectin and collagen. Human alveolar epithelial cells and human pulmonary microvascular endothelial cells were cultured on opposite sides of the membrane. Air was introduced to the epithelia side, forming an air-liquid interface and mimicking the lining of alveolar air space when cells become confluent. To simulate breathing motions, two lateral microchambers were incorporated—when vacuum was applied to the chambers, elastic deformation occurred, causing the PDMS membrane and adherent cells to stretch. The system successfully mimicked key lung functions, including pathogen-induced inflammatory responses and response to cytokine exposure [14]. As *in vitro* models, they offer major advantages such as miniaturization, integration, low consumption, and accurate control of parameters such as fluid shear stress [15].

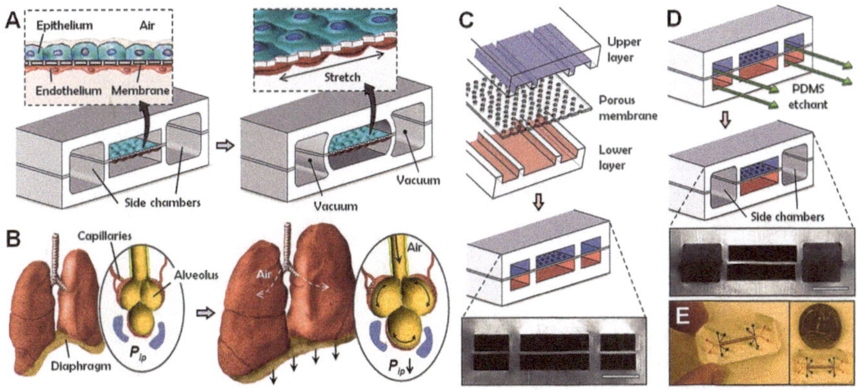

Fig. 14.8 Biologically inspired design of a human breathing lung-on-a-chip microdevice. (**a**) The microfabricated lung mimic device uses compartmentalized poly(dimethyl siloxane) (PDMS) microchannels to form an alveolar-capillary barrier on a PDMS membrane coated with fibronectin and collagen. The device recreates physiological breathing movements by applying vacuum to the side chambers and causing mechanical stretching of the PDMS membrane. (**b**) During inhalation in the living lung, contraction of the diaphragm causes a reduction in intrapleural pressure (Pip), leading to distension of the alveoli and physical stretching of the alveolar-capillary interface. (**c**) Three PDMS layers are aligned and irreversibly bonded to form two sets of three parallel microchannels separated by a 10-μm-thick PDMS membrane containing an array of 10 μm through-holes. Scale bar, 200 μm. (**d**) After permanent bonding, PDMS etchant is flowed through the side channels. Selective etching of the membrane layers in these channels produces two large side chambers to which vacuum is applied to cause mechanical stretching. Scale bar, 200 μm. (**e**) Images of an actual lung-on-a-chip microfluidic device viewed from above. (Adapted from [14] with permission from the American Association for the Advancement of Science)

Organs-on-chips technology integrates cell biology, engineering and material sciences to simulate native tissues [15]. PDMS is most commonly used material for fabricating organs-on-chips due to its ease of fabrication, optical transparency, low autofluorescence, high elasticity, excellent oxygen permeability, and good deformability. To enhance performance, researchers have explored alternative materials such as poly(methyl methacrylate) and gelatin-methacryloyl, as well as surface coatings. Common fabrication techniques for microfluidic devices include soft lithography, photolithography, injection molding, and 3D printing [16].

To date, a variety of organs-on-chips have been developed, including kidney on a chip, lung on a chip, heart on a chip, skin on a chip, pancreas on a chip, brain on a chip, placenta on a chip, intestine on a chip, and blood brain barrier on a chip [15, 17, 18]. For instance, a placenta-on-a-chip system was fabricated using soft lithography. It consisted of two PDMS microfluidic channels separated by a thin extracellular matrix membrane, with human trophoblasts and human umbilical vein endothelial cells (HUVECs) seeded on the opposite sides (Fig. 14.9). Over time, this model formed the placental barrier, which was then quantitatively assessed for its barrier properties [19].

Different organ chips can be fluidically linked to create multi-organ chip systems or even a human-on-a-chip [20, 21]. For example, an automated Interrogator

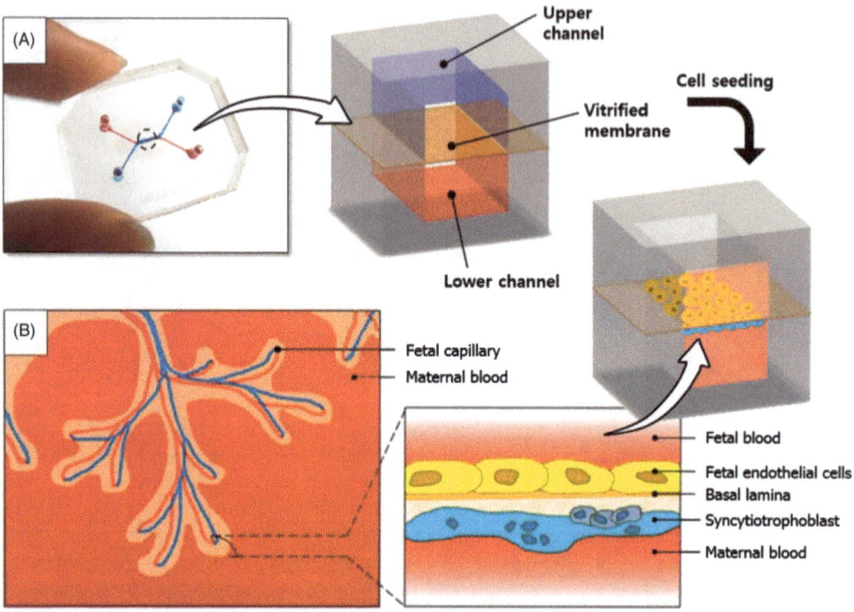

Fig. 14.9 A placenta-on-a-chip microdevice: (**a**) The microengineered device is composed of the upper (blue) and lower (red) poly(dimethyl siloxane) (PDMS) chambers separated by a vitrified collagen membrane. (**b**) Endothelial cells and trophoblasts are co-cultured in close apposition on the opposite sides of the membrane. (Adapted from [19] with permission from Taylor & Francis)

platform successfully maintained the viability and organ-specific functions of eight vascularized, two-channel organs-on-chips (intestine, liver, kidney, heart, lung, skin, blood-brain-barrier, and brain) for 3 weeks in culture (Fig. 14.10) [22]. Organs-on-chips, multi-organ chip systems in particular, hold great potential for disease modeling and personalized medicine. Drug administration methods such as aerosolized, oral, and intravenous delivery can be simulated by introducing compounds into the airspace of a lung chip, the lumen of an intestine chip, or directly into the vascular channel, respectively. Linked liver and kidney chips can be used to quantify drug metabolism and clearance, respectively, while other organ chips can be added to assess efficacy and potency (Fig. 14.11). For personalized medicine, organ chips generated using patient-derived cells can model rare genetic disorders, identify toxicities difficult to study clinically, and compare drug responses across different subpopulations (Fig. 14.12) [17].

A recent global literature meta-analysis of organoid and organ-on-chip research after 2010 was conducted and the algorithm identified and tracked 149 organs or organ substructures modeled as organoids and 107 for organ-on-chip, along with various stem cell sources, 130 diseases, and applications across 16 groups of non-human and non-mouse organisms. Over the past decade, annual publication in these areas reached 41% and 50%, respectively [23]. Additionally, Google search trends

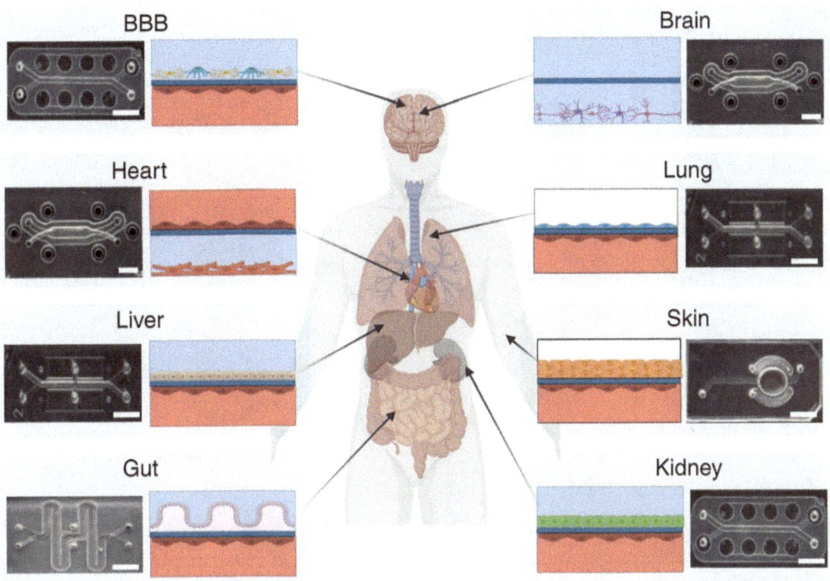

Fig. 14.10 Representative photograph and sectional schematic of each organ-on-a-chip. BBB: lower channel brain microvascular endothelial cells, upper channel brain pericytes and cortical astrocytes. Brain: differentiated human primary neural stem cells forming networks of neurons and astrocytes. Lung: lower channel human umbilical cord vascular endothelial cells, upper channel, lung epithelial cells. Skin: lower channel containing dermal microvascular endothelial cells, upper channel containing keratinocytes and dermal fibroblasts. Kidney: lower channel kidney derived endothelial cells, upper channel containing proximal tubule epithelial cells. Gut: lower channel human umbilical cord vascular endothelial cells, upper channel containing villi-like structures of gut epithelial cells, Liver: lower channel containing liver sinusoidal endothelial cells, upper channel containing hepatocytes. Heart: lower channel human umbilical cord vascular endothelial cells, upper channel cardiomyocytes differentiated from cardiomyocytes. Scale bar: 5 mm. *BBB* blood-brain-barrier. (Adapted from [22] with permission from permission from Springer Nature)

further confirm the rapid rise of both technologies (Fig. 14.13). Several companies specialize in organs-on-chips technology. For example, Hesperos, Inc., founded in 2015, aims to accelerate drug discovery by leveraging their Human-on-a-Chip® system (Fig. 14.14) [24]. As interest in alternatives to animal testing continues to grow, regulatory agencies and research institutions are moving toward validating and standardizing new *in vitro* technologies, particularly organoids and organs-on-chips [25]. In 2025, FDA released its *Roadmap to Reducing Animal Testing in Preclinical Safety Studies*, specifying both organoids and organs-on-chips (microphysiological systems) as scientifically validated new approach methodologies (NAMs) for reducing animal testing in preclinical safety studies (i.e., through *in vitro* human-derived systems) [26]. The Joint Research Centre of the European Commission also made a news announcement—*Setting out a roadmap for standardisation of organ-on-chip technology* [27]. Importantly, a multidisciplinary team of stakeholders, including

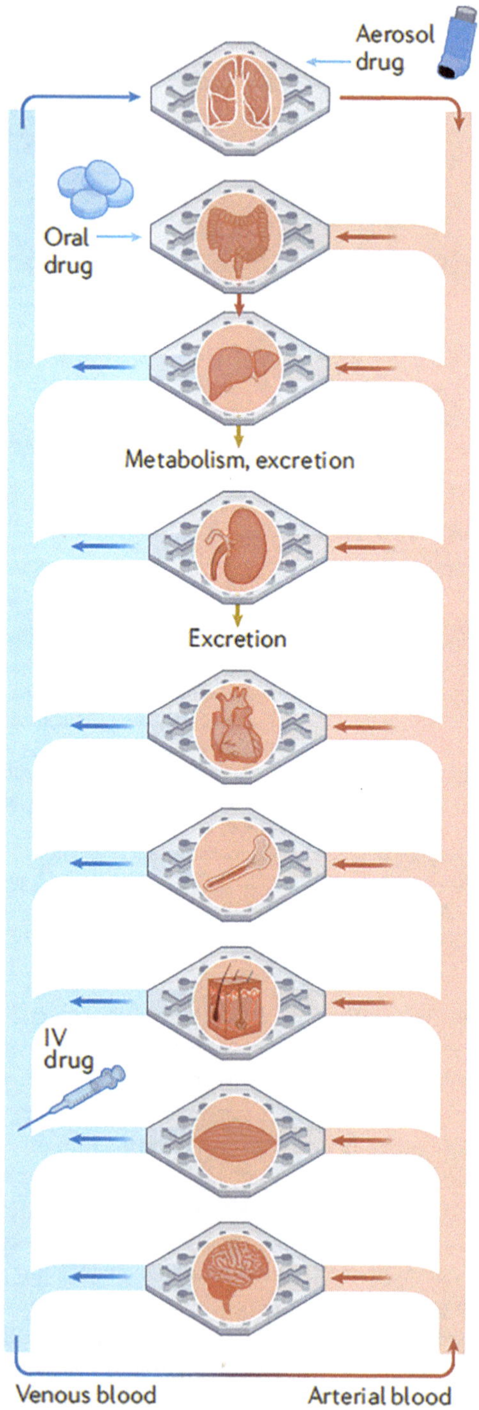

Fig. 14.11 Multi-organ chip systems linked by flow channels can mimic the physiological linking of organs in our bodies to study drug absorption, distribution, metabolism and excretion occur in the human body. *IV* intravenous. (Adapted from [17] with permission from Springer Nature)

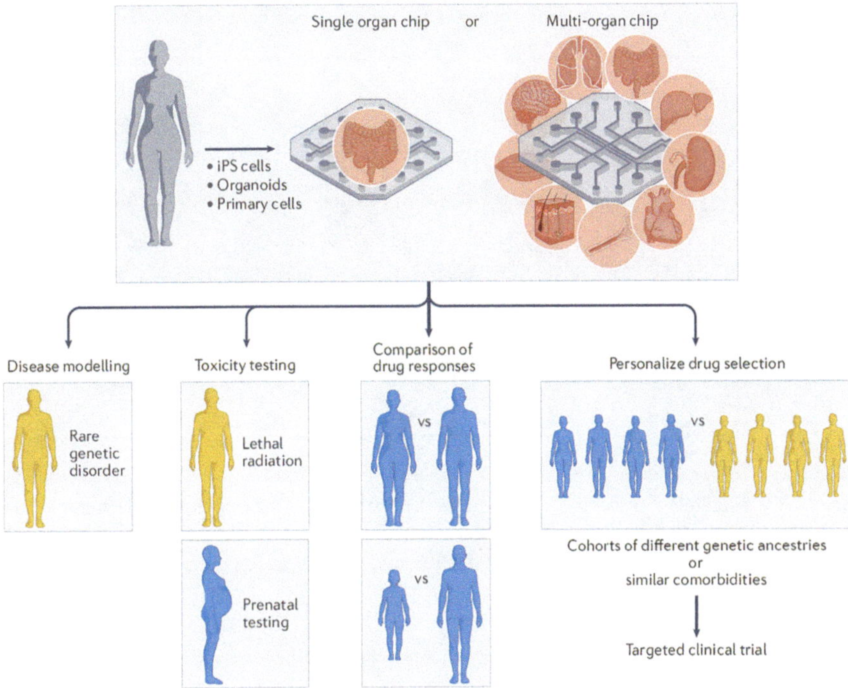

Fig. 14.12 Human organ chip applications for personalized medicine. *iPS cells* induced pluripotent stem cells. (Adapted from [17] with permission from Springer Nature)

Fig. 14.13 Increasing Google search hits over time for the terms "organoid/organoids" and "organ-on-a-chip/organs-on-chips"

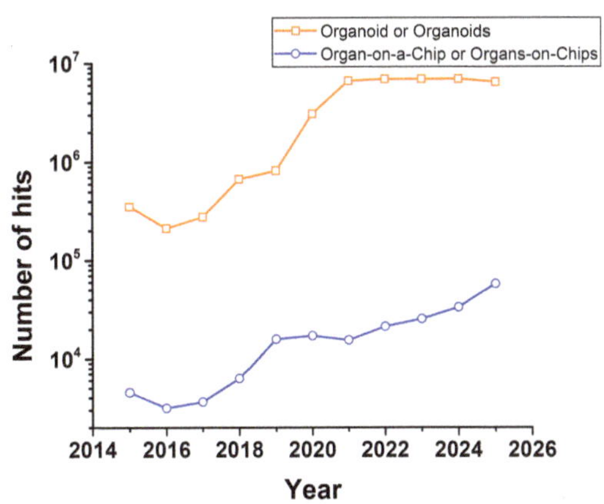

representatives from the FDA, the National Institute of Standards and Technology (NIST), the European Union, academia, and industry, was established to provide a framework for future development of guidelines/standards governing engineering concepts of organ-on-a-chip models [28].

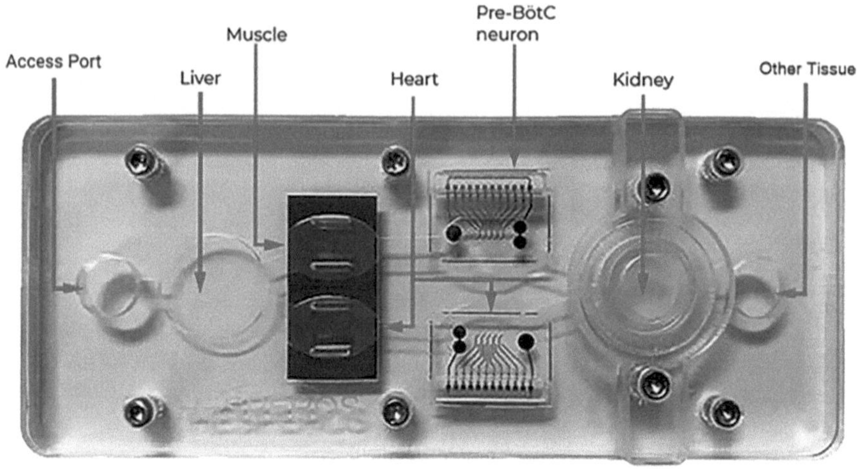

Fig. 14.14 Example 5-organ model funded by the National Institute of Health's Helping to End Addiction Long-term® Initiative (NIH HEAL Initiative®) to study the effects of an overdose. (Credit: Hesperos, Inc. (https://hesperosinc.com/))

Review and Practice Questions

1. What are organoids?
2. What is organoidgenesis? And what are the three key design considerations in narrative engineering?
3. What are organs-on-chips, and what are the core principles behind their design?
4. Read the article entitled "Reconstituting Organ-Level Lung Functions on a Chip" and answer the following questions:
   (a) How did the authors fabricate the biomimetic microsystem that reproduces the key structural, functional, and mechanical properties of the human alveolar-capillary interface?
   (b) How did the authors mimic the subatmospheric, pressure-driven stretching of the alveoli?
   (c) How did the authors demonstrate that this microsystem could recapitulate the pulmonary inflammation?
   (d) How did the authors demonstrate this that microsystem could mimic the innate cellular response to pulmonary infection of bacterial origin?
5. Conduct a literature search on organs-on-chips and write a short essay on an organ-on-a-chip system not covered in this chapter.
6. How can human organ chips be utilized to advance personalized medicine?

References

1. de Souza, N., *Organoids*. Nature Methods, 2018. **15**(1):23–23.
2. Yang, S., et al., *Organoids: The current status and biomedical applications*. MedComm, 2023. **4**(3):e274.

3. *Method of the Year 2017: Organoids*. Nature Methods, 2018. **15**(1):1.
4. Takebe, T. and J.M. Wells, *Organoids by design*. Science, 2019. **364**(6444):956–959.
5. Tang, X.-Y., et al., *Human organoids in basic research and clinical applications*. Signal Transduction and Targeted Therapy, 2022. **7**(1):168.
6. Liu, X., et al., *Recent progress on the organoids: Techniques, advantages and applications*. Biomedicine & Pharmacotherapy, 2025. **185**:117942.
7. Verstegen, M.M.A., et al., *Clinical applications of human organoids*. Nature Medicine, 2025. **31**(2):409–421.
8. Wei, X., et al., *A human organoid drug screen identifies α2-adrenergic receptor signaling as a therapeutic target for cartilage regeneration*. Cell Stem Cell, 2024. **31**(12):1813–1830.e8.
9. Wang, M., et al., *Morphological diversification and functional maturation of human astrocytes in glia-enriched cortical organoid transplanted in mouse brain*. Nature Biotechnology 2025. **43**(1):52–62.
10. Carolina, E., et al., *Generation of human iPSC-derived 3D bile duct within liver organoid by incorporating human iPSC-derived blood vessel*. Nature Communications, 2024. **15**(1): 7424.
11. Kuse, Y., et al., *Placenta-derived factors contribute to human iPSC-liver organoid growth*. Nature Communications, 2025. **16**(1):2493.
12. Cho, J., et al., *Self-assembled organoid-tissue modules for scalable organoid engineering: Application to chondrogenic regeneration*. Acta Biomaterialia, 2025. **197**:152–166.
13. Park, S.E., A. Georgescu, and D. Huh, *Organoids-on-a-chip*. Science, 2019. **364**(6444):960–965.
14. Huh, D., et al., *Reconstituting Organ-Level Lung Functions on a Chip*. Science, 2010. **328**(5986):1662–1668.
15. Singh, D., et al., *Journey of organ on a chip technology and its role in future healthcare scenario*. Applied Surface Science Advances, 2022. **9**:100246.
16. Cao, U.M.N., et al., *Microfluidic Organ-on-A-chip: A Guide to Biomaterial Choice and Fabrication*. International Journal of Molecular Sciences, 2023. **24**(4):3232.
17. Ingber, D.E., *Human organs-on-chips for disease modelling, drug development and personalized medicine*. Nature Reviews Genetics, 2022. **23**(8):467–491.
18. Lermant, A., et al., *Development of a human iPSC-derived placental barrier-on-chip model*. iScience, 2023. **26**(7):107240.
19. Lee, J.S., et al., *Placenta-on-a-chip: a novel platform to study the biology of the human placenta*. The Journal of Maternal-Fetal & Neonatal Medicine, 2016. **29**(7):1046–1054.
20. Huang, Y., et al., *From Organ-on-a-Chip to Human-on-a-Chip: A Review of Research Progress and Latest Applications*. ACS Sensors, 2024. **9**(7):3466–3488.
21. Srivastava, S.K., et al., *Organ-on-chip technology: Opportunities and challenges*. Biotechnology Notes, 2024. **5**:8–12.
22. Novak, R., et al., *Robotic fluidic coupling and interrogation of multiple vascularized organ chips*. Nature Biomedical Engineering, 2020. **4**(4):407–420.
23. Shoji, J.-y., et al., *Global Literature Analysis of Organoid and Organ-on-Chip Research*. Advanced Healthcare Materials, 2024. **13**(21):2301067.
24. Hesperos. *About Us*. Available from: https://hesperosinc.com/about/.
25. Reardon, S., *Beyond lab animals*. Science, 2025. **389**(6761):676–679.
26. U.S. Food and Drug Administration. *Roadmap to Reducing Animal Testing in Preclinical Safety Studies*. Available from: https://www.fda.gov/files/newsroom/published/roadmap_to_reducing_animal_testing_in_preclinical_safety_studies.pdf.
27. Joint Research Centre, European Commision. *Setting out a roadmap for standardisation of organ-on-chip technology*. Available from: https://joint-research-centre.ec.europa.eu/jrc-news-and-updates/setting-out-roadmap-standardisation-organ-chip-technology-2025-01-13_en.
28. Reyes, D.R., et al., *From animal testing to in vitro systems: advancing standardization in microphysiological systems*. Lab on a Chip, 2024. **24**(5):1076–1087.

Scaffold Design and Fabrication: Bioprinting

15

Learning Objectives

- Interpret a typical bioprinting workflow
- Analyze and compare different types of bioprinters—underlying mechanisms, advantages, and limitations
- Evaluate and contrast different types of bioinks—composition and suitability for specific tissue engineering applications

15.1 Introduction to Bioprinting

(Three-dimensional) Bioprinting is an emerging additive manufacturing technology that integrates living cells/spheroids/organoids, scaffolds, and bioactive molecules to fabricate biological structures that closely mimic native tissue. This innovative approach holds great promise for advancing personalized medicine, tissue regeneration, and pharmaceutical testing [1–3]. Compared to other tissue fabrication techniques, bioprinting offers several advantages including greater flexibility, customization, scalability, reliability, durability, and relatively high speed [2]. Bioprinting is built upon three central approaches: biomimicry, autonomous self-assembly, and mini-tissue building blocks. Biomimicry involves replicating specific cellular and extracellular components of tissues, such as the branching architecture of the vascular system. Autonomous self-assembly draws on an in-depth understanding of embryonic tissue genesis and organogenesis, requiring environmental manipulation to drive natural developmental processes in bioprinted constructs. Mini-tissues, functional building blocks of tissues and organs, can be fabricated and assembled into larger constructs [4, 5].

The history of bioprinting dates back to 1988, when Klebe first demonstrated cell deposition using a standard Hewlett-Packard (HP) inkjet printer. In 1999, Odde and Renn applied laser-assisted bioprinting to deposit living cells for constructing

W. Zhang, *Cell Culture and Tissue Engineering*,
https://doi.org/10.1007/978-3-032-32096-4_15

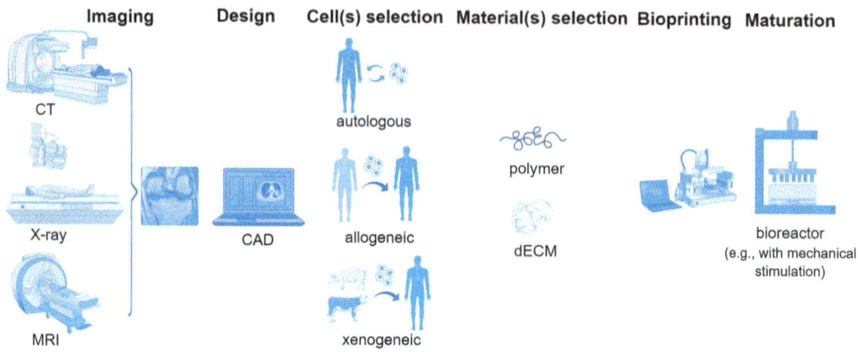

Fig. 15.1 A typical workflow for bioprinting. *CT* computed tomography, *MRI* magnetic resonance imaging, *CAD* computer-aided design, *dECM* decellularized extracellular matrix

anatomically complex models. A scaffold seeded with human cells in the shape of a bladder was bioprinted in 2001. In 2002, the first extrusion-based bioprinting technology, later commercialized as "3D-Bioplotter", was introduced. This was followed in 2003 by the development of the first inkjet bioprinter, created by modifying an HP standard inkjet printer. In 2004, cell-loaded bioprinting was implemented using a commercial stereolithography (SLA) printer. *In situ* bioprinting was attempted in 2012 [6]. More recently, the exploration of 4D bioprinting has begun, which incorporates "time" as the fourth dimension. The 4D bioprinting refers to the combination of cells, biological molecules, and stimuli-response biomaterials to develop 3D constructs that can transform structurally and/or functionally over time in response to external stimuli [5, 7, 8].

The overall bioprinting workflow is illustrated in Fig. 15.1. The process begins with structural design using computational tools (e.g., computer-aided design, CAD), often based on imaging data from medical devices, such as computed tomography (CT) scanning, X-rays, and magnetic resonance imaging (MRI). This is followed by the selection of appropriate cell(s) and biomaterials. Once the construct is fabricated using a bioprinter, tissue maturation is typically carried out in bioreactors, particularly those capable of delivering a particular stimulation (e.g., mechanical or electrical) to support functional development [4, 6, 9–11].

15.2 Bioprinter Types

There are various types of bioprinters. Representative examples, shown in Fig. 15.2, include extrusion-based, inkjet, laser-assisted and stereolithography bioprinters [12]. A comparison of extrusion-based, inkjet, and laser-assisted bioprinters is provided in Table 15.1. Extrusion-based bioprinters are the most widely used and cost-effective. They can be further classified based on the extrusion mechanism: pneumatic or mechanical (e.g., piston or screw) systems (Fig. 15.2). Mechanical extrusion generally offers improved spatial control but involves a more complex design [4]. In extrusion-based systems, bioinks are dispensed through nozzles in a

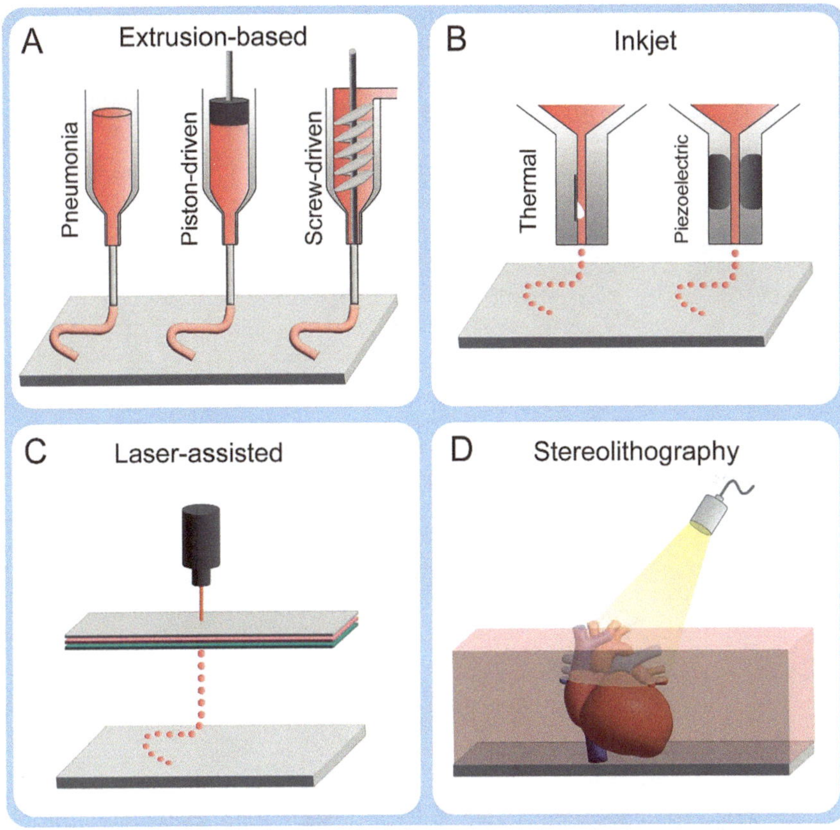

Fig. 15.2 Representative 3D bioprinting techniques. (**a**) Extrusion-based bioprinting; (**b**) Inkjet bioprinting; (**c**) Laser-assisted bioprinting; (**d**) Stereolithography bioprinting. (Adapted from [12] with permission from Elsevier)

controlled manner and deposited layer by layer to form 3D structures [12]. Extrusion-based systems support a broad range of bioinks and are particularly suited for high cell density applications. However, common drawbacks include lower printing resolution, slower printing speeds, and reduced cell viability due to shear stress during extrusion [4, 12]. Inkjet bioprinters, by contrast, utilize thermal or acoustic forces to eject droplets of bioink onto a substrate. However, the physical stress required for droplet generation limits the range of usable bioinks in terms of viscosity (<10 centipoise), physical form (generally liquid to allow droplet formation), and cell density (typically <10^7 cells/mL) [4]. Despite these limitations, inkjet bioprinting offers significant advantages in resolution and deposition accuracy, as it can generate droplets in the picoliter range [12, 16, 17]. Laser-assisted bioprinting, though less common than extrusion-based and inkjet methods, operates based on laser-induced forward transfer principles. A typical laser-assisted bioprinter consists of a pulsed laser beam, a 'ribbon' that has a donor transport support usually

Table 15.1 Comparison of bioprinters[a]

	Bioprinters		
	Extrusion-based	Inkjet	Laser-assisted
Bioink viscosities	30 - > 6 × 10^7 mPa s	3.5–12 mPa s	1–300 mPa s
Printing speed	Slow	Fast	Medium-fast
Resolution	Low-moderate (μm to mm scale)	High (50–300 μm)	High (microscale)
Vertical structure quality	Good	Poor	Fair
Cell viability	Varies depending on the literature		
Cell density	High	Low	Medium
Advantages	1. Simple process 2. Can print a wide range of bioinks 3. Suitable for high cell density printing	1. High resolution 2. High throughput	1. High resolution 2. Nozzle-free system
Disadvantages	1. Low resolution 2. Not suitable for low-viscosity bioinks	1. Limited to low-viscosity (liquid) bioinks 2. Not suitable for high cell densities	1. High cost 2. Risk of thermal damage

[a]Note: Summarized from references [4, 13–15]

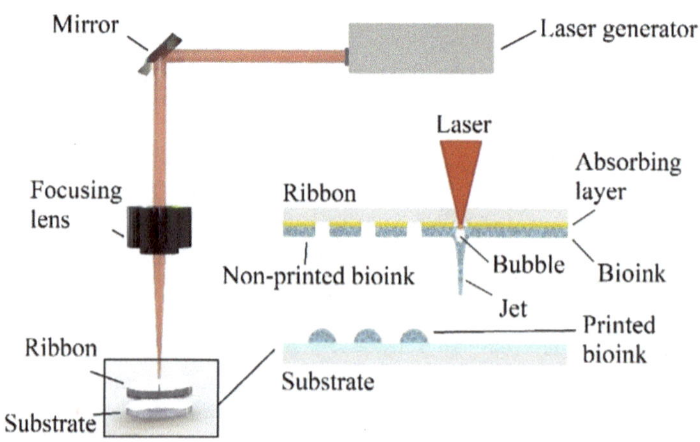

Fig. 15.3 Structure and main components of a laser-assisted bioprinter. (Adapted from [20] with permission from John Wiley and Sons)

made from glass that is covered with a laser-energy-absorbing layer (e.g., gold) and a layer of biological material prepared in a liquid solution, and a receiving substrate placed opposite the ribbon. Focused laser pulses create a high-pressure bubble (vaporization-induced) in the absorbing layer, propelling the bioink from the ribbon toward the substrate (Fig. 15.3) [4, 18–20]. Laser-assisted bioprinting enables high-resolution patterning due to its precise control and high printing frequency, and it avoids nozzle-related issues. However, it is crucial that the cells used in

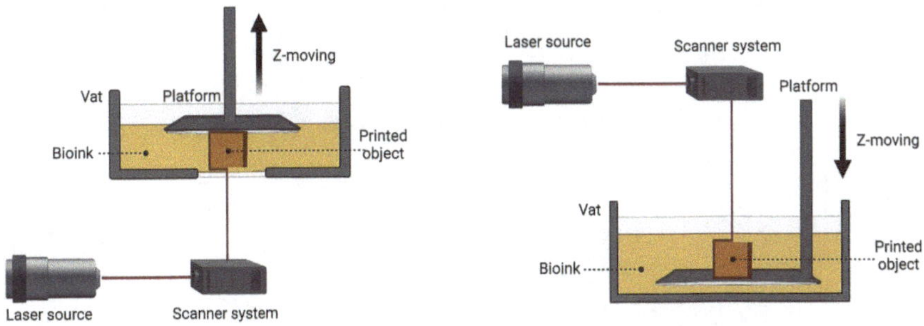

Fig. 15.4 Schematic stereolithography-based bioprinting: laser provides light for the photopolymerization of bioink in vat through a computer-controlled moving laser beam. When a layer is completely bioprinted *via* point-by-point curing, the build platform will lift up (left) or move down (right) in the z-direction until the bioprinted object is fished line-by-line and layer-by-layer. (Adapted from [21] with permission from Elsevier)

laser-assisted bioprinting can withstand both the applied laser energy and the potentially disruptive effects of the laser-induced transfer process [12].

Stereolithography bioprinting utilizes digital micromirror arrays to precisely control the light intensity in each pixel, allowing the polymerization of light-sensitive (ultraviolet (UV) or visible light) materials in a line-by-line and layer-by-layer fashion (Figs. 15.2d and 15.4). This technology enables fast and high-resolution fabrication of large, complex structures while avoiding shear stress and viscosity-related limitations common in other bioprinting methods. However, challenges remain—most notably the potential cytotoxicity associated with the UV exposure and photoinitiators used to initiate polymerization. These factors limit the selection of suitable photosensitive biomaterials. Hence, biomaterials for stereolithography must balance biocompatibility with the mechanical integrity required during and after printing. Additionally, post-polymerization steps, such as further photocuring, may be necessary to fully stabilize the printed constructs [12, 21, 22].

15.3 Bioink Types

Bioinks can be defined in various ways, with the primary distinction being whether or not cells are considered as a required component. A comprehensive definition describes bioink as "*a formulation of cells suitable for processing by an automated biofabrication technology that may also contain biologically active components and biomaterials*". Cells in bioinks can exist in different forms—such as single cells, cell aggregates (e.g., spheroids), organoids, cells seeded onto microcarriers, or encapsulated cells [23]. Common cell sources for bioprinting include stem cells, primary cells, transdifferentiated cells, and cell lines [12].

This section's focus is on acellular components of bioinks. Biomaterials used in bioink formulation include natural polymers, synthetic polymers, decellularized extracellular matrix (dECM), and various combinations of these. Natural polymers

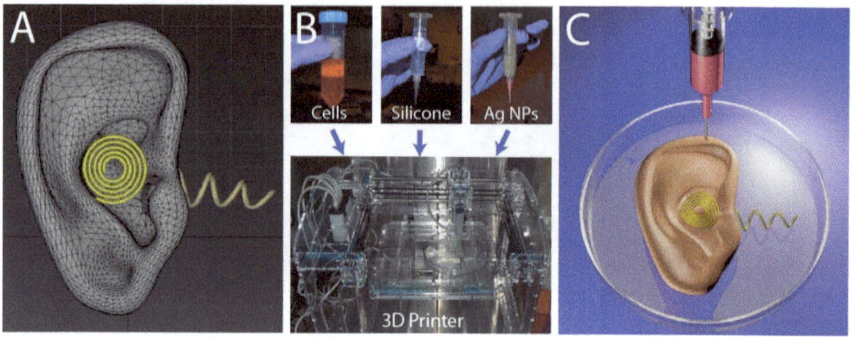

Fig. 15.5 Three-dimensional (3D) interweaving of biology and electronics *via* additive manufacturing to generate a bionic ear. (**a**) CAD drawing of the bionic ear. (**b**) (top) Optical images of the functional materials, including biological (chondrocytes), structural (silicone), and electronic (AgNP-infused silicone) used to form the bionic ear. (bottom) a 3D printer used for the printing process. (**c**) Illustration of the 3D printed bionic ear. *CAD* computer-aided design. (Adapted with permission from [31]. Copyright © 2013 American Chemical Society)

are typically more biocompatible and biodegradable than synthetic ones but tend to be less tunable [4, 24, 25]. dECM-based bioinks provide native structural elements and functional cytokines that can promote cell development; however, their poor mechanical properties and printability remain significant challenges [26, 27].

In addition to structural biomaterials, non-structural additives are often incorporated into bioinks to enhance functionality. These include microspheres, metallic nanoparticles, magnetic particles, carbon nanotubes, and bioactive molecules such as growth factors [28–30]. For instance, electrically conductive silver nanoparticles (AgNPs) were added in an alginate-based bioink containing chondrocytes to bioprint a 3D bionic ear (Fig. 15.5) [31]. It is worth noting that cellulose, the most abundant natural polymer, and its derivatives have been extensively studied for bioprinting applications, serving as matrix, binder, rheology modifier, reinforcement, and excipient [32]. Figure 15.6 summarizes the various types of biomaterials and functional additives commonly used in bioink formulations.

When designing a bioink, several key material properties (Fig. 15.7) must be considered to ensure suitability for bioprinting applications [4, 33, 34]:

1. Printability—the material should be easily dispensable/extrudable and capable of forming stable structures during the printing process (shape fidelity). Optical properties should also be taken into account for certain bioprinting technologies.
2. Mechanical properties—the bioink should maintain its shape after printing and exhibit mechanical characteristics appropriate for the target tissue.
3. Biomimicry—the material should mimic the native extracellular matrix (ECM), including tissue-specific composition and architecture.
4. Biocompatibility—the bioink should not elicit undesirable local or systemic host responses, and it must support cell survival, proliferation, and differentiation, thereby promoting tissue maturation.

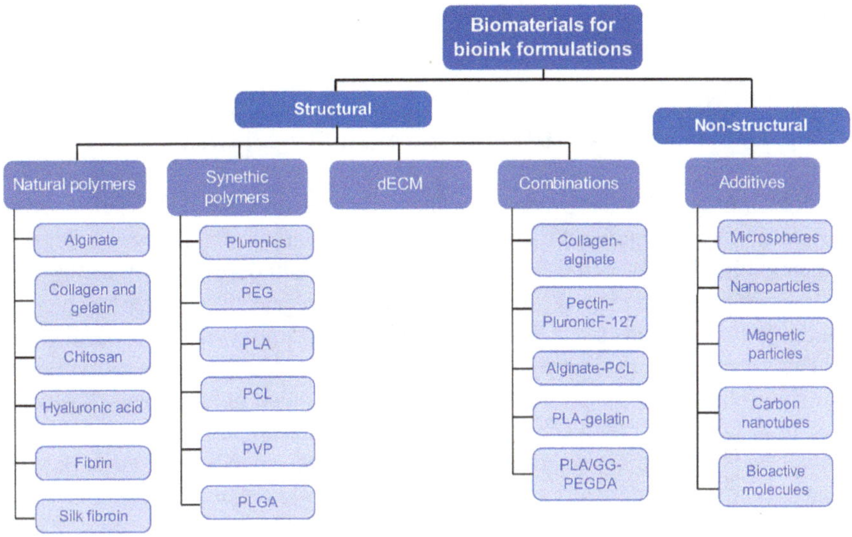

Fig. 15.6 Types of biomaterials and functional additives commonly used in bioink formulations. *PEG* polyethylene glycol, *PLA* polylactic acid, *PCL* polycaprolactone, *PVP* polyvinylpyrrolidone, *PLGA* poly(lactic-*co*-glycolic) acid, *GG-PEGDA* gellan gum-poly (ethylene glycol) diacrylate

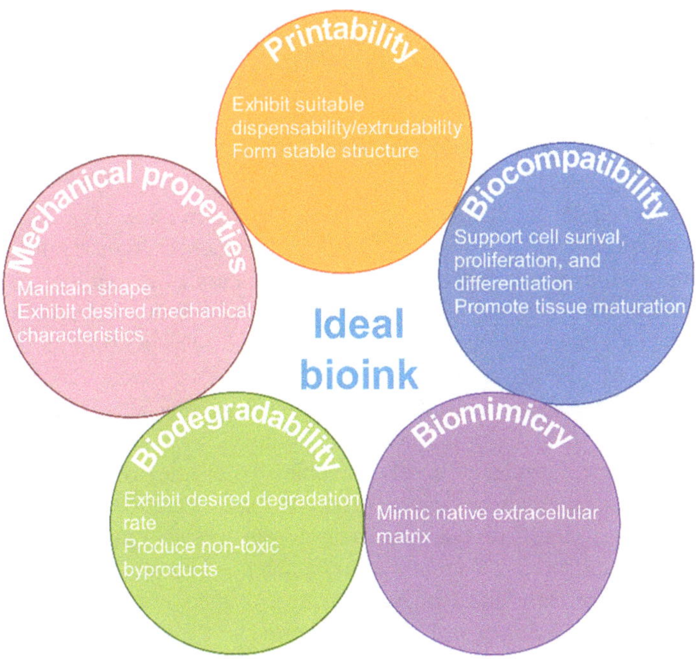

Fig. 15.7 Schematic of deal bioink properties

5. Biodegradability—the degradation rate should align with the rate at which cells synthesize their own ECM, and the degradation byproducts must be non-toxic.

15.4 Representative Advancements in Bioprinting

The field of bioprinting has evolved significantly in recent years. This section showcases some representative advancements in bioprinting. First, the concept of "dynamic bioink" emerged. Dynamic bioinks have been developed by incorporating reversible dynamic molecular network building blocks into conventional static bioink. Compared to the static bioink, dynamic bioink, mimicking the natural extracellular matrix, possesses reversible dynamic molecular networks that provide cellular activity and growth and thus enhances the maturation of bioprinted organ-like constructs. Dynamic bioinks fall into two main categories based on the molecular design principles: supramolecular interactions (hydrogen bonds, host-guest interactions, electrostatic interactions, π-π stacking, and hydrophobic interactions) and reversible chemical bonds (such as reversible Diels-Alder reactions, Schiff base bonds, and disulfide bonds). Besides improving the printability and shape stability, dynamic bioinks facilitate the precise construction of cellular microenvironments through their unique reversible supramolecular interactions and dynamic chemical bonding—providing cells with essential biomechanical and biochemical cues which are critical in regulating cell behavior (including spreading, expansion, migration, and differentiation) and facilitating subsequent tissue maturation (Fig. 15.8) [33].

Additives have also been incorporated in bioink for non-structural purposes, as discussed in Sect. 15.3. Among these, microspheres serve various roles, such as supporting vascularization and enabling controlled release of bioactive molecules. For instance, incorporation of microspheres, prepared through various methods, into pectin/Pluronic F-127 bioinks have been investigated [28, 29, 35]. This system has been successfully applied in endometrium repair in mice [36]. As shown in Fig. 15.9, β-estradiol (E2), a hormone essential for maintaining the microenvironment of the endometrium and stimulating regeneration of the endometrium, was encapsulated into the chitosan-coated pectin-based microspheres to provide a long-term, sustained release of E2 for endometrial repair. The pectin/Pluronic F-127 hydrogel scaffold offered a structural matrix supporting the attachment, growth, and migration of bone marrow–derived mesenchymal stem cells (BMSCs).

In bioprinting, the Freeform Reversible Embedding of Suspended Hydrogels (FRESH) approach, introduced in 2015, enables the printing of complex 3D scaffolds, particularly from soft materials. A typical FRESH platform involves extruding bioinks into a yield-stress support bath that temporarily holds the bioinks in place until they are cured. This method is characterized by three key features: (i) a viscoelastic support bath with Bingham plastic-like rheological behavior for freeform fabrication, (ii) a customizable aqueous phase in the support bath compatible with the multiplexing of gelation mechanisms, and (iii) the ability to liquefy the support bath for nondestructive release of printed constructs under biologically compatible conditions [37]. A. Lee et al. demonstrated a FRESH method for

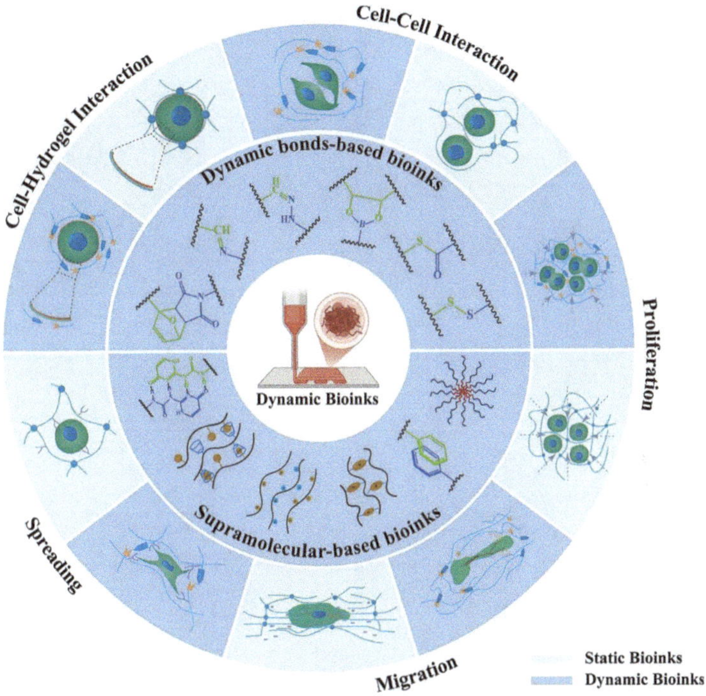

Fig. 15.8 Types of dynamic bioinks and a comparison of their effects on cell behavior versus static bioinks. (Adapted from [33] with permission from Elsevier)

3D-bioprinting collagen to engineer components of the human heart at various scales—from capillaries to the full organ. In this method, a gelatin microparticle slurry served as the support bath, which can be melted away at 37 °C after bioprinting. Acidic collagen solution was extruded into the support bath buffered to pH 7.4, where rapid neutralization induced gelation and filament formation [38]. The FRESH 3D bioprinting of collagen types II and III has also been recently explored [39]. Additionally, Maedeh Rahimnejad et al. investigated the FRESH bioprinting of chitosan and chitosan-gelatin hydrogels using a Pluronic-based support bath with shear thinning behavior and rapid self-healing properties. An acidic chitosan solution mixed with a combination of β-glycerophosphate (BPG) and sodium hydrogen carbonate (SHC) produced a physiologically compatible mixture suitable for incorporating cells and bioactive factors. Upon incubation at 37 °C, the mixture rapidly gelled into a strong, porous hydrogel without the need for chemical crosslinkers or non-physiological conditions (Fig. 15.10) [40].

As discussed in Sect. 15.1, 4D bioprinting aims to generate intricate tissue constructs capable of transforming their shape and characteristics in response to physical, chemical, or biological stimuli. These constructs are designed to support tissue/organ regeneration, repair, or replacement by mimicking native tissue/organ functions and integrating seamlessly into the human body (Fig. 15.11) [8, 41]. 4D bioprinting is especially promising for developing vascularization tissues due to its

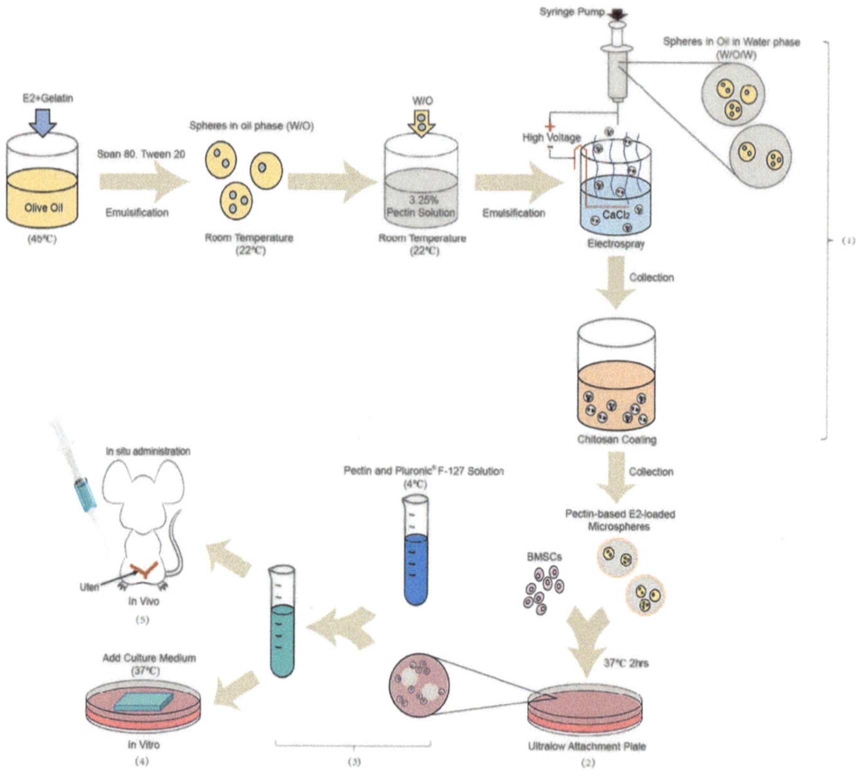

Fig. 15.9 A schematic illustration of the procedure for fabricating pectin-based E2-loaded microspheres and pectin/Pluronic F-127 scaffolds for endometrial regeneration: (1) preparing microsphere by electrospray and coating with chitosan; (2) incubating cells with E2-loaded microspheres; (3) mixing with pectin-Pluronic F-127 solution; (4) evaluating the therapeutic effects of E2 MPs composite scaffolds *in vitro*; and (5) evaluating the BMSCs/E2 MPs/scaffolds system for *in vivo* endometrial regeneration in mice. *BMSCs* bone marrow–derived mesenchymal stem cells, *E2 MPs* β-estradiol-loaded microspheres. (Adapted from [36]. Copyright © 2024 Mary Ann Liebert, Inc)

ability to produce deformable architectures that respond to external cues [41]. A recent study introduced an innovative platform that enables the bioprinting of tissues with programmable and predicted 4D shape-morphing, driven by cell-generated forces. In this approach, collagen-hyaluronic acid bioinks are deposited within a yield-stress granular support hydrogel composed of agarose microgels. The viscoelastic properties of the support hydrogel allow and regulate 4D shape-morphing behavior. The shape transformation is influenced by key factors including the initial print geometry, bioink ECM composition, cell phenotype, and viscoelasticity of the support hydrogel (Fig. 15.12) [42].

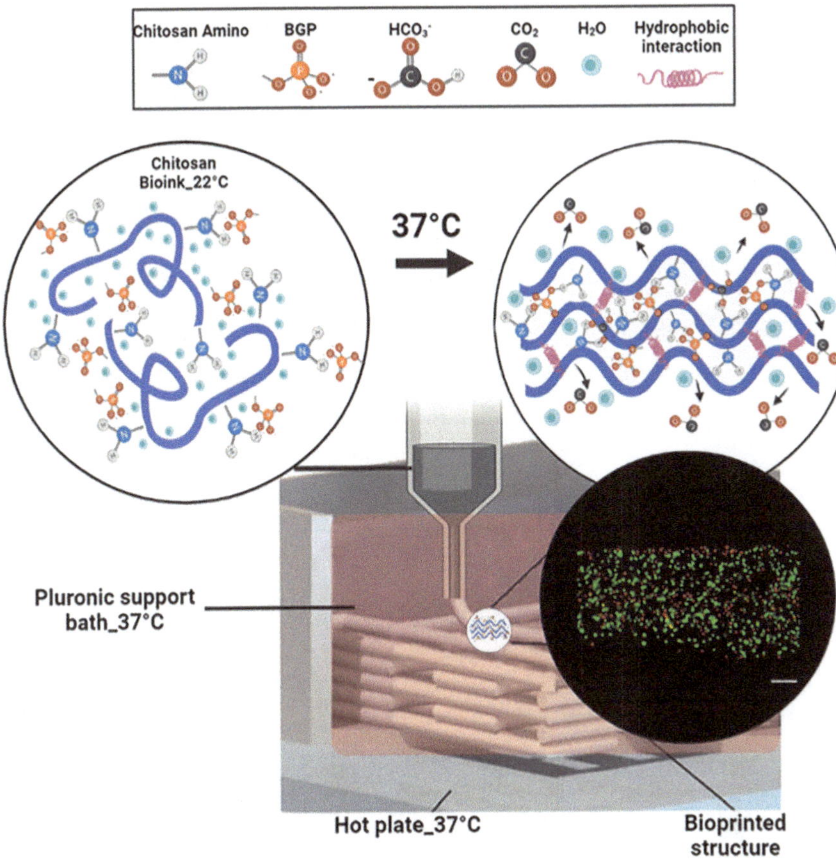

Fig. 15.10 Mechanism of FRESH bioprinting of chitosan and chitosan-gelatin hydrogels using a Pluronic-based support bath. *BGP* β-glycerophosphate. (Adapted from [40] with permission from Elsevier)

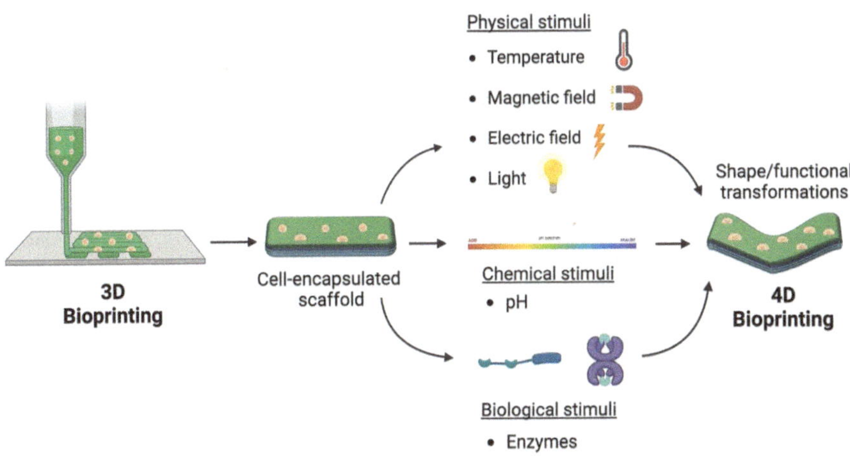

Fig. 15.11 Graphical summary of 3D and 4D bioprinting of cell-encapsulated structures. (Adapted from [8]. Image created using Biorender.com)

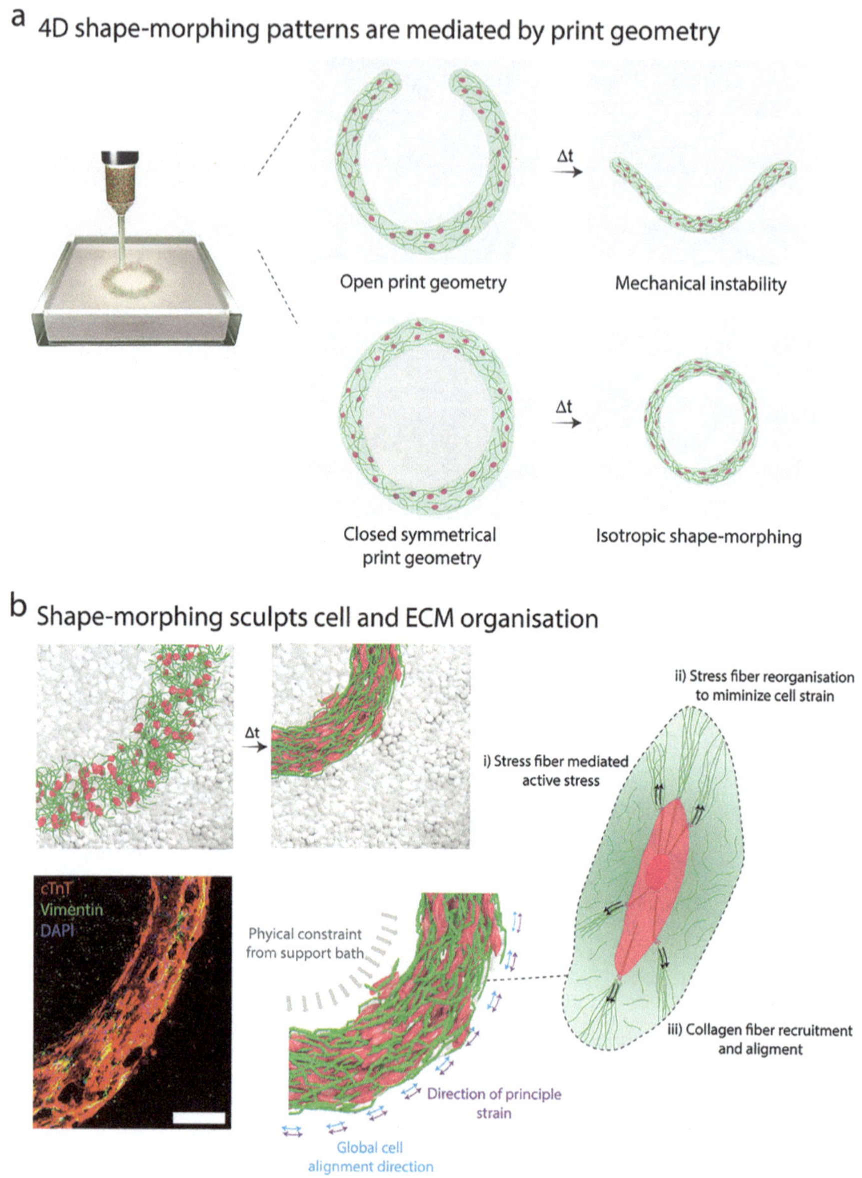

Fig. 15.12 Overview of 4D bioprinting platform for programmed tissue shape-morphing during their maturation: (**a**) Initial print geometry influences the shape-morphing under cell-generated contraction forces—open geometries show mechanical instabilities, while closed symmetrical geometries undergo isotropic shrinkage. (**b**) As cells exert active contractile forces on the collagen network and against the viscoelastic support bath, stress fiber and collagen remodeling occurs, aligning cells along the principal tissue axis. This guided remodeling can be harnessed to enhance the maturation of bioprinted induced pluripotent stem cell (iPSC)-derived heart tissues (scale bar: 100 μm). *cTNT* cardiac-specific troponin T, *DAPI* 4′,6-diamidino-2-phenylindole. (Adapted from [42])

Review and Practice Questions

1. What is bioprinting?
2. What are the common types of bioprinters, and how do their printing mechanisms differ?
3. What is a bioink? What types of biomaterials are used in bioink formulations, and what are the general characteristics for an ideal bioink?
4. What is a dynamic bioink, and how do dynamic bioinks influence cell behavior?
5. How does the FRESH bioprinting technique work?
6. What is the "fourth dimension" refer to in 4D bioprinting, and what is the primary objective of this approach?
7. Conduct a literature review and write a short essay summarizing recent advancements in bioprinting not covered in this chapter.

References

1. Hull, S.M., et al., *3D bioprinting of dynamic hydrogel bioinks enabled by small molecule modulators.* Science Advances, 2023. **9**(13):eade7880.
2. Yilmaz, B., et al., *Bioprinting: A review of processes, materials and applications.* Bioprinting, 2021. **23**:e00148.
3. Huang, M.S., et al., *Organoid bioprinting: from cells to functional tissues.* Nature Reviews Bioengineering, 2025. **3**(2):126–142.
4. Murphy, S.V. and A. Atala, *3D bioprinting of tissues and organs.* Nature Biotechnology, 2014. **32**(8):773–785.
5. Mendoza-Cerezo, L., et al., *Evolution of bioprinting and current applications.* International Journal of Bioprinting, 2023. **9**(4):742.
6. Gu, Z., et al., *Development of 3D bioprinting: From printing methods to biomedical applications.* Asian Journal of Pharmaceutical Sciences, 2020. **15**(5):529–557.
7. Gao, B., et al., *4D Bioprinting for Biomedical Applications.* Trends in Biotechnology, 2016. **34**(9):746–756.
8. Faber, L., A. Yau, and Y. Chen, *Translational biomaterials of four-dimensional bioprinting for tissue regeneration.* Biofabrication, 2024. **16**(1):012001.
9. Xie, Z., et al., *3D Bioprinting in Tissue Engineering for Medical Applications: The Classic and the Hybrid.* Polymers, 2020. **12**(8):1717.
10. Mirshafiei, M., et al., *Advancements in tissue and organ 3D bioprinting: Current techniques, applications, and future perspectives.* Materials & Design, 2024. **240**:112853.
11. Zhang, W., *Recent Progress in Bioprinting: From Bioink Design to Applications.* Bioengineering, 2022. **9**(12):785.
12. Ma, Y., et al., *Advancements of 3D bioprinting in regenerative medicine: Exploring cell sources for organ fabrication.* Heliyon, 2024. **10**(3):e24593.
13. Derakhshanfar, S., et al., *3D bioprinting for biomedical devices and tissue engineering: A review of recent trends and advances.* Bioactive Materials, 2018. **3**(2):144–156.
14. Aljohani, W., et al., *Bioprinting and its applications in tissue engineering and regenerative medicine.* International Journal of Biological Macromolecules, 2018. **107**:261–275.
15. Gao, G., et al., *Organ Bioprinting: Are We There Yet?* Advanced Healthcare Materials, 2018. **7**(1):1701018.
16. Barui, S., *3D inkjet printing of biomaterials: Principles and applications.* Medical Devices & Sensors, 2021. **4**(1):e10143.
17. Zhao, D.-k., et al., *Inkjet 3D bioprinting for tissue engineering and pharmaceutics.* Journal of Zhejiang University-SCIENCE A, 2022. **23**(12):955–973.

18. Ventura, R.D., *An Overview of Laser-assisted Bioprinting (LAB) in Tissue Engineering Applications.* Medical Lasers, 2021. **10**(2):76–81.
19. Guillemot, F., et al., *Laser-Assisted Cell Printing: Principle, Physical Parameters Versus Cell Fate and Perspectives in Tissue Engineering.* Nanomedicine, 2010. **5**(3):507–515.
20. Dou, C., et al., *A State-of-the-Art Review of Laser-Assisted Bioprinting and its Future Research Trends.* ChemBioEng Reviews, 2021. **8**(5):517–534.
21. Li, W., et al., *Stereolithography apparatus and digital light processing-based 3D bioprinting for tissue fabrication.* iScience, 2023. **26**(2):106039.
22. Kačarević Ž, P., et al., *An Introduction to 3D Bioprinting: Possibilities, Challenges and Future Aspects.* Materials, 2018. **11**(11):2199.
23. Groll, J., et al., *A definition of bioinks and their distinction from biomaterial inks.* Biofabrication, 2019. **11**(1):013001.
24. Vanaei, S., et al., *An Overview on Materials and Techniques in 3D Bioprinting Toward Biomedical Application.* Engineered Regeneration, 2021. **2**:1–18.
25. Jose, J., et al., *Recent advances in the design and development of bioink formulations for various biomedical applications.* Results in Engineering, 2024. **22**:102060.
26. Liu, H., et al., *Recent Advances in Decellularized Matrix-Derived Materials for Bioink and 3D Bioprinting.* Gels, 2023. **9**(3):195.
27. Lian, L., et al., *Rapid Volumetric Bioprinting of Decellularized Extracellular Matrix Bioinks.* Advanced Materials, 2024. **36**(34):2304846.
28. Zieman, J., et al., *Development of Gelatin-Coated Hydrogel Microspheres for Novel Bioink Design: A Crosslinker Study.* Pharmaceutics, 2023. **15**(1):90.
29. Kanungo, M., et al., *Development of Gelatin-Coated Microspheres for Novel Bioink Design.* Polymers, 2021. **13**(19):3339.
30. Gungor-Ozkerim, P.S., et al., *Bioinks for 3D bioprinting: an overview.* Biomaterials Science, 2018. **6**(5):915–946.
31. Mannoor, M.S., et al., *3D Printed Bionic Ears.* Nano Letters, 2013. **13**(6):2634–2639.
32. Sharma, C., et al., *Cellulose-based 3D printing bio-inks for biomedical applications: A review.* International Journal of Biological Macromolecules, 2025. **305**:141174.
33. Ma, D., et al., *Dynamic bioinks for tissue/organ bioprinting: Principle, challenge, and perspective.* Progress in Materials Science, 2026. **155**:101527.
34. Schwab, A., et al., *Printability and Shape Fidelity of Bioinks in 3D Bioprinting.* Chemical Reviews, 2020. **120**(19):11028–11055.
35. Johnson, D.L., et al., *Design of pectin-based bioink containing bioactive agent-loaded microspheres for bioprinting.* Biomedical Physics & Engineering Express, 2019. **5**(6):067004.
36. Wu, Y., et al., *E2-Loaded Microcapsules and Bone Marrow–Derived Mesenchymal Stem Cells with Injectable Scaffolds for Endometrial Regeneration Application.* Tissue Engineering Part A, 2023. **30**(3-4):115–130.
37. Shiwarski, D.J., et al., *Emergence of FRESH 3D printing as a platform for advanced tissue biofabrication.* APL Bioengineering, 2021. **5**(1):010904.
38. Lee, A., et al., *3D bioprinting of collagen to rebuild components of the human heart.* Science, 2019. **365**(6452):482–487.
39. Moss, S.P., D.J. Shiwarski, and A.W. Feinberg, *FRESH 3D Bioprinting of Collagen Types I, II, and III.* ACS Biomaterials Science & Engineering, 2025. **11**(1):556–563.
40. Rahimnejad, M., et al., *FRESH bioprinting of biodegradable chitosan thermosensitive hydrogels.* Bioprinting, 2022. **27**:e00209.
41. Arif, Z.U., et al., *A review on four-dimensional (4D) bioprinting in pursuit of advanced tissue engineering applications.* Bioprinting, 2022. **27**:e00203.
42. Pramanick, A., et al., *4D Bioprinting Shape-Morphing Tissues in Granular Support Hydrogels: Sculpting Structure and Guiding Maturation.* Advanced Functional Materials, 2025. **35**(5):2414559.

Scaffold Design and Fabrication: Decellularization

16

Learning Objectives

- Define decellularization and evaluate its significance in tissue engineering
- Compare and assess various decellularization methods
- Explore and assess the applications of decellularized extracellular matrix (dECM) materials in tissue engineering

16.1 Introduction to Decellularization

Decellularization is the process of removing cellular and nuclear components from tissues or organs using various agents/techniques, while preserving the native ultrastructure of the extracellular matrix (ECM) components such as collagen, elastin, fibronectin, laminin, proteoglycans, glycosaminoglycans, and various growth factors [1]. The material obtained through this process is termed as decellularized extracellular matrix (dECM). Nearly every human tissue type has been successfully decellularized (Fig. 16.1). Compared to other scaffold types, dECM offers several unique advantages, making it a gold-standard for scaffolds [2–4]:

1. dECM-derived scaffolds closely resemble the native environment in both biochemical composition and anatomical architectures;
2. Acellular matrices support cellular growth and functional differentiation without triggering an immune response, partly due to the highly conserved proteins and ligands in the ECM across species;
3. Decellularization provides a relatively straightforward method to obtain scaffolds that replicate the complex 3D structure of an organ while preserving the extracellular matrix.

W. Zhang, *Cell Culture and Tissue Engineering*,
https://doi.org/10.1007/978-3-032-32096-4_16

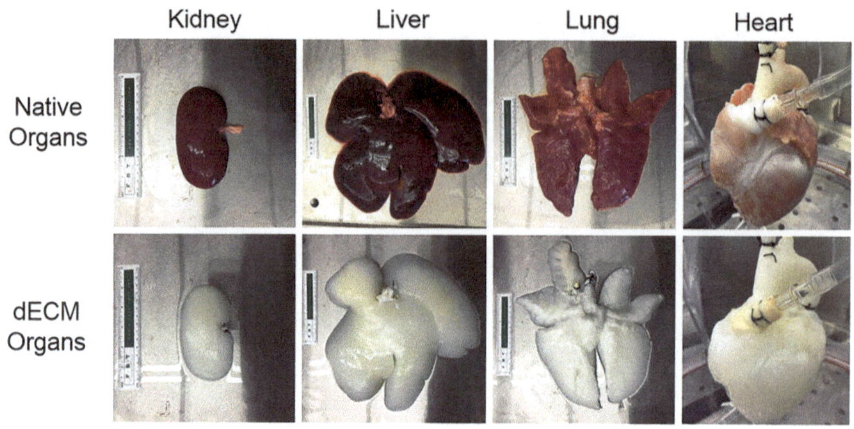

Fig. 16.1 Examples of decellularized organs. (Adapted from [4] with permission from Elsevier)

16.2 Methods of Decellularization

Decellularization methods are generally classified into three major categories: physical, chemical, and biological (Fig. 16.2 and Table 16.1) [5–12]. Physical methods include freeze-thaw cycles, pressure-based techniques, and uses of superficial fluids. Freeze-thaw cycles cause cell death by forming ice crystals that rupture cell membranes and introduce osmotic stress, which help detach cells from the ECM. Pressure can mechanically disrupt cell membranes. When combined with chemical or biological decellularization agents, pressure gradients enhance the delivery of agents and removal of cellular debris. Supercritical fluids, such as CO_2, offer low viscosity and high transport diffusivity, enabling rapid and efficient decellularization. While physical approaches facilitate mechanical disruption of cells, they are typically insufficient alone and are often combined with chemical or biological treatments. Chemical methods use acids/bases and detergents. Acids and based dissolve cytoplasmic components and remove nucleic acids. Non-ionic detergents like Triton X-100 can penetrate the lipid bilayer of the cell membrane, targeting the lipid-lipid and lipid-protein interactions, disrupting the membrane structure and releasing cell contents. In contrast, ionic detergents such as sodium dodecyl sulfate (SDS) effectively lyse cells but may also denature proteins and compromise ECM components and microstructure. While chemical methods are highly effective at removing cellular components, they carry a risk of ECM damage and loss of bioactivity. Biological methods involve enzymes, chelators and toxins. Enzymes such as phospholipase A2, proteases (e.g., trypsin), and nucleases target specific biological substrates to break down cellular components. However, enzymatic activity may be inhibited by natural protease inhibitors released from lysed cells, and prolonged exposure of some enzymes, such as trypsin, can degrade the ECM. Chelating agents like ethylenediaminetetraacetic acid (EDTA) disrupt cell adhesion by binding divalent metal cations (e.g., Ca^{2+}) at

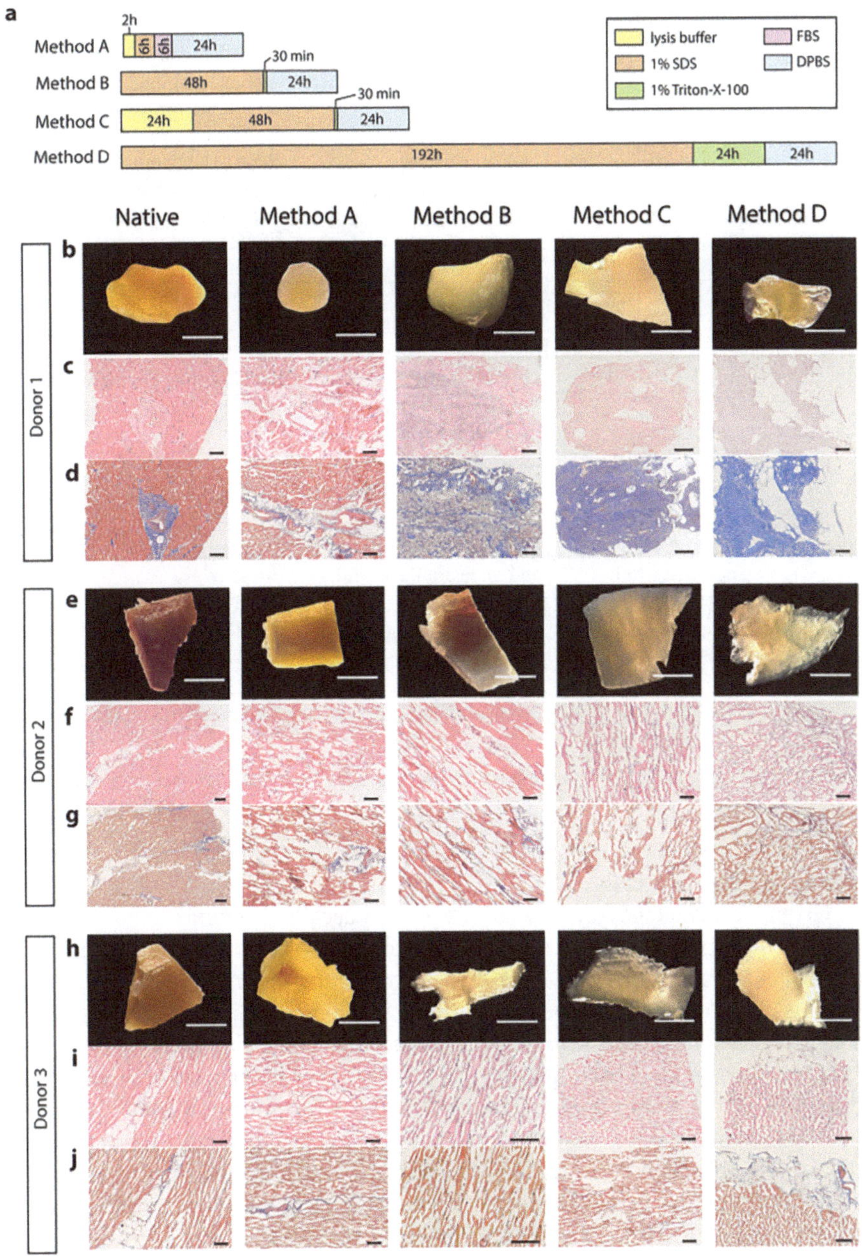

Fig. 16.2 Histological assessment of the different chemical-based decellularization methods applied to human cardiac tissue. (**a**) Chemical-based decellularization methods tested, including the solutions and exposure duration used in each step. For each donor: (**b**), (**e**), (**h**) Gross tissue morphology. Scale bars: 1 mm; (**c**), (**f**), (**i**) Hematoxylin & Eosin staining. Scale bars: 100 μm; (**d**), (**g**), (**j**) Masson's Trichome staining. Scale bars: 100 μm. *SDS* sodium dodecyl sulfate, *FBS* fetal bovine serum, *DPBS* Dulbecco's phosphate-buffered saline. (Adapted from [13])

Table 16.1 Comparison of various decellularization approaches

	Agent/technique	Mechanism	Advantages and disadvantages
Physical	Freeze-thaw	Ice formation leads to cell damage	Advantages: Easy to perform; no residue Disadvantages: Need to be combined with other approaches for efficient decellularization; may damage ECM; may need specialized equipment
	Pressure	Damage cell and aid in cell detachment	
	Supercritical fluid (e.g., CO_2)	Pressure can damage cells; improve transport properties	
Chemical	Acids and bases	Solubilize cellular components; remove nucleic acids; denature proteins	Advantages: Effective Disadvantages: Residue; may damage ECM
	Non-ionic detergents (e.g., Triton X-100)	Disrupt lipid-lipid and lipid-protein interactions	
	Ionic detergents (e.g., SDS)	Disrupt lipid membranes; denature proteins	
Biological	Enzymes (e.g., nucleases and trypsin)	Specifically remove cells and undesirable acellular materials	Advantages: Effective Disadvantages: Residue; may damage ECM; may be costly
	Chelators (e.g., EDTA)	Bind metallic ions (e.g., Ca^{2+}) and release cells from ECM	
	Toxins (e.g., latrunculin B)	Produce specific cytotoxic effects	

ECM extracellular matrix, *SDS* sodium dodecyl sulfate, *EDTA* ethylenediaminetetraacetic acid

cell-ECM interfaces. Toxins, such as latrunculin B, may be used to selectively damage and remove cells.

Given the advantages and limitations of each decellularization method, combining multiple approaches is often necessary to maximize cell and nuclear material removal while minimizing damage to the extracellular matrix [6]. There is no one-size fits-all strategy; the choice of decellularization method should consider factors such as tissue/organ type, size, thickness, intended application and other variables [5, 13]. Interindividual heterogeneity can also influence the success of human tissue decellularization. For example, a study evaluating four different methods for decellularizing human cardiac tissue found that both SDS and Triton X-100 were essential for effective DNA removal. However, these agents also altered the biochemical composition of the ECM. Notably, the preservation of ECM architecture varied depending on the tissue donor (Fig. 16.2) [13].

To further enhance the decellularization outcomes—especially for whole organs or large tissues such as human limbs—perfusion decellularization can be employed. Unlike immersion and agitation methods, perfusion decellularization involves circulating detergents through the native vasculature. This technique leverages the existing vascular network, ensuring more uniform and deeper tissue exposure, leading to more efficient removal of cellular components from large 3D-tissue compartments [14, 15]. Figure 16.3 illustrates a setup for pressure-controlled perfusion decellularization of whole organs.

Following decellularization, a visual inspection is often the first step—decellularized tissues/organs should appear translucent white, indicating successful removal of cellular components. Beyond visual confirmation, it's essential to ensure the removal of residual detergents, preservation of tissue architecture (such as vasculature), and, critically, the absence of cellular and nuclear materials that could trigger adverse remodeling responses. Recommended minimal criteria include [12, 15]:

(a) <50 ng dsDNA per mg ECM dry weight;
(b) <200 bp DNA fragment length;
(c) lack of visible nuclear material in tissue sections after staining.

A sterilization process is typically required, particularly for clinical applications and long-term storage. Sterilization methods include UV irradiation, radiation, and chemical sterilants such as ethylene oxide or peracetic acid [7, 9]. Despite these measures, dECM products still face immunologic challenges (Fig. 16.4). Factors such as the decellularization technique (e.g., apoptosis-assisted techniques), ECM alteration, cytotoxicity residues, graft source, and implantation site can all influence immunogenicity. Common strategies to mitigate these risks include: selective antigen removal; crosslinking to mask antigenic sites; solubilization to eliminate soluble protein antigens; and sterilization to remove pathogen-associated immunogenicity [16]. A recent study proposed comprehensive criteria for the safe clinical application of dECM derived from both human and animal sources, covering aspects from decellularization efficacy to aseptic processing of decellularization to ensure sterility (Fig. 16.5) [17].

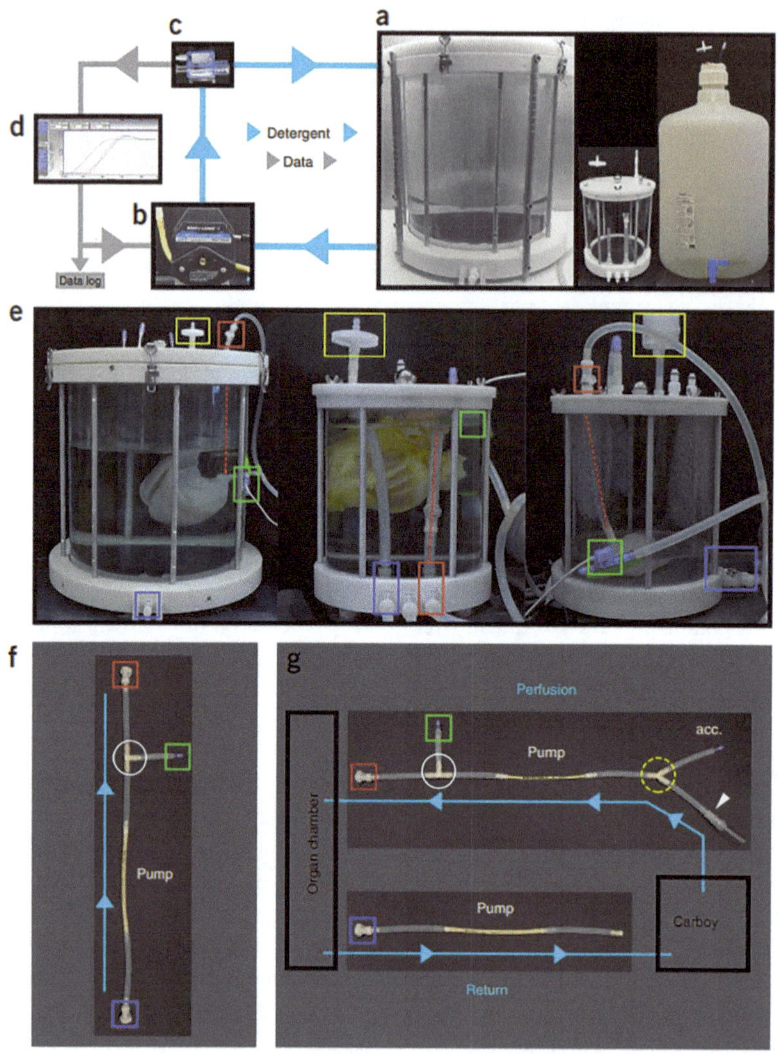

Fig. 16.3 Overview of large organ perfusion decellularization setup. (**a**–**c**) Detergent (blue path) flows from the reservoir (**a**; left to right: large-organ chamber; medium-organ chamber, or 50-L carboy) by means of a pump (**b**), past a pressure sensor (**c**) and into the perfusion line of the organ chamber (**a**). (**d**) The pressure read by the sensor is sent to the pump controller. (**e**) Decellularization chamber setups for large lungs (left), heart (middle) and kidney (right). Perfusate inlets are outlined in red, drains in blue, pressure sensors in green and air filters in yellow. (**f**) Dashed red line indicates the tubing path from the chamber perfusate input to the organ cannula. Tubing for large-organ chambers transports perfusate from the chamber drain connector (blue outline) through a section of L/S 17 pump tubing, past a T-connector (white circle) with a pressure sensor connector (green outline) to the chamber input connector (red outline). (**g**) The return tubing set for medium-organ chambers (bottom) transports perfusate from the chamber drain connection (blue outline) through a section of L/S 17 pump tubing to the carboy reservoir. The perfusion set for medium-organ chambers (top) transports the perfusate from the carboy, past a Y-connector (dashed yellow circle) with auxiliary connector, through a section of L/S 16 pump tubing, past a T-connector (white circle) with a pressure sensor connector (green outline) to the chamber input connector (red outline). (Adapted from [15] with permission from Springer Nature)

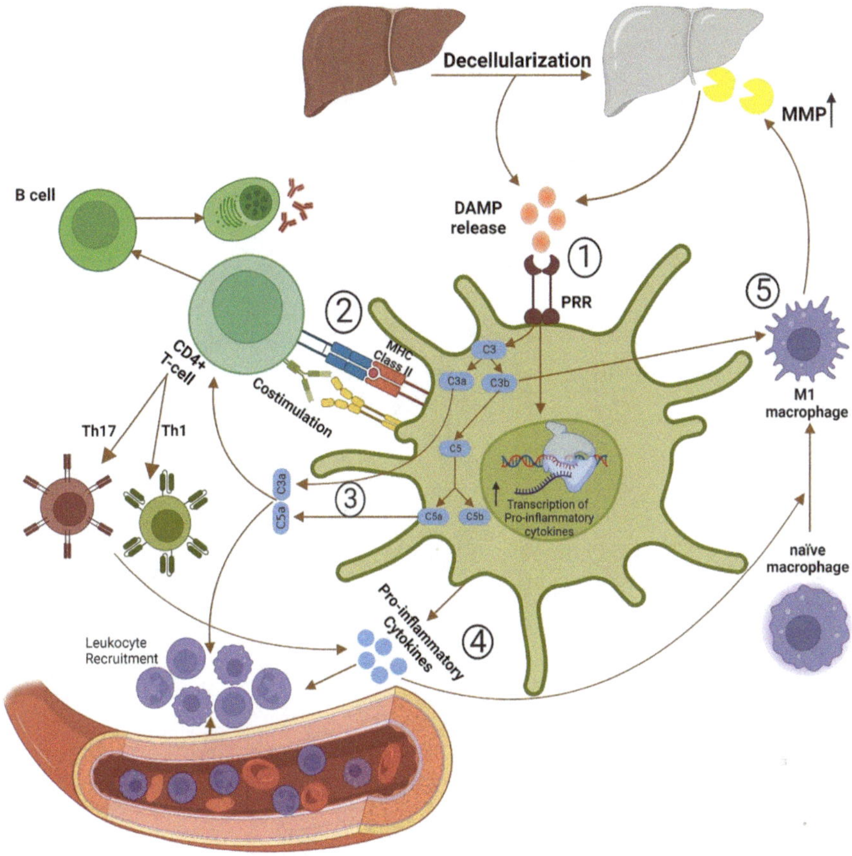

Fig. 16.4 Innate and adaptive immune response against dECM scaffolds. (1) Decellularization strategies inevitably produce various amounts of damage-associated molecular patterns (DAMPs), which trigger pattern recognition receptors (PRRs) on innate immune cells. (2) PRR stimulation in antigen-presenting cells provides prerequisites of T cell activation, including pro-inflammatory cytokine and major histocompatibility complex and co-stimulatory molecules, as well as triggering a complement cascade that produces C3a, C3b, and C5a. T cell activation is the essential trigger for B cells, which mediate antibody production against remaining antigens within dECM scaffolds. Antigen-antibody interactions further trigger complement cascades and implement a vicious cycle. (3) C3a and C5a recruit immune cells and induce T helper type 1 (TH1) and TH17 polarization in helper $CD4^{+}$ T cells, increasing the production of pro-inflammatory cytokines. (4) Elevated cytokines at the implantation site attract additional immune cells and induces M1 polarization in naïve macrophages. (5) C3b complements enhance matrix metalloproteinases (MMP) production in M1 macrophages, accelerating the decellularized extracellular matrix (dECM) scaffold degradation and failure. *dECM* decellularized extracellular matrix. (Adapted from [16])

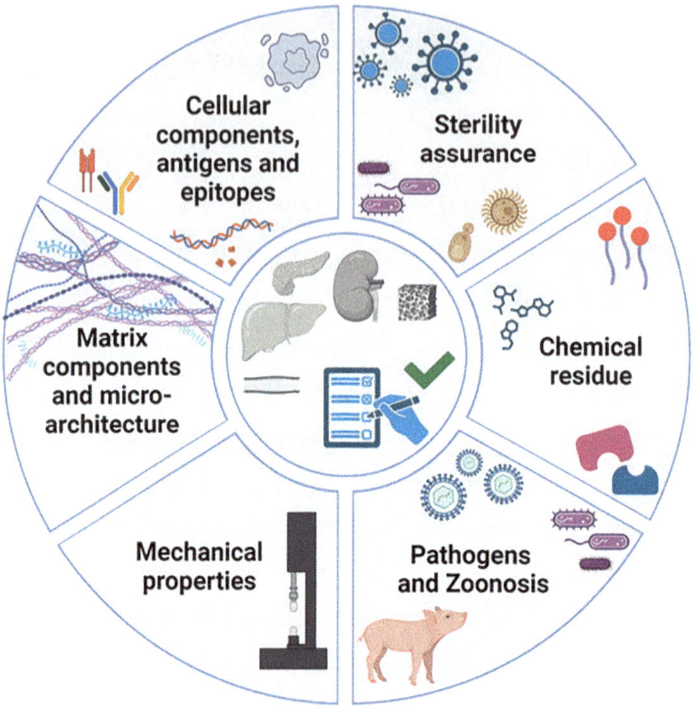

Fig. 16.5 Quality and safety criteria for translation for clinical translation of decellularized scaffolds. Key factors to assess include: cellular residue and DNA fragments; matrix components; effects of decellularization on mechanical properties; presence of microorganisms; residual detergents and other chemicals; and, for animal-derived tissues or organs, the presence of pathogens and animal-specific antigens or epitopes. Zoonosis, infectious diseases transmitted from a non-human animal to humans must also be carefully evaluated. (Adapted from [17])

16.3 Utilization of dECM Materials

The major application of dECM is in the fabrication of organoids and engineered organs [7]. Beyond therapeutic use, dECM is also applied in coating biomaterials, organoid cultures, disease modeling, and as a promising drug delivery platform for the controlled release of bioactive molecules [5, 18–21]. The dECM can be utilized in various forms: directly used for recellularization or as injectable gel or as solutions for bioprinting or electrospinning [4, 22–24].

Recellularization refers to repopulating a cellular ECM scaffold with specific cell types [25]. Major challenges associated with recellularization include identifying optimal cell sources, controlling the cell distribution, and promoting vascularization [25, 26]. Michael S. Xu et al. demonstrated a perfusion-recellularization strategy using a co-culture of human umbilical vein endothelial and mesenchymal stromal cells, which enhanced the cell engraftment and neovascularization in decellularized flaps compared to monocultures (Fig. 16.6) [27]. Additionally, recent research showed that modifying dECM surfaces with chemokines and cytokines

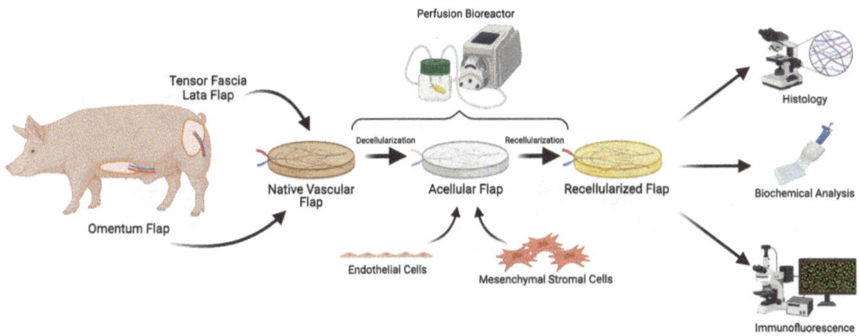

Fig. 16.6 Schematic of tissue perfusion-decellularization and recellularization of porcine omentum and tensor fascia lata flaps. During recellularization, a co-culture of human umbilical vein endothelial and mesenchymal stem cells was manually perfused in the arterial inlet. (Adapted from [27])

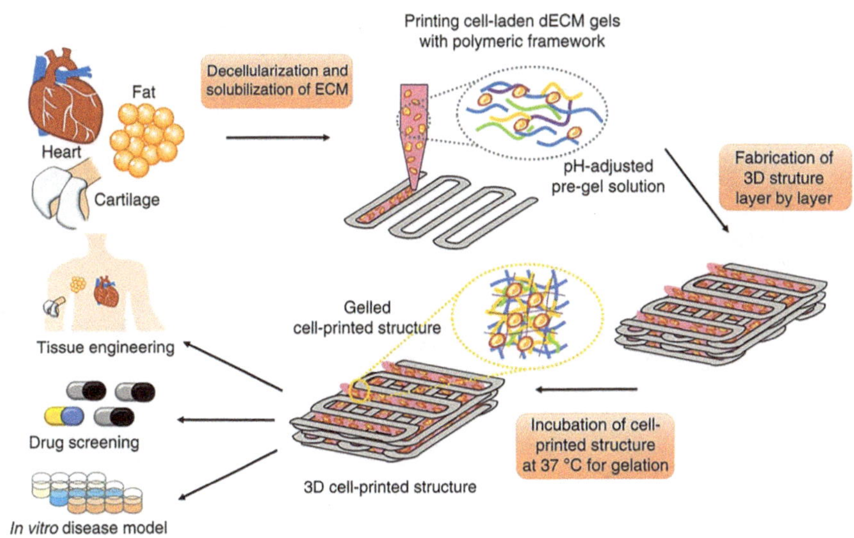

Fig. 16.7 Schematic illustrating the tissue printing process using dECM bioink. *dECM* decellularized extracellular matrix. (Adapted from [29] with permission from Springer Nature)

could improve recellularization efficiency and promote targeted cell differentiation [28].

Bioinks derived from dECM offer significant potential for creating a supportive microenvironment conducive to 3D tissue growth. Typically, dECM is solubilized in acidic conditions and then adjusted to physiological pH to form a printable bioink (Fig. 16.7) [29]. Nima Tabatabaei Rezaei et al. developed a novel hepatic tissue bioink by combining liver dECM methacrylate with gelatin methacrylate. The resulting solution demonstrated rapid crosslinking under visible light and tunable mechanical, rheological, and physiological properties (Fig. 16.8) [30].

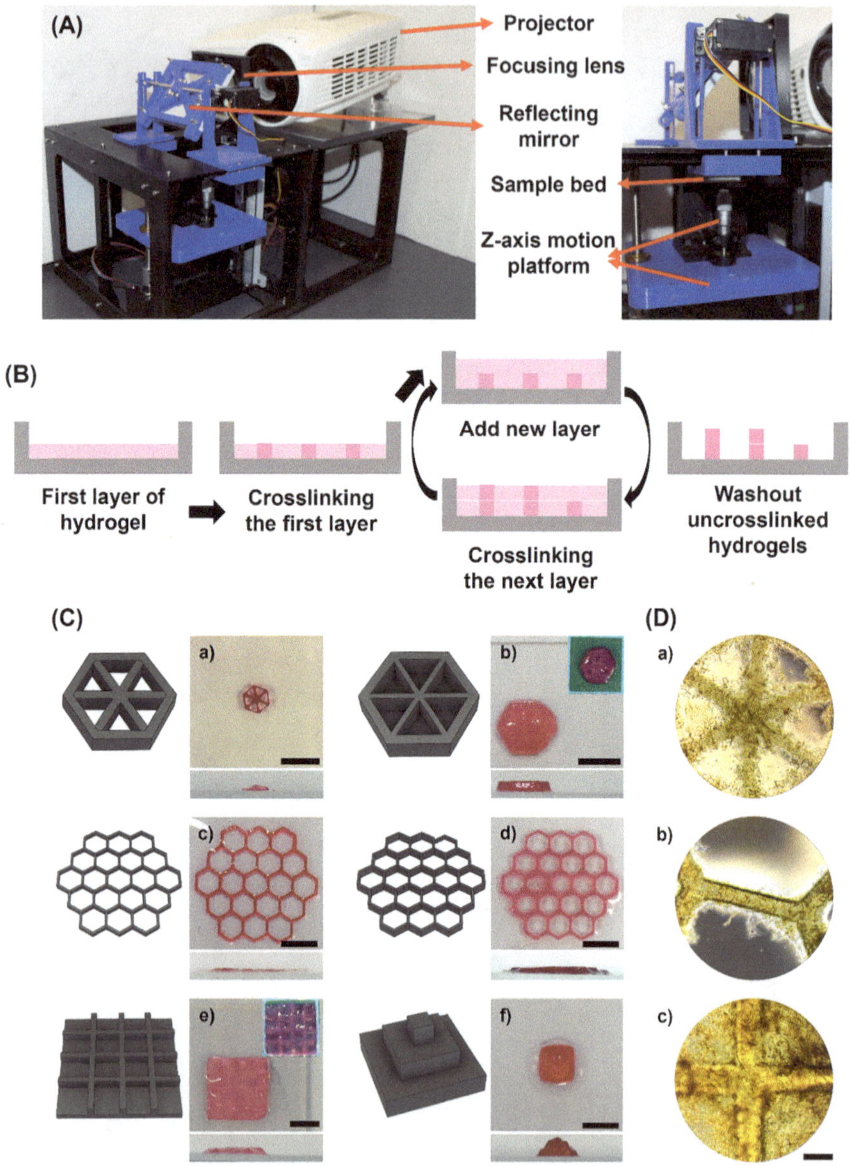

Fig. 16.8 3D bioprinting of 5G0.5LdMA hydrogel. (**a**) DLP-based bioprinting setup used for layer-by-layer additive fabrication of 3D structures, highlighting the system's key components. (**b**) Schematic representation of the DLP bioprinting process. (**c**) 3D printing of various shapes with different thicknesses, showing the design, stained printed hydrogels, and side views indicating structure thickness: (**a**) 0.7 mm, (**b**) 1.5 mm, (**c**) 0.5 mm, (**d**) 1 mm, (**e**) 1.2 mm, and (**f**) 3 mm (scale bar = 5 mm). (**d**) Brightfield images of the printed structures: **a**, **c**, and **e**. Scale bar = 500 μm. *G* gelatin methacrylate, *LdMA* liver decellularized matrix methacrylate, *DLP* digital light processing. (Adapted from [30])

Review and Practice Questions

1. What is decellularization, and what are the advantages of using dECM for scaffolding?
2. What are the three main decellularization approaches? What are their respective advantages and disadvantages? Provide two examples and mechanisms for each approach.
3. What is perfusion decellularization?
4. Describe the mechanism of innate and adaptive immune response against dECM scaffolds.
5. How can immunogenicity be reduced post-decellularization?
6. What are the primary applications of dECM?
7. What is recellularization, and challenges are associated with it?
8. Conduct a literature review and summarize a recent study on the application of injectable dECM-based hydrogels.

References

1. Choudhury, D., et al., *Decellularization systems and devices: State-of-the-art.* Acta Biomaterialia, 2020. **115**:51–59.
2. Pellegata, A.F., A.M. Tedeschi, and P. De Coppi, *Whole Organ Tissue Vascularization: Engineering the Tree to Develop the Fruits.* Frontiers in Bioengineering and Biotechnology, 2018. **6**:56.
3. McInnes, A.D., M.A.J. Moser, and X. Chen, *Preparation and Use of Decellularized Extracellular Matrix for Tissue Engineering.* Journal of Functional Biomaterials, 2022. **13**(4):240.
4. Taylor, D.A., et al., *Decellularized matrices in regenerative medicine.* Acta Biomaterialia, 2018. **74**:4–89.
5. Neishabouri, A., et al., *Decellularization in Tissue Engineering and Regenerative Medicine: Evaluation, Modification, and Application Methods.* Frontiers in Bioengineering and Biotechnology, 2022. **10**:805299.
6. Zhang, X., et al., *Decellularized extracellular matrix scaffolds: Recent trends and emerging strategies in tissue engineering.* Bioactive Materials, 2022. **10**:15–31.
7. Guo, X., et al., *Decellularized extracellular matrix for organoid and engineered organ culture.* Journal of Tissue Engineering, 2024. **15**:20417314241300386.
8. Mendibil, U., et al., *Tissue-Specific Decellularization Methods: Rationale and Strategies to Achieve Regenerative Compounds.* International Journal of Molecular Sciences, 2020. **21**(15):5447.
9. Zhao, J., et al., *Decellularization techniques pave the way for tissue engineering and regenerative medicine: a narrative review.* Regenerative Medicine Reports, 2024. **1**(2):117–130.
10. Biazar, E., *Decellularized Scaffolds for Tissue Regeneration: Techniques and Applications*, in *Advances in Regenerative Medicine and Tissue Engineering*, M.K. Paul, B. Bisht, and B.N. Singh, Editors. 2024, IntechOpen.
11. Nicholls, D.L., et al., *Perfusion decellularization for vascularized composite allotransplantation.* SAGE Open Med, 2022. **10**:20503121221123893.
12. Crapo, P.M., T.W. Gilbert, and S.F. Badylak, *An overview of tissue and whole organ decellularization processes.* Biomaterials, 2011. **32**(12):3233–3243.
13. Tenreiro, M.F., et al., *Interindividual heterogeneity affects the outcome of human cardiac tissue decellularization.* Scientific Reports, 2021. **11**(1):20834.

14. Gerli, M.F.M., et al., *Perfusion decellularization of a human limb: A novel platform for composite tissue engineering and reconstructive surgery.* PLOS One, 2018. **13**(1):e0191497.
15. Guyette, J.P., et al., *Perfusion decellularization of whole organs.* Nature Protocols, 2014. **9**(6):1451–1468.
16. Kasravi, M., et al., *Immunogenicity of decellularized extracellular matrix scaffolds: a bottleneck in tissue engineering and regenerative medicine.* Biomaterials Research, 2023. **27**(1):10.
17. van Hengel, E.V.A., et al., *Towards Safety and Regulation Criteria for Clinical Applications of Decellularized Organ-Derived Matrices.* Bioengineering, 2025. **12**(2):136.
18. Li, J., et al., *Pulmonary decellularized extracellular matrix (dECM) modified polyethylene terephthalate three-dimensional cell carriers regulate the proliferation and paracrine activity of mesenchymal stem cells.* Frontiers in Bioengineering and Biotechnology, 2023. **11**:1324424.
19. Giobbe, G.G., et al., *Extracellular matrix hydrogel derived from decellularized tissues enables endodermal organoid culture.* Nature Communications, 2019. **10**(1):5658.
20. Li, C., et al., *Enhancing organoid culture: harnessing the potential of decellularized extracellular matrix hydrogels for mimicking microenvironments.* Journal of Biomedical Science, 2024. **31**(1):96.
21. Kolliopoulos, V. and A.G. Mikos, *Decellularized extracellular matrix as a drug delivery carrier.* Journal of Controlled Release, 2025. **382**:113661.
22. Zhang, S., et al., *The considerations on selecting the appropriate decellularized ECM for specific regeneration demands.* Materials Today Bio, 2024. **29**:101301.
23. Gadre, M., et al., *Decellularization and Their Significance for Tissue Regeneration in the Era of 3D Bioprinting.* ACS Omega, 2024. **9**(7):7375–7392.
24. Liu, W., et al., *Decellularized extracellular matrix materials for treatment of ischemic cardiomyopathy.* Bioactive Materials, 2024. **33**:460–482.
25. Ahmed, E., T. Saleh, and M. Xu, *Recellularization of Native Tissue Derived Acellular Scaffolds with Mesenchymal Stem Cells.* Cells, 2021. **10**(7):1787.
26. Stoian, A., et al., *Two Decades of Advances and Limitations in Organ Recellularization.* Current Issues in Molecular Biology, 2024. **46**(8):9179–9214.
27. Xu, M.S., et al., *Bioengineering of vascularized porcine flaps using perfusion-recellularization.* Scientific Reports, 2024. **14**(1):7590.
28. Choi, M., et al., *Surface modification of decellularized kidney scaffold with chemokine and AKI-CKD cytokine juice to increase the recellularization efficiency of bio-engineered kidney.* Biomaterials, 2025. **316**:123007.
29. Pati, F., et al., *Printing three-dimensional tissue analogues with decellularized extracellular matrix bioink.* Nature Communications, 2014. **5**(1):3935.
30. Tabatabaei Rezaei, N., et al., *Bioprinting of hepatic tissue model using photocrosslinkable dECM-containing composite hydrogel.* Materials Today Bio, 2025. **32**:101824.

Tissue Fabrication Technology: Vascularization and Innervation

17

Learning Objectives

- Define vascularization and analyze its importance in tissue development and regeneration
- Evaluate and design strategies that promote vascularization in tissue engineering applications
- Define innervation and assess its significance in tissue function
- Evaluate and design approaches that promote innervation in tissue engineering applications

17.1 The Human Cardiovascular System and Strategies for Promoting Vascularization in Tissue Engineering

The cardiovascular system delivers nutrients and oxygen to all cells in the body. As shown in Fig. 17.1, it consists of the heart and blood vessels. The pumping action of the heart creates the pressure needed to propel blood through the *pulmonary circulation* (to the lungs) and the *systemic circulation* (to the rest of the body). Arteries carry blood away from the heart, while veins return the blood to the heart. Blood passes from the arterial to the venous system through microscopic capillaries [1, 2]. Nearly all tissues in the body are supplied with nutrients and oxygen by the vascular system [3]. Functional blood vessels are also essential for effective removal of metabolic waste.

Vascularization is the process of growing blood vessels into tissue to improve oxygen and nutrient delivery. It remains a limiting factor in engineering large 3D tissue constructs, whether during *in vitro* preculture or after *in vivo* implantation [4]. Ideally, engineered vasculature should be integrated into the host vasculature *via* anastomosis, allowing for immediate nutrient supply to all cells following surgery [5]. Two fundamental processes involved in the formation of new blood vessels are

W. Zhang, *Cell Culture and Tissue Engineering*,
https://doi.org/10.1007/978-3-032-32096-4_17

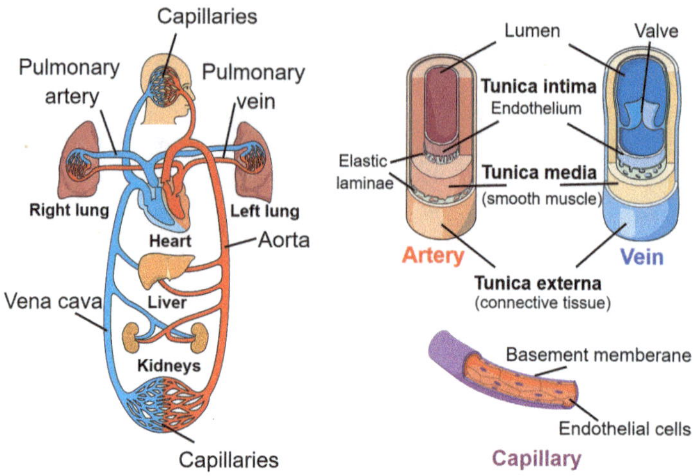

Fig. 17.1 Human cardiovascular system: pulmonary circulation and systemic circulation (left); and structures of blood vessels (right)

vasculogenesis and angiogenesis. Vasculogenesis refers to the differentiation of endothelial progenitor cells (angioblasts) into endothelial cells and the *de novo* formation of a primitive vascular network during early embryogenesis. These endothelial progenitor cells are derived from mesodermal cells. In parallel, extraembryonic vasculogenesis occurs in the yolk sac, where vasculature is formed from endothelial cells that originate from haemangioblasts (Fig. 17.2a). Angiogenesis is the formation of new capillaries—blood vessels that lack a fully developed tunica media—from preexisting blood vessels, either through sprouting or intussusception. Sprouting is executed by tip cells, which respond to vascular endothelial growth factor (VEGF) *via* VEGF receptors (VEGFRs) (Fig. 17.2b) [6–8]. Other key regulators of angiogenesis include hypoxia-inducible factor (HIF), platelet-derived growth factor (PDGF), angiopoietin (Ang), transforming growth factor (TGF), and epithelia growth factor (EGF) [9].

As shown in Fig. 17.3, there are two primary strategies for vascularization: cell-based strategy, which relies on the ability of endothelial cells to form new vessels (i.e., neoangiogenesis), and scaffold-based strategy, which focuses on the creation of vascular structures using either biologically-derived or synthetic materials [3]. Alternatively, vascularization strategies can be categorized based on post-implantation behavior: angiogenesis, which involves the ingrowth of vascular sprouts from the host microvasculature, and inosculation, which involves the integration of a pre-formed vascular network with the host [10]. There are five key approaches to stimulate angiogenic processes translating the basic principles of neoangiogenesis into practices. As shown in Fig. 17.4, these approaches include *in vitro* and *in vivo* prevascularization (e.g., using macoporous scaffold), addition of angiogenic growth factors, immobilization of bioactive molecules (such as peptides) on biomaterials to enhance neoangiogenesis, and co-cutlure of multiple types of cells [3, 11].

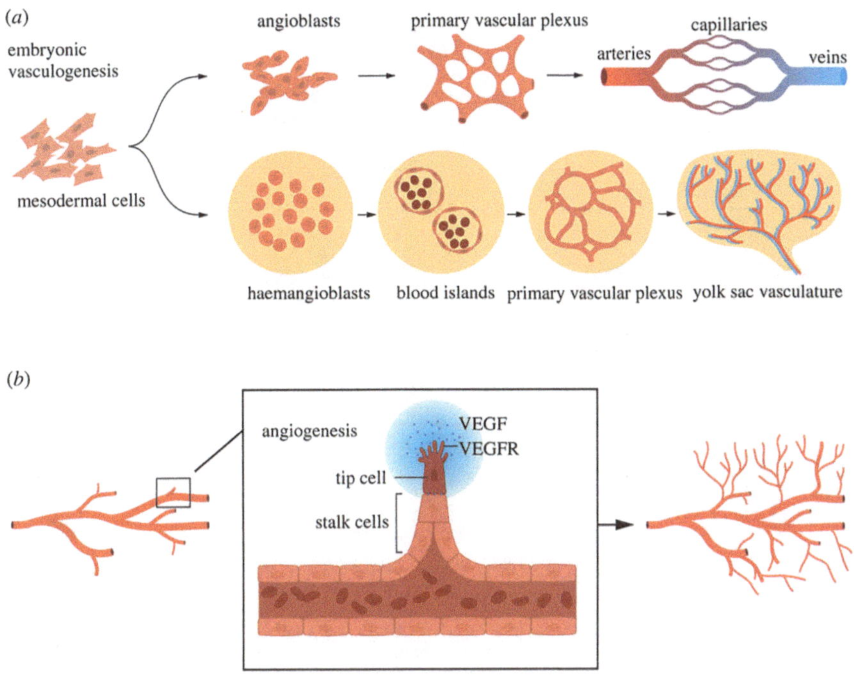

Fig. 17.2 Vasculogenesis (**a**) and angiogenesis (**b**). *VEGF* vascular endothelial growth factor, *VEGFR* VEGF receptor. (Adapted from [7] with permission from The Royal Society)

Fig. 17.3 Two main vascularization strategies: cell-based strategy (**a**) and scaffold-based strategy (**b**). *ECs* endothelial cells

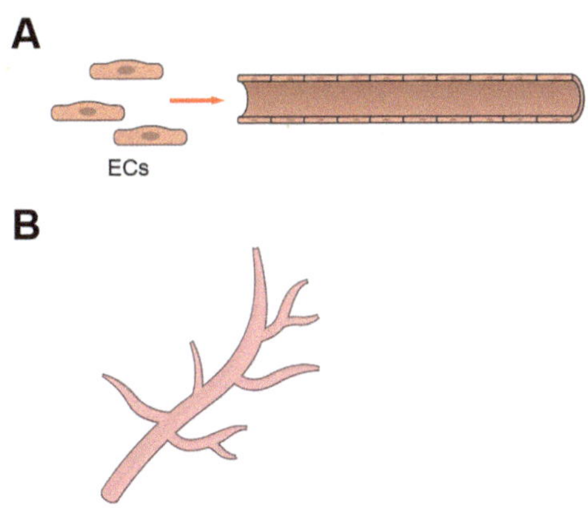

VEGF is the most commonly used growth factor to promote vascularization. However, angiogenesis is a dynamic, multistep process involving the activation, migration, proliferation, and tube formation of endothelial cells, followed by the stabilization of the newly formed vessels through perivascular cells (Fig. 17.5). Therefore, it is often beneficial to combine various growth factors to achieve synergistic effects, especially for generating larger and more complex vascular structures.

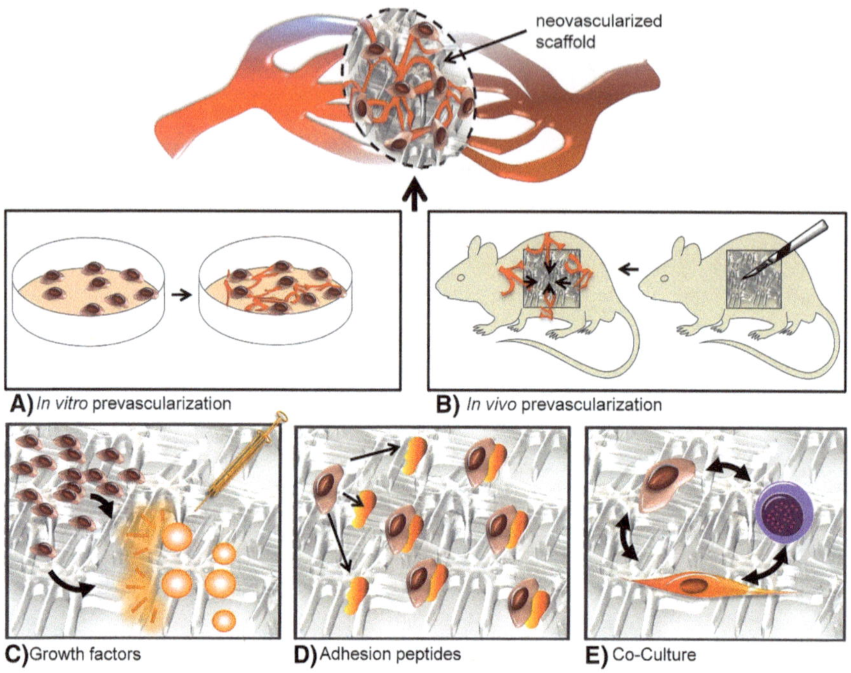

Fig. 17.4 Five different approaches for inducing vascularization by stimulation of angiogenic processes. (Adapted from [3] with permission from Elsevier)

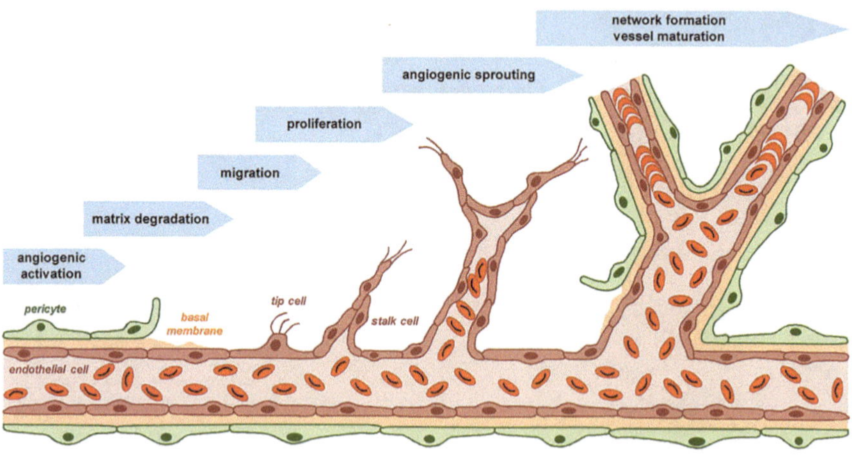

Fig. 17.5 The process of angiogenesis: (i) angiogenic activation of microvessels by growth factors, (ii) detachment of stabilizing pericytes and degradation of the basal membrane by matrix metalloproteinases (MMPs), (iii) migration of endothelial tip cells towards an angiogenic stimulus and (iv) proliferation of following endothelial stalk cells, (v) formation of angiogenic sprouts, (vi) sprouts develop a lumen and interconnect to form blood-perfused microvascular networks, which are stabilized by basement membrane formation and recruitment of perivascular cells. (Adapted from [13])

For instance, factors promoting angiogenesis (e.g., VEGF, PDGF, Ang-2) can be combined with those promoting vessel maturation (e.g., PDGF, TGF-β, and Ang-1) [9, 12, 13]. Among the cells used for vascularization, endothelial cells, such as human umbilical vein endothelial cells (HUVECs), are the most common. Other promising sources include endothelial progenitor cells derived from postnatal peripheral blood or bone marrow [9]. Studies have shown that co-culturing endothelial cells with supporting cells can enhance vascularization. Supporting cells that have been investigated include fibroblasts, pericytes, and mesenchymal stem cells (MSCs). A recent study explored the effects of the temporal and sequential delivery of multiple growth factors and support cells on vascularization within 3D polyethylene glycol (PEG) hydrogels (Fig. 17.6). The hydrogel was formed through enzymatic crosslinking using PEG-K (K: K-peptide; sensitive to degradation by matrix metalloproteinases) and PEK-Q (Q: Q-peptide; a transglutaminase substrate). The results identified an optimal growth factor cocktail consisting of Ang1, PDGF-BB, and Ang-2 in endothelial growth medium (EGM-2) for *in vitro* studies. Additionally, both fibroblasts and MSCs, co-cultured with HUVECs supported the vessel-like structure formation, separately, with MSCs lead to more branching points. When TGF-β was added at the end to mimic fibrosis, smaller vessels and more collagen secretion were observed [12].

Successful immobilization of peptides/proteins on biomaterials/scaffolds have also been studied to enhance neoangiogenesis. Short peptide adhesion sequences, such as RGD (derived from fibronectin), are well known for supporting cell adhesion. In addition, short peptides can facilitate the attachment and migration of endothelial cells. Controlled release of growth factors can be achieved by immobilizing glycosaminoglycan heparin or heparin mimic molecules, which contain growth factor binding sites [3]. Beyond peptide/protein immobilization, biomaterials/scaffolds can be modified to promote vascularization through adjustment in chemical composition, degradation properties, architecture, porosity, and surface topography. For instance, scaffolds with large pore sizes and high interconnectivity support vascular ingrowth. Scaffolds with embedded microchannels may guide vascular ingrowth [9, 11, 14, 15]. Interestingly, the incorporation of microspheres/capsules into scaffolds has also been shown to promote vasculature formation [16, 17].

Current *in vitro* prevascularization approaches, as illustrated in Fig. 17.7, include cell seeding, spheroids generation, and cell sheet technology. The cell seeding approach is the incorporation of vessel-forming cells into scaffolds. Mural cells have been utilized to support the development and maintenance of performed microvascular networks. Spheroids can be generated by co-culturing vessel-forming cells with tissue-specific cells using techniques such as the hanging drop method. These spheroids can subsequently develop networks of vessel-like structures. In cell sheet technology, monolayer cell sheets are first generated and then stacked to create thicker vascularized constructs [9, 18]. The *in vivo* prevascularization strategy typically involves temporarily implanting non-vascularized constructs at highly vascularized host sites, and then explanting, transplanting them to the targeted site [19].

To fabricate scaffolds/tissues with pre-defined vascular networks, various techniques have been developed. These include decellularization (discussed further in

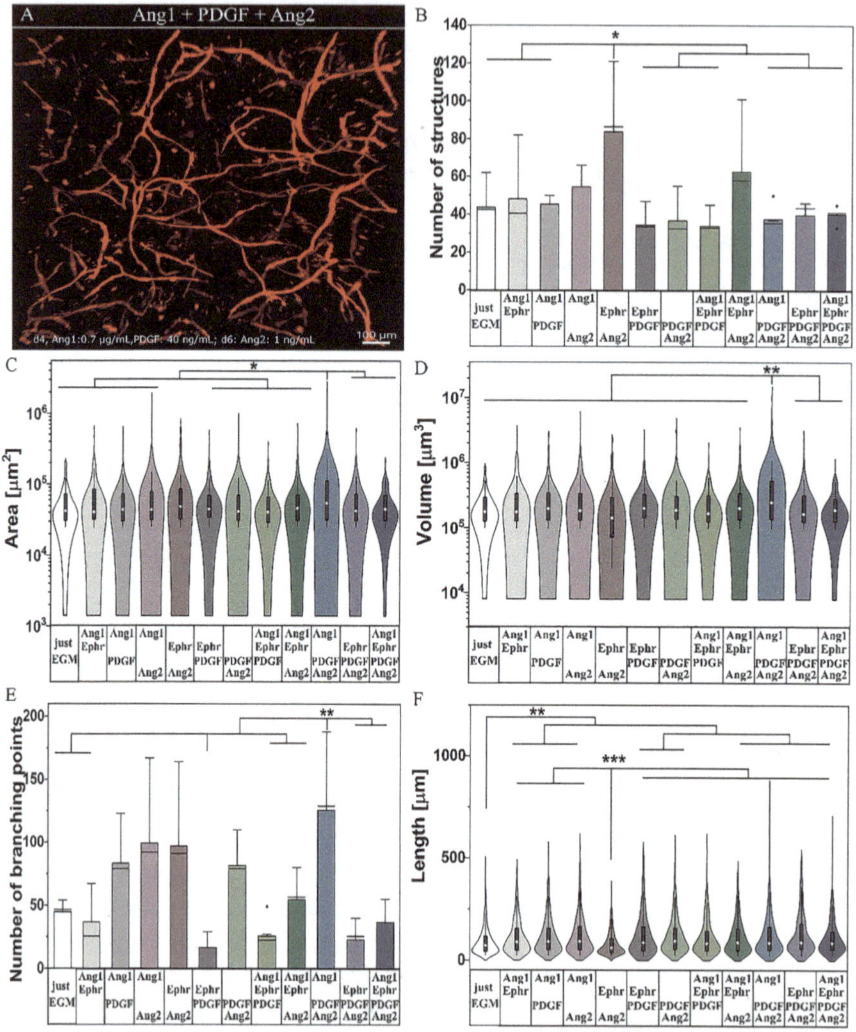

Fig. 17.6 Effects of growth factor combinations. (**a**) Immunofluorescence staining of PECAM/CD-31 (red) after 7 days of culture of normal human dermal fibroblasts (NHDF) and human umbilical vein endothelial cells (HUVEC) mixtures (3:1 ratio) in endothelial growth medium (EGM-2) with the optimal growth factor combination, scale bar: 100 μm; Quantification of vascular structures: number of structures (**b**), area (**c**), volume (**d**), number of branching points (**e**), and length (**f**). n: 6 replicates per condition ($p^* < 0.05$; $p^{**} < 0.01$; $p^{***} < 0.001$). Platelet endothelial cell adhesion molecule (PECAM-1), also known as cluster of differentiation 31 (CD31), is commonly expressed on endothelial cells. *Ang1* angiopoietin 1, *PDGF* platelet-derived growth factor, *Ang2* angiopoietin 2, *Ephr* Ephrin-B2. (Adapted from [12] with permission from John Wiley and Sons)

Chap. 16), 3D printing, electrospinning, molding, spatial laser degradation, layer-by-layer fabrication, and microfluidics approaches [3, 9, 20–22]. An interesting example is the use of a microreplication method to transfer the microvascular structure of leaf venation into synthetic matrices, including biomaterial hydrogel

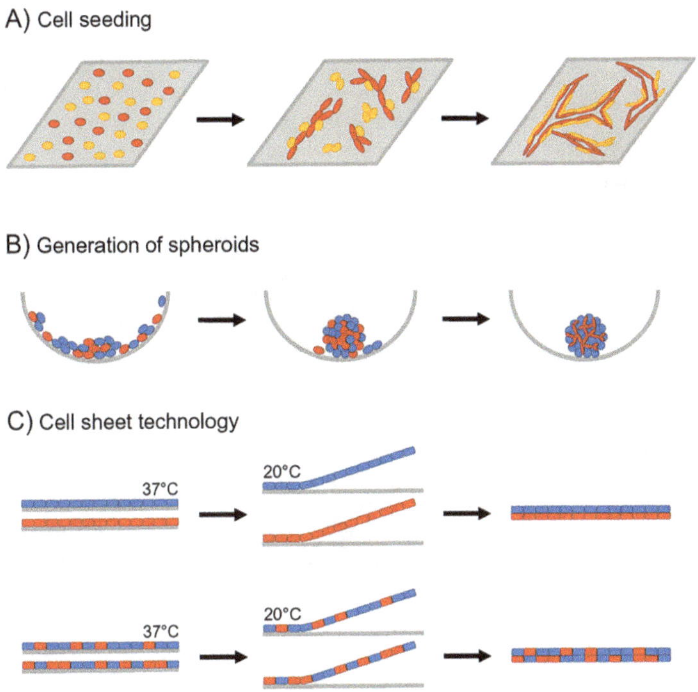

Fig. 17.7 Current *in vitro* prevascularization approaches. *Red* endothelial cells, *orange* mural cells, *blue* tissue-specific cells. (Adapted from [18] with permission from Elsevier)

(Fig. 17.8). This method is based on the observation that fluid transport in leaf venation follows the same Murry's law as blood vasculature in animals. It was demonstrated that the endothelial cells were able to grow and spread within this nature-inspired microfluidic network [23].

Another strategy for vascularization is the use of microvascular fragments (MVFs), such as those derived from adipose tissue. MVFs are small, functional vascular units that exhibit 3D capillary structures composed of endothelial and stromal cells. These fragments exhibit high angiogenic and regenerative potential [22, 24]. MVFs can be isolated from tissues *via* mechanical mincing and short-term enzymatic digestion. Upon implantation, they rapidly reassemble into mature blood-perfused microvascular networks by connecting with one another and integrating with the host vasculature [22]. The primary molecular mechanisms by which MVF's promote tissue vascularization post-implantation include implant paracrine action, indirect activation of host cells, activation of local immune cells, and hypoxia-inducing factor signaling pathway [25]. MVFs have been applied in various tissue engineering applications. For example, transplantation of adipose tissue-derived MVFs has been shown to enhance therapy for critical limb ischemia (Fig. 17.9). Other applications include dental pulp regeneration [26], treatment of full-thickness skin defects [27], lymphangiogenesis [28], and more.

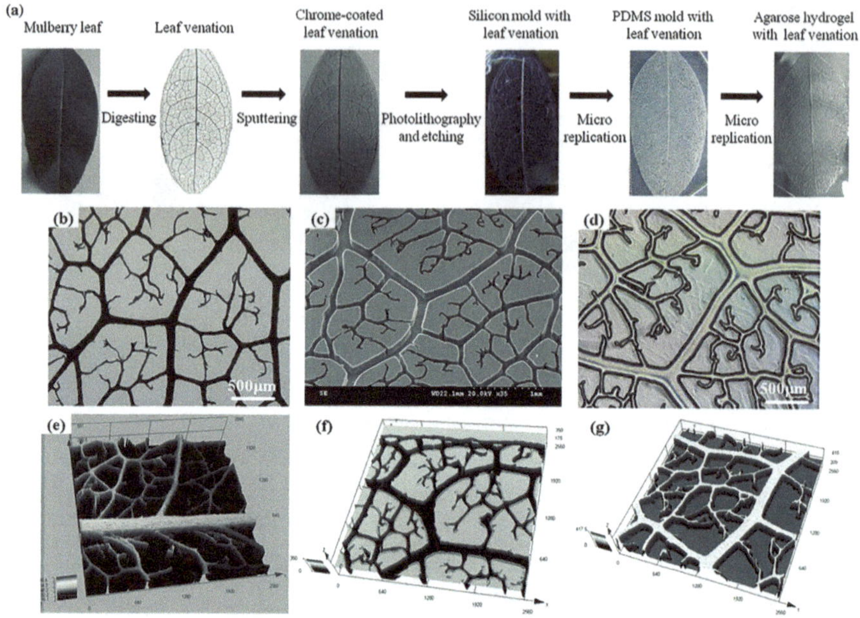

Fig. 17.8 Fabrication of nature-inspired microfluidic network. (**a**) Schematic illustration of the microreplication process from a natural leaf to synthetic materials, (**b**) native leaf venation network, (**c**) dried-etched microfluidic network in a silicon mold, (**d**) negative pattern of the microfluidic network in a (PDMS) mold, (**e–f**) 3D profiles of the microfluidic networks in leaf venation, silicon mold and PDMS mold. (Adapted from [23] with permission from John Wiley and Sons)

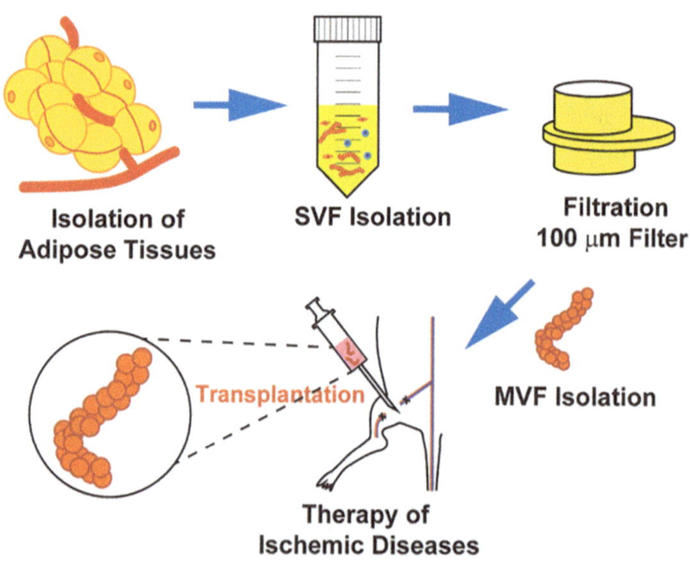

Fig. 17.9 Adipose tissue-derived MVF isolation and transplantation for therapy of ischemic diseases. *SVF* stromal vascular fractions, *MVF* microvascular fragments. (Adapted from [24])

Currently, significant attention is focused on the development of vascularized, particularly pre-vascularized tissue constructs. However, creating vascular networks that not only integrate seamlessly with the host circulation but also effectively support engrafted cells within engineered tissue remains a critical challenge [29]. Looking ahead, immune compatibility of cell-based implants requires intensive investigation. In terms of cell sources for supporting vascular development, stem cells present considerable promise due to their superior proliferative and synthetic capabilities [30]. It is also important to note that functional blood vessels are composed not only of endothelial cells but smooth muscle cells, which are vital for mechanical strength and barrier function, as well as fibroblasts, which help to recreate the extracellular matrix for maturing vascular tissue. Therefore, constructs containing only an endothelial layer are not sufficient to fully replicate the structure and function of native vasculature [20].

17.2 The Human Nervous System and Strategies for Promoting Innervation in Tissue Engineering

The nervous system is composed of neurons, which generate and conduct electrochemical impulses, and supporting cells. It is divided into central nervous system (CNS)—comprising the brain and spinal cord—and peripheral nervous system (PNS), which includes the cranial nerves arising from the brain and the spinal nerves arising from the spinal cord (Fig. 17.10). Neurons consist of three major regions: the cell body, dendrites, and an axon. The supporting cells (neuroglia) differ between CNS and PNS. In the CNS, they include oligodendrocytes, microglia, astrocytes, and ependymal cells. The PNS contains Schwann cells and satellite cells (Fig. 17.10) [31].

The process by which nerves grow and form connections with target tissues or organs is known as innervation [32]. Innervation plays a pivotal role in the functional control

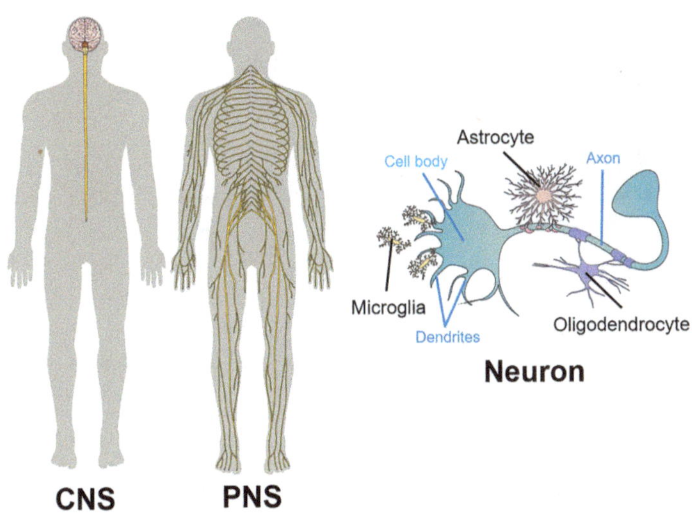

Fig. 17.10 Central nervous system (CNS), peripheral nervous system (PNS), and the structure of a neuron with supporting cells

and modulation of tissues and organs. Unfortunately, it has often been overlooked in most non-neural tissue engineering applications, largely due to the complexity of replicating the heterogeneous native cell types and structures present in innervated organs [33]. Several strategies have been explored to incorporate/promote innervation in tissue engineered constructs: the incorporation of neural (stem) cells, use of scaffolds designed with channels/conduits to guide nerve innervation, sustained delivery of bioactive molecules, such as neurotrophins and neurotrophic factors, and pre-innervation [32–35].

Ji Hyun Kim et al. investigated the integration of human neural stem cells into bioprinted skeletal muscle constructs, demonstrating the neural cell integration facilitated rapid innervation and accelerated muscle function restoration [34]. A biomimetic nerve guidance conduit with sustained insulin-like growth factor-1 (IGF-1) delivery capacity—enabled by microcapsules - was recently developed and shown to enhance peripheral nerve regeneration. IFG-1 was selected due to its multifaceted roles in supporting neuronal health and functions. As shown in Fig. 17.11, the biomimetic conduit with interconnected 3D porous network was made of chitosan and gelatin microcapsules, offering excellent mechanical

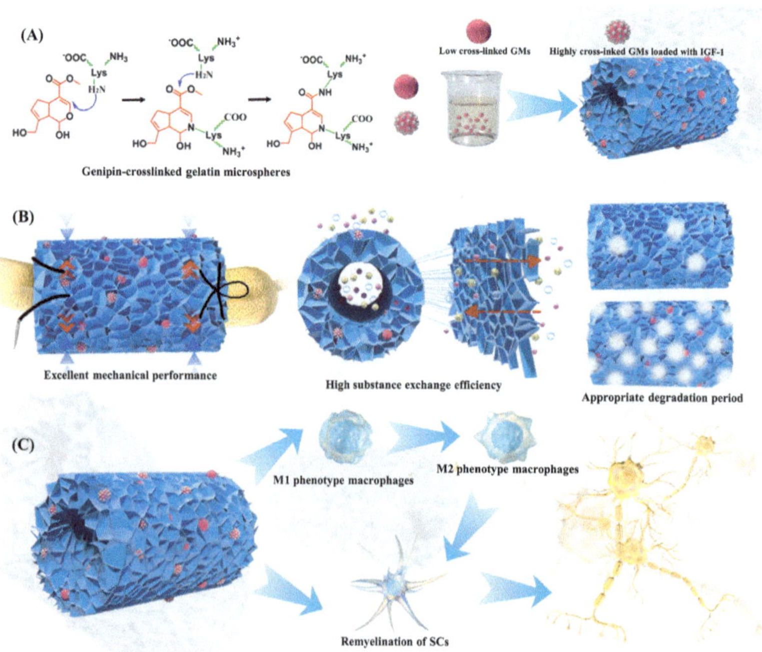

Fig. 17.11 Schematic illustrates the preparation, properties, and regenerative potential of the biomimetic extracellular matrix nerve guide conduit (NGC). (**a**) Multifunctional gelatin microcapsules (GMs) with varying degrees of crosslinking enable the sustained delivery of IGF-1 and the formation of an interconnected porous network. (**b**) The NGC demonstrates superior tensile strength, elongation at break, radial compression strength, and material exchange efficiency, along with a degradation rate aligned with the peripheral nerve regeneration process. (**c**) The NGC promotes Schwann cell (SC) remyelination and nerve axon regrowth, modulates macrophage immune phenotypes, and provides a favorable immunological microenvironment for peripheral nerve regeneration, facilitating repair. *IGF-1* insulin-like growth factor-1 gelatin microcapsules. (Adapted from [36] with permission from Elsevier)

performance, high transport property, and an appropriate degradation rate. The system promoted early survival, proliferation, and remyelination potential of Schwann cells, while supporting neurite outgrowth [36]. Poly (lactic-*co*-glycolic acid) (PLGA) microspheres have also been employed for neurotrophic factors delivery and were shown support neurite growth over a longer period compared to the control group [37]. In terms of pre-innervation, one approach involves transplantable miniature constructs consisting of living neurons and axonal tracts internalized within hydrogel tubes. Specifically, the system utilized hollow agarose tubes with a relatively stiff outer casing to guide the unidirectional outgrowth of axons through a soft collagen matrix. The construct supported the seeded dorsal root ganglia neuron survival and axonal extension [38]. A recent study introduced a wrinkled coating (made from silk fibroin to introduce topographical cues) to improve neuron elongation and alignment within conduits besides better cell viability and adhesion (Fig. 17.12) [39]. Stretch-grown neural constructs (i.e., TENGs) have been employed to achieve directed innervation of biofabricated tissues and organs. The strategy leverages the natural axonal growth mechanism of stretch growth, which enables axons to extend at rapid rates without the aid of chemical cues, physical guides or growth cones (Fig. 17.13) [33, 40–42].

When it comes to non-traditional approaches, conductive materials have been utilized to fabricate nerve conduits due to their ability to promote nerve regeneration through their electrical properties [43]. A recent study incorporated exosomes

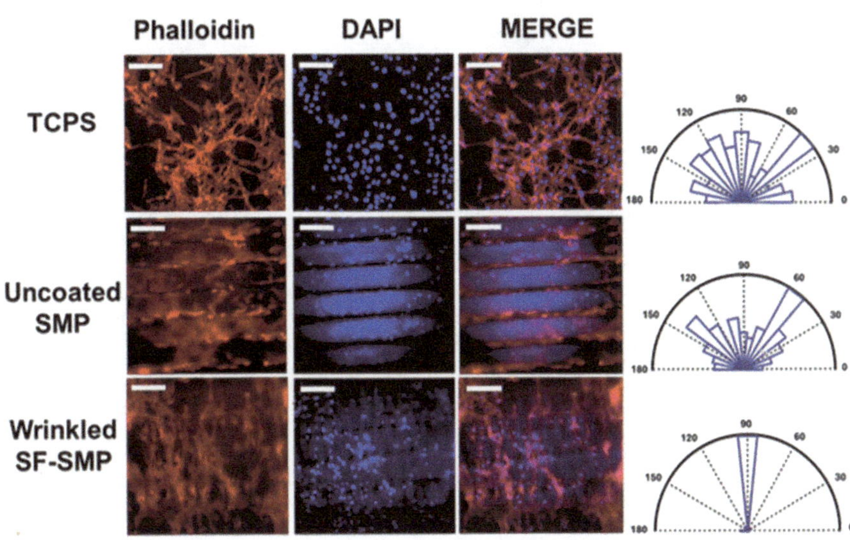

Fig. 17.12 Nuclear alignment and neurite extension of 50B11 cells (dorsal root ganglions cell line) on conduits after 72 h of culture. Angular histograms generated from fluorescence micrographs revealed significantly greater nuclear alignment on wrinkled conduits compared to uncoated conduits and tissue culture polystyrene (TCPS) (scale bar = 100 μm). Actin cytoskeleton and nuclei were stained using phalloidin and DAPI, respectively. *SF* silk fibroin, *SMP* shape-memory polymer. (Adapted from [39] with permission from John Wiley and Sons)

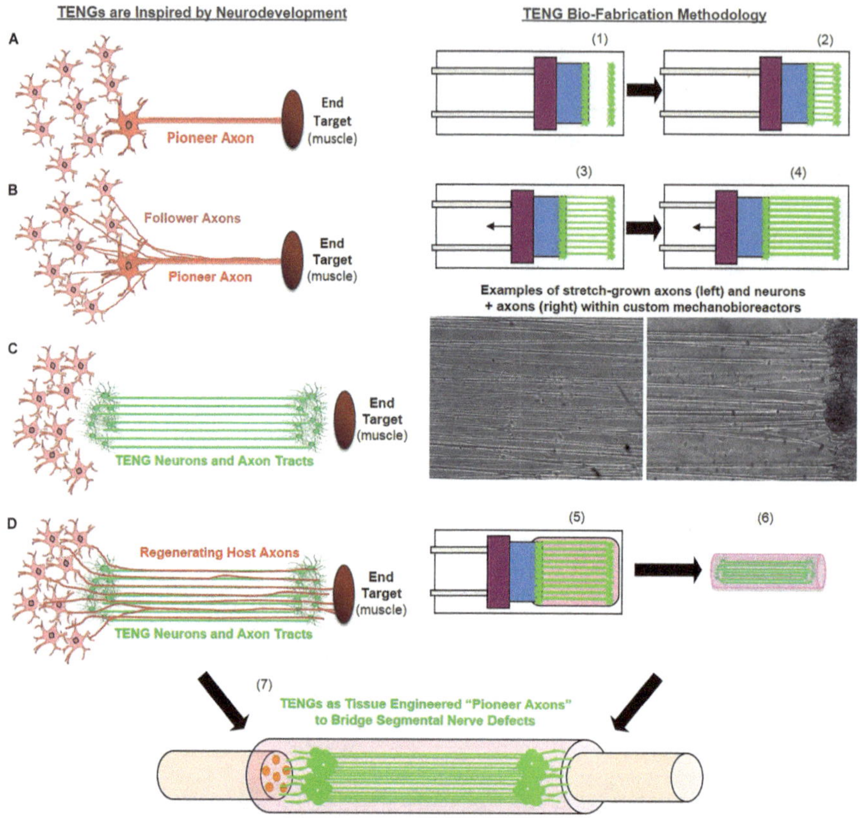

Fig. 17.13 Tissue Engineered Nerve Graft (TENG): inspiration, biofabrication, and surgical implementation. LEFT: TENGs are inspired by axonal pathfinding during nervous system development, where (**a**) "pioneer axons" first reach their target and (**b**) subsequently guide "follower axons". In TENGs, axons are (**c**) tissue engineered analogs of "pioneer axons," functioning as a (**d**) living scaffold to direct and target regenerating host axons across segmental nerve defects. RIGHT: TENGs are biofabricated in custom mechanobioreactors *via* the process of axon "stretch-growth." The process includes: (1) plating Primary DRG neurons in mechanobioreactors. (2) allowing natural axon outgrowth to integrate two neuron populations. (3) using a computer-controlled micro-stepper motor to gradually separate the two neuron populations, applying mechanical tension. (4) Inducing axon "stretch-growth," leading to increased length, diameter, and fasciculation over days to weeks (at 1–10 mm/day). (5) Immediately prior to implant, neurons and stretch-grown axons are embedded in extracellular matrix (ECM) for stabilization. (6) The ECM containing neurons and stretch-grown axons is "rolled" and transferred into a nerve guide tube (NGT). (7) The NGT is then sutured to sciatic nerve to bridge the excised segment. (Adapted from [40])

into a 3D printed hierarchical porous scaffold, which demonstrated neuro-promotive properties that enhanced bone regeneration. As shown in Fig. 17.14, these exosomes were derived from nerve growth factor (NGF)-preconditioned mesenchymal stem cells [44]. Looking forward, integrating innervation into various processes during tissue/organ engineering will be increasing important. It is anticipated the stem

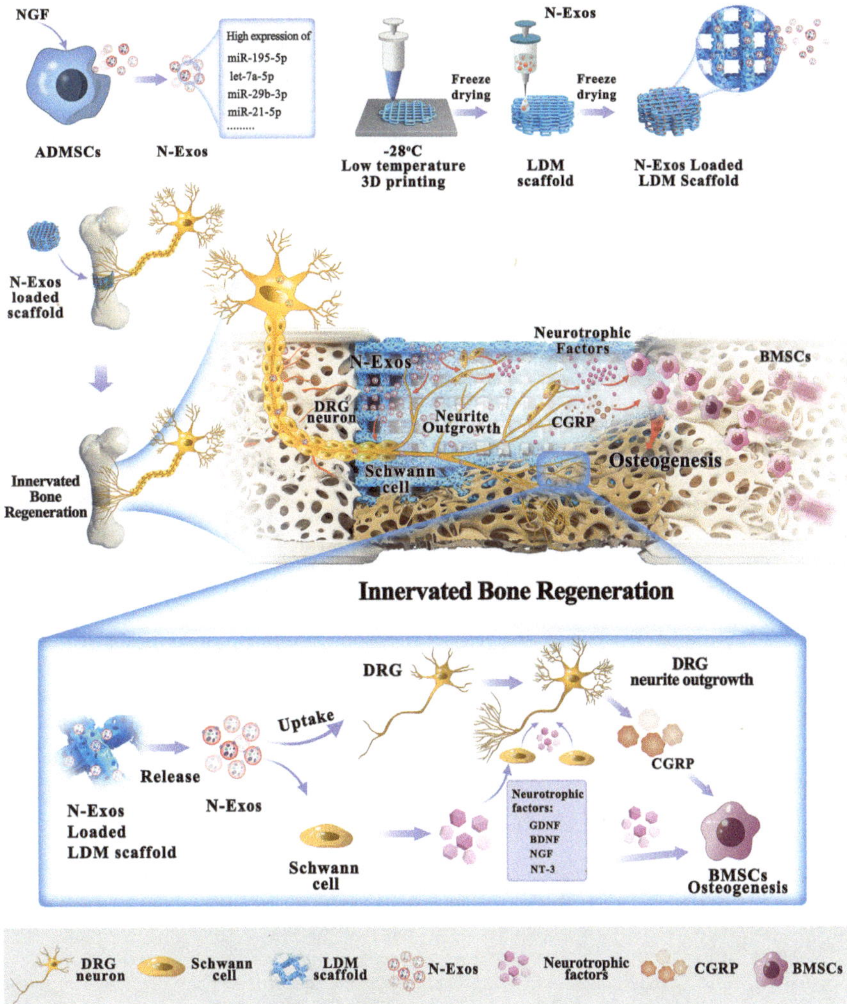

Fig. 17.14 (**a**) N-Exos-functionalized hierarchical porous scaffolds were fabricated using the LDM printing technique, followed by N-Exos loading. (**b**) The N-Exos-functionalized LDM-printed hierarchical porous scaffolds could promote the axonal growth and CGRP expression in DRG neurons as well as enhance the regenerative potential and neurotrophic effects of SCs by transferring multicomponent functional miRNAs, and consequently, synergistically stimulate the osteogenic differentiation capacity of BMSCs to achieve innervated bone regeneration. *N-exosomes* NGF-stimulated MSCs-derived exosomes, *NFG* nerve growth factor, *ADMSCs* adipose tissue-derived mesenchymal stem cells, *LDM* low-temperature deposition modeling, *DRGs* dorsal root ganglions, *CGRP* calcitonin gene related peptide, *BMSCs* bone marrow-derived MSCs, *GDNF* glial cell line-derived neurotrophic factor, *BDNF* brain-derived neurotrophic factor, *NT-3*: neurotrophin-3. (Adapted with permission from [44] Copyright © 2024 American Chemical Society)

cell-derived, multilineage constructs (including neural phenotypes) created with complex cellular structures and guidance patterns would be the ultimate approach [33]. Furthermore, achieving a fine balance between complexity and controllability is essential to develop constructs that are both biomimetic and manufacturable/scalable. These constructs must be intricate enough to replicate native tissue architecture, yet reproducible and practical for large-scale fabrication [35].

Review and Practice Questions

1. What are vascularization, vasculogenesis, and angiogenesis, respectively?
2. Why is vascularization critical in tissue engineering?
3. What are the five key approaches to stimulate angiogenesis?
4. Describe the process of angiogenesis.
5. How can immobilization of peptides/proteins be used to promote vascularization?
6. How can biomaterials/scaffolds be modified to promote vascularization?
7. What are the three main *in vitro* prevascularization strategies?
8. What is the typical *in vivo* prevascularization strategy?
9. How can microvascular fragments be utilized for vascularization?
10. What is innervation?
11. What are the common approaches for incorporating/promoting innervation?
12. Explain the theory and methodology behind Tissue Engineered Nerve Grafts (TENGs).

References

1. Kimball, J.W., *15.3A: Anatomy of Human Circulatory System*, in *Biology*. 2025, Tufts University & Harvard: LibreTexts. Available from: https://bio.libretexts.org/@go/page/5393.
2. Fox, S. and K. Rompolski, *13. Blood, Heart and Circulation*, in *Human Physiology*. 2025, McGraw Hill.
3. Novosel, E.C., C. Kleinhans, and P.J. Kluger, *Vascularization is the key challenge in tissue engineering*. Advanced Drug Delivery Reviews, 2011. **63**(4):300–311.
4. Chau, L.T., et al., *10 - Microfluidic devices for developing tissue scaffolds*, in *Microfluidic Devices for Biomedical Applications*, X. Li and Y. Zhou, Editors. 2013, Woodhead Publishing.
5. Yang, G., et al., *Vascularization in tissue engineering: fundamentals and state-of-art*. Progress in Biomedical Engineering, 2020. **2**(1):012002.
6. Kolte, D., J.A. McClung, and W.S. Aronow, *Chapter 6 - Vasculogenesis and Angiogenesis*, in *Translational Research in Coronary Artery Disease*, W.S. Aronow and J.A. McClung, Editors. 2016, Academic Press.
7. Del Gaudio, F., D. Liu, and U. Lendahl, *Notch signalling in healthy and diseased vasculature*. Open Biology, 2022. **12**(4):220004.
8. Dudley, A.C. and A.W. Griffioen, *Pathological angiogenesis: mechanisms and therapeutic strategies*. Angiogenesis, 2023. **26**(3):313–347.
9. Filippi, M., et al., *Chapter 14 - Strategies to promote vascularization, survival, and functionality of engineered tissues*, in *Tissue Engineering (Third Edition)*, J. De Boer, et al., Editors. 2023, Academic.

10. Laschke, M.W. and M.D. Menger, *Vascularization in Tissue Engineering: Angiogenesis versus Inosculation.* European Surgical Research, 2012. **48**(2):85–92.
11. Lopes, S.V., et al., *Vascularization Approaches in Tissue Engineering: Recent Developments on Evaluation Tests and Modulation.* ACS Applied Bio Materials, 2021. **4**(4):2941–2956.
12. Bastard, C., et al., *How Does Temporal and Sequential Delivery of Multiple Growth Factors Affect Vascularization Inside 3D Hydrogels?* Advanced Therapeutics, 2024. **7**(1):2300091.
13. Laschke, M.W., Y. Gu, and M.D. Menger, *Replacement in angiogenesis research: Studying mechanisms of blood vessel development by animal-free in vitro, in vivo and in silico approaches.* Frontiers in Physiology, 2022. **13**:981161.
14. Herrmann, M., et al., *Chapter 14 - Vascularization, Survival, and Functionality of Tissue-Engineered Constructs*, in *Tissue Engineering (Second Edition)*, C.A.V. Blitterswijk and J. De Boer, Editors. 2014, Academic Press.
15. Zhao, W., et al., *Biomaterials to promote vascularization in tissue engineering organs and ischemic fibrotic diseases.* MedComm – Biomaterials and Applications, 2022. **1**(1):e16.
16. Zhang, W., J.K. Choi, and X. He, *Engineering Microvascularized 3D Tissue Using Alginate-Chitosan Microcapsules.* Journal of Biomaterials and Tissue Engineering, 2017. **7**(2):170–173.
17. Agarwal, P., et al., *Microfluidics Enabled Bottom-Up Engineering of 3D Vascularized Tumor for Drug Discovery.* ACS Nano, 2017. **11**(7):6691–6702.
18. Laschke, M.W. and M.D. Menger, *Prevascularization in tissue engineering: Current concepts and future directions.* Biotechnology Advances, 2016. **34**(2):112–121.
19. Masson-Meyers, D.S. and L. Tayebi, *Vascularization strategies in tissue engineering approaches for soft tissue repair.* Journal of Tissue Engineering and Regenerative Medicine, 2021. **15**(9):747–762.
20. Devillard, C.D. and C.A. Marquette, *Vascular Tissue Engineering: Challenges and Requirements for an Ideal Large Scale Blood Vessel.* Frontiers in Bioengineering and Biotechnology, 2021. **9**:721843.
21. Chen, J., et al., *Current Strategies for Engineered Vascular Grafts and Vascularized Tissue Engineering.* Polymers, 2023. **15**(9):2015.
22. Iqbal, M.Z., et al., *Breathing new life into tissue engineering: exploring cutting-edge vascularization strategies for skin substitutes.* Angiogenesis, 2024. **27**(4):587–621.
23. He, J., et al., *Fabrication of Nature-Inspired Microfluidic Network for Perfusable Tissue Constructs.* Advanced Healthcare Materials, 2013. **2**(8):1108–1113.
24. Park, G.T., et al., *Transplantation of adipose tissue-derived microvascular fragments promotes therapy of critical limb ischemia.* Biomaterials Research, 2023. **27**(1):70.
25. Gao, Y., et al., *Application and Mechanism of Adipose Tissue-Derived Microvascular Fragments in Tissue Repair and Regeneration.* Biomolecules, 2025. **15**(3):422.
26. Xu, X., et al., *Adipose Tissue-derived Microvascular Fragments as Vascularization Units for Dental Pulp Regeneration.* J Endod, 2021. **47**(7):1092–1100.
27. Frueh, F.S., et al., *Adipose Tissue-Derived Microvascular Fragments Improve Vascularization, Lymphangiogenesis, and Integration of Dermal Skin Substitutes.* Journal of Investigative Dermatology, 2017. **137**(1):217–227.
28. Frueh, F.S., et al., *Adipose tissue-derived microvascular fragments promote lymphangiogenesis in a murine lymphedema model.* Journal of Tissue Engineering, 2022. **13**: 20417314221109957.
29. Brady, E.L., et al., *Engineered tissue vascularization and engraftment depends on host model.* Scientific Reports, 2023. **13**(1):1973.
30. Song, H.H.G., et al., *Vascular Tissue Engineering: Progress, Challenges, and Clinical Promise.* Cell Stem Cell, 2018. **22**(3):340–354.
31. Fox, S. and K. Rompolski, *7. The Nervous System: Neurons and Synapses*, in *Human Physiology*. 2025, McGraw Hill.
32. Karvinen, J. and M. Kellomäki, *Design aspects and characterization of hydrogel-based bioinks for extrusion-based bioprinting.* Bioprinting, 2023. **32**:e00274.
33. Das, S., et al., *Innervation: the missing link for biofabricated tissues and organs.* npj Regenerative Medicine, 2020. **5**(1):11.

34. Kim, J.H., et al., *Neural cell integration into 3D bioprinted skeletal muscle constructs accelerates restoration of muscle function.* Nature Communications, 2020. **11**(1):1025.
35. Newman Frisch, A., et al., *Advances in vascularization and innervation of constructs for neural tissue engineering.* Current Opinion in Biotechnology, 2022. **73**:188–197.
36. Wan, T., et al., *Biomimetic ECM nerve guidance conduit with dynamic 3D interconnected porous network and sustained IGF-1 delivery for enhanced peripheral nerve regeneration and immune modulation.* Materials Today Bio, 2025. **30**:101403.
37. Poerio, A., et al., *Towards innervation of bioengineered muscle constructs: Development of a sustained neurotrophic factor delivery and release system.* Bioprinting, 2022. **27**:e00220.
38. Cullen, D.K., et al., *Microtissue Engineered Constructs with Living Axons for Targeted Nervous System Reconstruction.* Tissue Engineering Part A, 2012. **18**(21-22):2280–2289.
39. Oguntade, E., et al., *A 3D Nerve Guidance Conduit Surface With a Wrinkled Protein Coating Toward Peripheral Nerve Injury Repair.* Polymers for Advanced Technologies, 2024. **35**(11):e70003.
40. Katiyar, K.S., et al., *Tissue Engineered Axon Tracts Serve as Living Scaffolds to Accelerate Axonal Regeneration and Functional Recovery Following Peripheral Nerve Injury in Rats.* Frontiers in Bioengineering and Biotechnology, 2020. **8**:492.
41. Smith, D.H., et al., *Tissue-engineered grafts exploit axon-facilitated axon regeneration and pathway protection to enable recovery after 5-cm nerve defects in pigs.* Science Advances, 2022. **8**(44):eabm3291.
42. Maggiore, J.C., et al., *Tissue engineered axon-based "living scaffolds" promote survival of spinal cord motor neurons following peripheral nerve repair.* Journal of Tissue Engineering and Regenerative Medicine, 2020. **14**(12):1892–1907.
43. Zarrintaj, P., et al., *Conductive biomaterials as nerve conduits: Recent advances and future challenges.* Applied Materials Today, 2020. **20**:100784.
44. Lian, M., et al., *Nerve Growth Factor-Preconditioned Mesenchymal Stem Cell-Derived Exosome-Functionalized 3D-Printed Hierarchical Porous Scaffolds with Neuro-Promotive Properties for Enhancing Innervated Bone Regeneration.* ACS Nano, 2024. **18**(10):7504–7520.

Part IV

Commercialization of Cellular and Tissue Engineered Products

Product and Process Design for Industrial Manufacturing of Cell and Tissue Engineered Products

18

Learning Objectives

- Evaluate the goals and core components/requirements of CGMP and CGTP
- Analyze and implement the QbD approach, including DoE
- Examine the applications and current challenges of integrating AI in cell and tissue engineering

18.1 CGMP and CGTP Regulations

CGMP stands for the Current Good Manufacturing Practice regulations enforced by the U.S. Food and Drug Administration (FDA), which provides systems to ensure proper design, monitoring, and control of manufacturing processes and facilities. The term "Current" signifies that companies must employ up-to-date technologies and systems to comply with these regulations. Adherence to CGMP includes establishing strong quality management systems, obtaining appropriate quality raw materials, establishing robust operating procedures, detecting and investigating product quality deviations, and maintaining reliable testing laboratories [1]. The CGMP requirements in 21 Code of Federal Regulations (CFR) Parts 210 and 211 govern the methods to be used in, and the facilities or controls to be used for, the manufacture, processing, packing, or holding of a drug to ensure that such drugs meets the requirements of the FD&C ACT as to safety, and has the identity and strength and meets the quality and purity characteristics that it purports or is represented to possess (Table 18.1). Due to the broader scope of these regulations, most of the CGMP regulations under Parts 210 and 211 apply to human cells, tissues, or cellular or tissue-based products (HCT/Ps) regulated as biological products [2]. The CGMP regulations in Parts 600 through 680 of 21 CFR pertain to drugs that are also biological products for human use. Additionally, 21 CFR Part 1271 applies to drugs that are also HCT/Ps (subject to review under an application submitted under

W. Zhang, *Cell Culture and Tissue Engineering*,
https://doi.org/10.1007/978-3-032-32096-4_18

Table 18.1 Definitions of key CGMP-related terms and their implications for HCT/Ps

Term	Definition	HCT/Ps implications
Safety	The relative freedom from harmful effect to persons affected, directly or indirectly, by a product when prudently administered, taking into consideration the character of the product in relation to the condition of the recipient at the time	Pathogens Endotoxins
Identity	A component's chemical structure and its physical form (e.g., polymorph, solvate, and appearance) including, if appropriate, its stereochemistry or immunochemistry	Biochemical markers of cells
Strength	(i) The concentration of the drug substance (for example, weight/weight, weight/volume, or unit dose/volume basis), and/or (ii) The potency, that is, the therapeutic activity of the drug product as indicated by appropriate laboratory tests or by adequately developed and controlled clinical data (expressed, for example, in terms of units by reference to a standard)	Number of cells Concentration of growth factors
Quality	The totality of features and characteristics that bear on the ability of a device to satisfy fitness-for-use, including safety and performance	Cell viability
Purity	The relative freedom from extraneous matter in the finished product, whether or not harmful to the recipient or deleterious to the product	Residual impurities Remaining feeder-layers cells

section 505 of the Federal Food, Drug and Cosmetic (FD&C) Act or under a biological product license application under section 351 of the Public Health Service Act) [3]. Clinical grade cell and tissue engineered products should be produced in a CGMP facility that is accredited by the regulatory body. The quality management system is the crucial part of CGMP, designed to control the manufacturing processes, facility management, process and method validations, personnel training and competency and also records. Environmental controls, monitoring and maintenance of the facility are critical to preventing contaminations and ensuring operational efficiency. Standard operating procedures (SOP) for all related processes and procedures involved in the production of cell and tissue-based products should be developed and validated before the products can be manufactured and released for therapeutic use [4]. An example of the manufacturing of human induced pluripotent stem cells under CGMP is shown in Fig. 18.1.

Current Good Tissue Practices (CGTP) are a comprehensive set of regulations governing HCT/Ps. Regulated under 21 CFR 1271 Subparts D and E, the primary goal of CGTP is to ensure that HCT/Ps are manufactured and processed in a manner designed to prevent the risk of introduction, transmission and spread of communicable diseases [5]. Core CGTP requirements are those requirements that directly relate to preventing the introduction, transmission, or spread of communicable disease of HCT/P. These include requirements for facilities; environmental controls; equipment; supplies and reagents; recovery; processing and process controls;

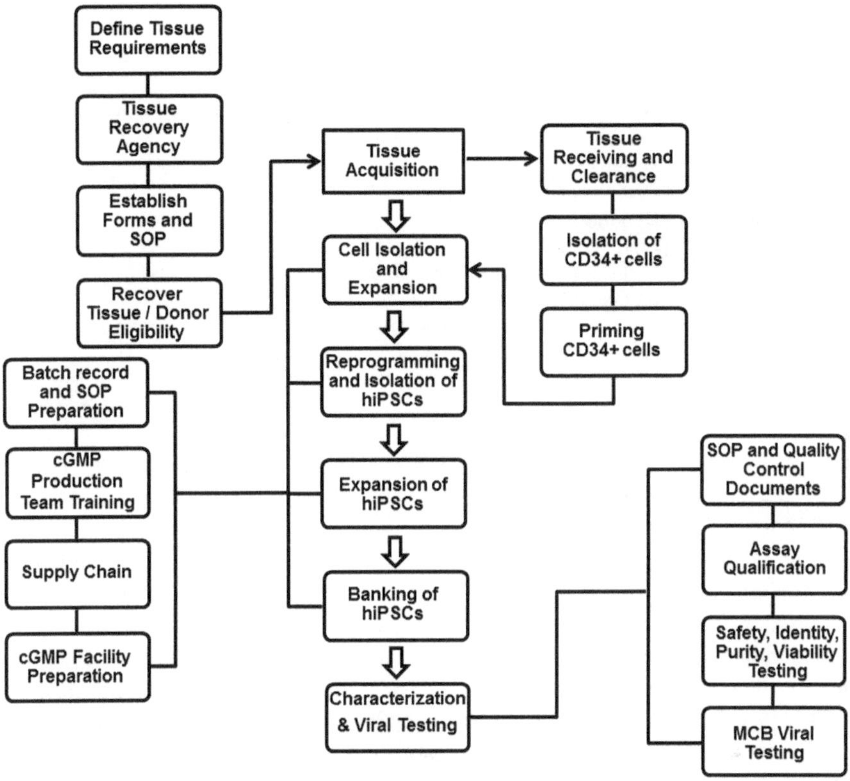

Fig. 18.1 An overview of the manufacturing of human induced pluripotent stem cells under cGMP. *CD* cluster of differentiation, *SOP* standard operating procedure, *cGMP* current Good Manufacturing Practice, *MCB* master cell bank. (Adapted from Baghbaderani, B.A. et al., cGMP-Manufactured Human Induced Pluripotent Stem Cells Are Available for Pre-clinical and Clinical Applications. Stem Cell Reports. 2015, 5(4): 647–659 with permission from Elsevier)

labeling controls; storage; receipt, redistribution shipment, and distribution of an HCT/P through; and donor eligibility determinations, donor screening, and donor testing [2, 6].

In general, the CGMP regulations and CGTP regulations require the same manufacturing practices, and compliance with the CGMP regulations results in compliance with the applicable CGTP requirements (Table 18.2). However, the following CGTP requirements necessitate additional manufacturing practices, as they are only partially or not at all covered by the corresponding CGMP regulations: all donor eligibility requirements; prevention of the introduction, transmission, or spread of communicable diseases; certain parts of manufacturing arrangements under a contract, agreement, or other arrangements; procedures for sharing with other establishments information pertaining to possible contamination or potential for transmission of communicable disease; audits; prohibition on pooling; predistribution shipment; HCT/P availability for distribution only after donor eligibility established;

Table 18.2 Comparison between Current Good Manufacturing Practice (CGMP) and Current Good Tissue Practice (CGTP)

	CGMP	CGTP
Goal	Assuring the safety, identity, strength, quality, and purity of a product	Prevention of the introduction, transmission, or spread of communicable diseases
Core components/ requirements	Organization and personnel Buildings and facilities Equipment Control of components and drug product containers and closures Production and process controls Packaging and labeling controls Holding and distribution Laboratory controls Records and reports	Facilities Environmental control Equipment Supplies and reagents Recovery Processing and process controls Labeling controls Storage Receipt, predistribution shipment, and distribution of an HCT/P Donor eligibility determinations, donor screening, and donor testing

packaging and shipping requirements; recordkeeping for 10 years (but facility cleaning and sanitation records for 3 years); and tracking. An example of a CGTP requirement not covered by CGMP regulations is the requirement for predistribution shipment. Under this regulation, if you ship an HCT/P within your establishment or between establishments (e.g., from a procurer to a processor), and the HCT/P is not available for distribution as described in § 1271.265(c), you must first determine and document whether preestablished criteria designed to prevent communicable disease transmission have been met, and the HCT/P must be shipped in quarantine [2]. European Union (EU)'s guidelines for GMP for medicinal products for human can be accessed at https://health.ec.europa.eu/document/download/7c6c5b3c-4902-46ea-b7ab-7608682fb68d_en. The EU and the USA signed a mutual recognition agreement (MRA) on GMP inspections, which allows EU authorities and their FDA counterparts to rely on each other's GMP inspections for some types of medicines, waive batch testing of products on entry into their territories, and share information on inspections and quality defects [7]. The main differences between FDA and EMA GMP regulations is that the FDA utilizes a rule-based, prescriptive approach, whereas the EMA adopts a principle-based directive focusing more on Quality Risk Management (QRM) and Quality Management Systems (QMS) [8]. Given the significant differences in GMP across regulatory agencies, harmonizing GMP practices based on globally applicable, up-to-date quality standards informed by cutting-edge science is critical [9, 10]. In addition to GMP, mandatory and recommended quality standards are also reflected in the quality management standards prescribed by the International Organization for Standardization (ISO), against which clinical or commercial product manufacturing facilities can achieve certification [11].

18.2 Quality by Design (QbD) Approach

The U.S. FDA defines quality by design (QbD) as a systematic approach to development that begins with predefined objectives and emphasizes product and process understanding and process control, based on sound science and quality risk management [12]. Similarly, European Medicines Agency (EMA) defines QbD as an approach that aims to ensure the quality of medicines by employing statistical, analytical and risk-management methodologies in the design, development and manufacturing of medicines [13]. As shown in Fig. 18.2, the QbD approach begins by defining the desired product characteristics (quality, safety and efficacy) from a patient and clinical perspective, i.e., the Quality Target Product Profile (QTPP) [14, 15]. The QTPP outlines the design criteria for the product and provides an understanding of what will ensure the quality, safety, and efficacy of a specific product for the patient. It therefore serves as the basis for the development of the critical quality attributes (CQA), critical process parameters (CPPs), and control strategy [16]. A CQA is a physical, chemical, biological, or microbiological property or characteristic of an output material (intermediate or finished product) that should be within an appropriate limit, range, or distribution to ensure the desired product quality. Critical material attributes (CMAs) refer to the physical, chemical, biological, or microbiological properties or characteristics of an input material. Critical process parameters (CPPs) are controlled or monitored parameters, such as speed and temperature, whose variability significantly influences product quality [17]. The design space is

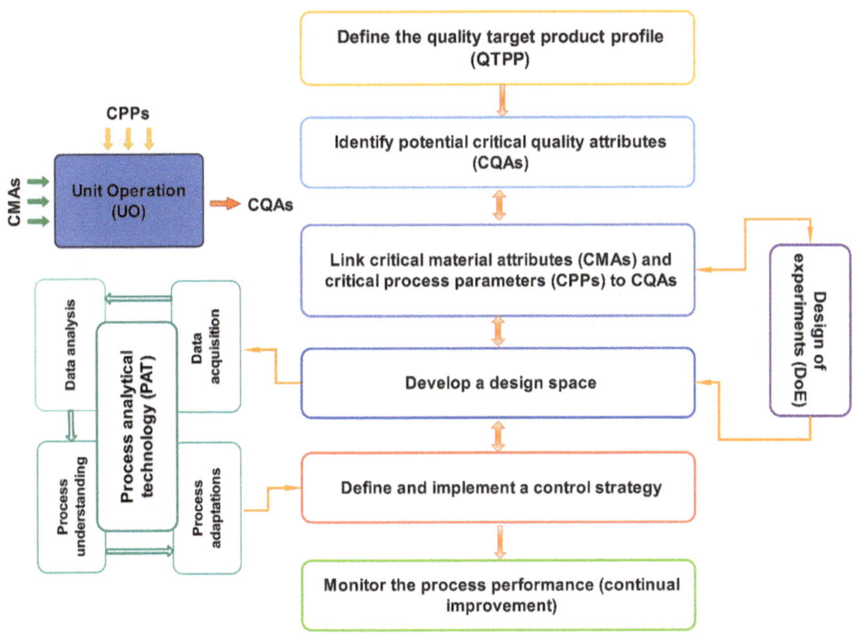

Fig. 18.2 A typical quality by design (QbD) scheme. (Figure is generated based on [15, 17])

defined as the multidimensional combination and interaction of input variables (e.g., material attributes) and process parameters that have been demonstrated to provide assurance of quality [18]. QbD enables the establishment of a design space comprising a combination of CPPs, which will ensure all CQAs are met. Operating within the design space should ensure consistent product quality [14, 19]. The control strategy is a planned set of controls, derived from current product and process understanding, that ensures performance and product quality. The application of process analytical technology (PAT) may be part of the control strategy to ensure the process remains within the established design space. PAT is a system for designing, analyzing, and controlling manufacturing through timely measurements (i.e., during processing) of critical quality and performance attributes of raw and in-process materials and processes with the goal of ensuring final product quality (Fig. 18.2) [17].

A manufacturing process for a cell and tissue engineering product can generally be broken down into multiple unit operations (UOs) [20]. It is essential to evaluate each unit operation or a combination of unit operations to identify CQAs, CMAs, and CPPs, and link CMAs and CPPs to CQAs: CQAs = f (CMA1, CMA2, CMA3, ... CPP1, CPP2, CPP3 ...) (Fig. 18.2) [17]. Considering the manufacturing process of an autologous skin graft (Fig. 18.3), the UOs include cell isolation, cell expansion, scaffold fabrication, cell seeding, and maturation [21]. For example, in the cell expansion unit operation, CPPs may include cell seeding density, medium composition, and culture time [20].

Design of experiments (DoE) is a powerful tool for optimizing bioprocesses, playing a critical role in the QbD strategy. It is valuable for exploring the experimental domain and providing statistics insights that help understand the impact of CPPs on CQAs [18]. DoE can help identify optimal conditions, CMAs, CPPs, and design space (Fig. 18.2) [17]. A DoE approach uses a controlled set of tests to model the relationships between factors (material attributes and process parameters) and observed responses (quality attributes), enabling effective experimental planning and analysis data. The core process of DoE is shown in Fig. 18.4. Typically, DoE is applied after a screening study has completed. Several DoE methods exist, including factorial experiments and response surface methodologies (RSM). In factorial

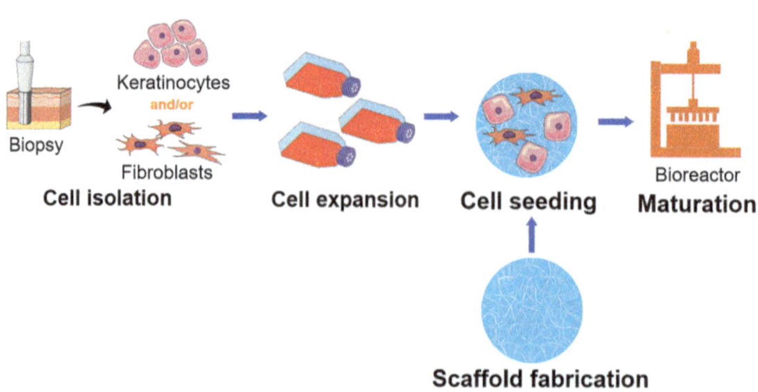

Fig. 18.3 Unit operations in the manufacturing process of an autologous skin graft

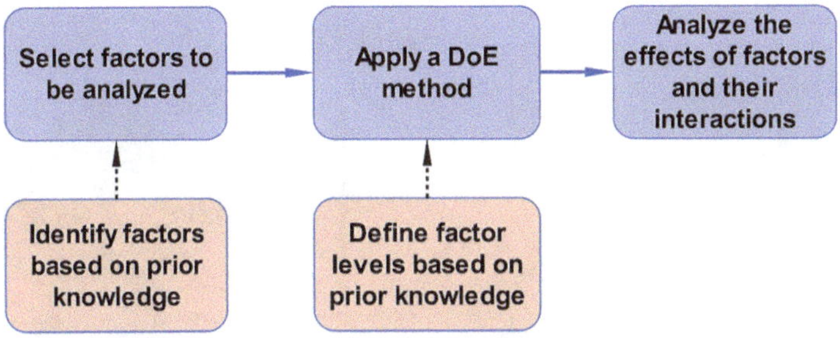

Fig. 18.4 Core design of experiments (DoE) process. (Figure is generated based on [22])

designs, all factors are assessed in all possible combinations of "high" and "low" levels. Fractional factorial designs can be used to reduce the number of experimental runs while still determining the main effects during screening experiments. RSM combines mathematical and statistical techniques with experimental designs to develop models relating responses and control factors, enabling prediction of response values and optimum control variables settings [22]. RSM has been applied in various cell and tissue engineering contexts. For example, Alireza Talaei et al. used RSM to efficiently optimize the composition of a hybrid gelatin-methacryloyl (GelMA) and hyaluronic acid methacryloyl (HAMA) hydrogel with respect to both mechanical and transport properties. The factors investigated were the concentrations of GelMA, HAMA, and Lithium phenyl-2,4,6-trimethylbenzoylphosphinate (LAP; photoinitiator). Mathematical models were fitted to the experimental data and used to predict the optimum hydrogel composition. As shown in Fig. 18.5, increasing GelMA and HAMA concentration resulted in enhanced hydrogel viscoelastic stiffness (measured *via* storage modulus) [23]. A recent study used RSM to build empirical models relating alginate-based hydrogel composition to its mechanical behavior [24].

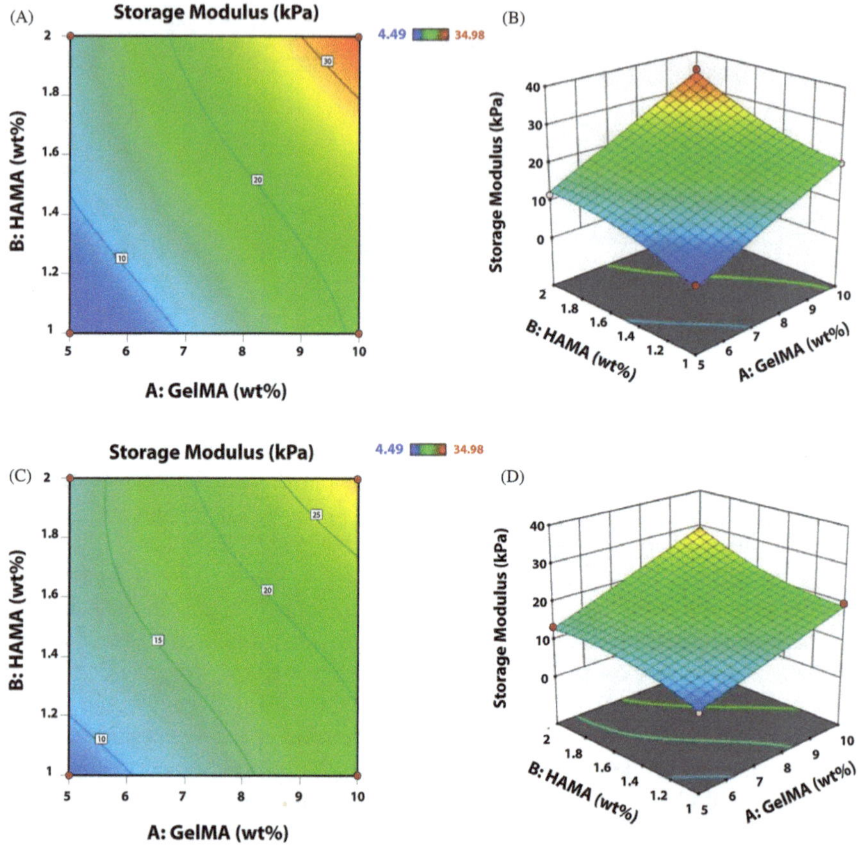

Fig. 18.5 2D and 3D response surface plots. (**a**, **b**) storage modulus (evaluating viscoelastic stiffness) as a function of GelMA/HAMA concentration at 0.1% (w/v) LAP. (**c**, **d**) Storage modulus as a function of GelMA/HAMA concentration at 0.2% (w/v) LAP. *GelMA* gelatin-methacryloyl, *HAMA* hyaluronic acid methacryloyl, *LAP* Lithium phenyl-2,4,6-trimethylbenzoylphosphinate. (Adapted from [23] with permission from John Wiely and Sons)

18.3 Artificial Intelligence in Cell and Tissue Engineering

Artificial intelligence (AI) is defined as the development of computer systems capable of performing tasks that routinely require human intelligence. A key subset of artificial intelligence (AI), machine learning (ML), is widely applied in healthcare. In cell therapy, AI has been utilized to improve the characterization and optimization of cell products by predicting their function and potency, as well as enabling non-invasive analysis of live cell cultures [25]. In tissue engineering, AI holds significant potential in optimizing biomaterial and scaffold design and fabrication (including 3D printing), predicting biological response to scaffolds, modeling scaffold *in vivo* performance, analyzing images, streamlining development processes, and enabling personalized medicine [22, 25–32].

Figure 18.6 illustrates the AI-aided workflow for tissue engineering scaffold design and prediction. The process begins with scaffold generation, followed by mechanical property simulation. Virtual topographies are then created and used to train a 3D Convolutional Neural Network (CNN). After training, the CNN is validated and tested on new scaffold designs, providing faster and comparable predictions to solutions, while also expanding scaffold libraries [25]. In predicting biological response to the scaffold, AI can assist in identifying correlations between material parameters and biological outcomes, determining the weight of scaffold characteristics, and analyzing correlations between scaffold parameters (Fig. 18.7) [26]. As for real-world applications, María Dolores Bermejillo Barrera et al. reported the use of 3D CNNs trained on digital tomographies obtained from the CAD models to predict the mechanical properties of innovative tissue engineering scaffolds [33]. Dina Ahmed Salem et al. explored the use of ML to predict the performance of nanofibrous scaffolds for skin tissue engineering, finding that the AdaBoost algorithm was the most effective at predicting the target variable [34]. An innovative foundation model, HydrogelFinder, has been developed for rational design, rapid production, and screening of self-assembling peptides. It comprises three key modules: Hydrogel-mining for literature data mining, HydrogelFinder-GPT employing a deep generative model, and HydrogelFinder-predict as a virtual

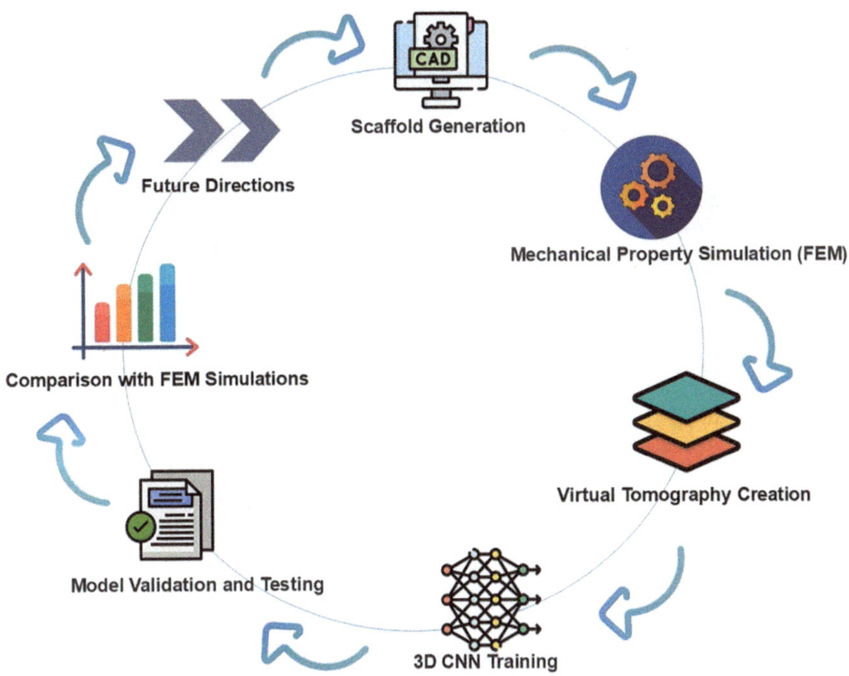

Fig. 18.6 Artificial intelligence (AI)-aided process for designing tissue engineering scaffolds. *CAD* computer-aided design, *FEM* finite element method, *CNN* convolutional neural network. (Adapted from [25] with permission from Elsevier)

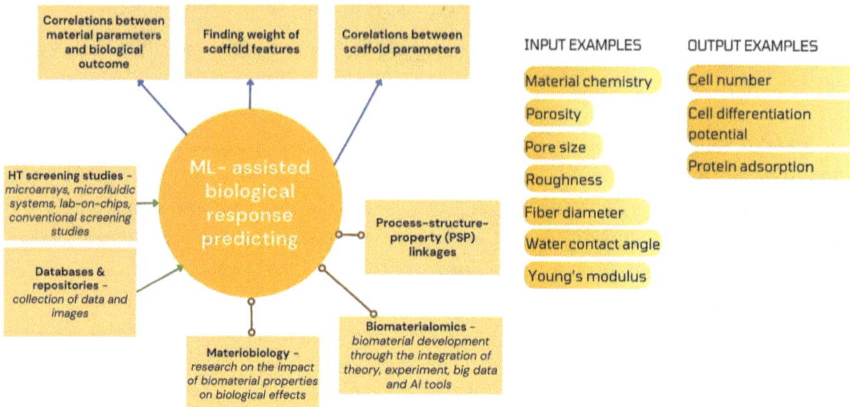

Fig. 18.7 The concept of machine learning (ML)-assisted prediction of biological response to scaffolds, along with associated components and examples of input and output items. *HT* high throughput. (Adapted from [26])

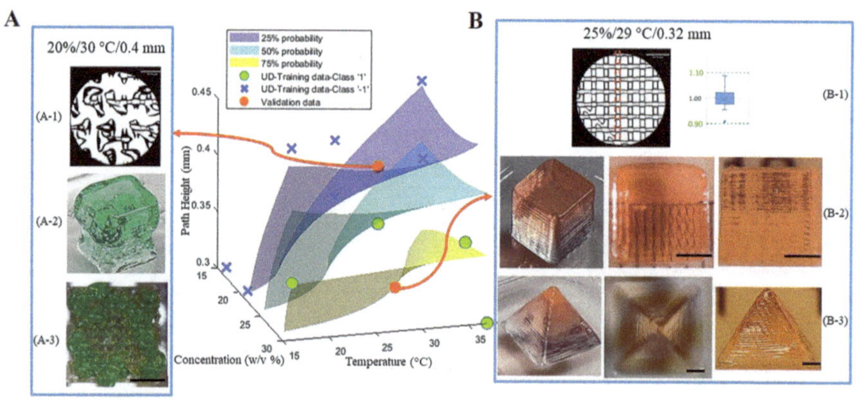

Fig. 18.8 Validation on probability space. (**a**) Low-printability validation: validation point selected at 20% (w/v) concentration, 30 °C nozzle temperature, and 0.4 mm path height. (A-1) Microscopic image evaluating the filament width. (A-2) Printed cube structure (5 × 5 × 5 mm^3). (A-3) Top view of the printed grid structure (5 × 5 × 5 mm^3). (**b**) High-printability validation: validation point at 25% (w/v) concentration, 29 °C nozzle temperature, and 0.32 mm path height. (B-1) Microscopic image evaluating the filament width index. (B-2) Printed cube structure (5 × 5 × 5 mm^3) with top view and side view. (B-3) Printed pyramid structure (10 mm × 10 mm × 8 mm) with top and side view. Scale bar for (A-1) and (B-1) is 1.87 mm and 2 mm for all other images. *UD* uniform design. (Adapted from [36])

screening tool [35]. Zhouquan Fu et al. investigated the key print parameters affecting printability and developed a machine learning model to optimize these parameters, achieving high quality prints with high probability (>75%) (Fig. 18.8) [36, 37].

Although AI shows great promises in cell and tissue engineering, several challenges—such as data quality, model interpretability, and standardization—must be addressed for widespread implementation [27]. The limited availability of large, well-curated, and standardized datasets hampers the development of robust AI models, as the use of large datasets is essential for achieving high identification and prediction precision [25, 26]. The inherent complexity and dynamic nature of biological processes in cell and tissue engineering contribute to data scarcity and quality. Furthermore, integrating AI into existing clinical and research frameworks is complicated by the vast heterogeneity of biological systems [25]. Establishing rigorous data collection methodologies and standardizing datasets to ensure their reliability has become increasingly critical. Standardization will also promote efficient data sharing and promote collaboration across the field [26].

Review and Practice Questions

1. Define key terms related to Current Good Manufacturing Practice (CGMP) and provide example implications of these terms for human cells, tissues, and cellular and tissue-based products (HCT/Ps).
2. Compare Current Good Manufacturing Practice (CGMP) and Current Good Tissue Practice (CGTP).
3. Describe the typical process of a quality by design (QbD) approach.
4. Identify the unit operations involved in the manufacturing process of a cell and tissue engineered product of your choice. For one unit operation, determine its critical material attributes (CMAs), critical process parameters (CPPs), and critical quality attributes (CQAs).
5. What is design of experiments (DoE) and how can it be applied within a quality by design (QbD) framework?
6. Conduct a literature review and write a short essay on a specific application of response surface methodology (RSM) in cell and tissue engineering that was not discussed in this chapter.
7. Conduct a literature review and write a short essay on a specific application of artificial intelligence (AI) in cell and tissue engineering that was not discussed in this chapter.
8. What are some of the key challenges in applying artificial intelligence (AI) to cell and tissue engineering?

References

1. U.S. Food and Drug Administration. *Facts About the Current Good Manufacturing Practice (CGMP)*. Available from: https://www.fda.gov/drugs/pharmaceutical-quality-resources/facts-about-current-good-manufacturing-practice-cgmp?_hsenc=p2ANqtz-_wTgMtFtJF4kYYDOPhT8Ax38FIGfL0LlnLicV-4mAemmVQLk0imQOROhBLPVuopltev0My.
2. U.S. Food and Drug Administration. *Current Good Tissue Practice (CGTP) and Additional Requirements for Manufacturers of Human Cells, Tissues, and Cellular and Tissue-Based Products (HCT/Ps)*. Available from: https://www.fda.gov/regulatory-information/search-

fda-guidance-documents/current-good-tissue-practice-cgtp-and-additional-requirements-manufacturers-human-cells-tissues-and.
3. U.S. National Archives. *PART 211—CURRENT GOOD MANUFACTURING PRACTICE FOR FINISHED PHARMACEUTICALS.* Avaialble from: https://www.ecfr.gov/current/title-21/chapter-I/subchapter-C/part-211.
4. Seet, W.T., et al., *Current Good Manufacturing Practice (cGMP) Facility and Production of Stem Cell*, in *Stem Cell Production: Processes, Practices and Regulations*, F.A. Khan, Editor. 2022, Springer.
5. Chaudhry, P. and M. Jamieson, *Good Tissue Practices*, in *Encyclopedia of Tissue Engineering and Regenerative Medicine*, R.L. Reis, Editor. 2019, Academic.
6. U.S. National Archives. *PART 1271—HUMAN CELLS, TISSUES, AND CELLULAR AND TISSUE-BASED PRODUCTS.* Available from: https://www.ecfr.gov/current/title-21/chapter-I/subchapter-L/part-1271.
7. European Medicines Agency. *Mutual recognition agreements (MRA).* Available from: https://www.ema.europa.eu/en/human-regulatory-overview/research-development/compliance-research-development/good-manufacturing-practice/mutual-recognition-agreements-mra.
8. Bradley, C. *FDA vs EMA: Key Differences in GMP Regulations Every Pharma Lab Must Know.* 2025. Available from: https://www.labmanager.com/fda-vs-ema-key-differences-in-gmp-regulations-every-pharma-lab-must-know-34062.
9. Petrelli, F., et al., *The requirements for manufacturing highly active or sensitising drugs comparing Good Manufacturing Practices.* Acta Biomedica, 2019. **90**(2):288–299.
10. Petrelli, F., et al., *Campaign manufacturing of highly active or sensitizing drugs: a comparison between the GMPs of various Regulatory Agencies.* La Clinica Terapeutica, 2020. **170**(1):e66-e73.
11. Patel, D., J. Reese Koç, and F. Otegbeye, *Creating a GMP cell processing program: A focus on quality and regulation.* Best Practice & Research Clinical Haematology, 2025. **38**(1):101614.
12. U.S. Food and Drug Administration. *Guidance for Industry Q8(R2) Pharmaceutical Development.* Available from: https://www.fda.gov/media/71535/download.
13. European Medicines Agency. *Quality by design.* Available from: https://www.ema.europa.eu/en/human-regulatory-overview/research-development/quality-design.
14. Maillot, C., et al., *Quality by design to define critical process parameters for mesenchymal stem cell expansion.* Biotechnology Advances, 2021. **50**:107765.
15. Van den Bergh, J. and E. Claes, *Cell by Design (CbD).* GMP Journal, 2021(32):23–26.
16. U.S. Food and Drug Administration. *Q8, Q9, & Q10 Questions and Answers -- Appendix: Q&As from Training Sessions (Q8, Q9, & Q10 Points to Consider).* Available from: https://www.fda.gov/regulatory-information/search-fda-guidance-documents/q8-q9-q10-questions-and-answers-appendix-qas-training-sessions-q8-q9-q10-points-consider#:~:text=The%20quality%20target%20product%20profile,%2C%20CPPs%2C%20and%20control%20strategy.
17. Zhang, W., *Commercialization of controlled release products*, in *Fundamentals and Design of Drug Delivery Systems: A Textbook with Lab Activities*. 2025, Academic Press.
18. Kasemiire, A., et al., *Design of experiments and design space approaches in the pharmaceutical bioprocess optimization.* European Journal of Pharmaceutics and Biopharmaceutics, 2021. **166**:144–154.
19. Udugama, I.A., et al., *The role of process systems engineering in applying quality by design (QbD) in mesenchymal stem cell production.* Computers & Chemical Engineering, 2023. **172**:108144.
20. Claes, E., et al., *Chapter 20 - Product and process design: scalable and sustainable tissue-engineered product manufacturing*, in *Tissue Engineering (Third Edition)*, J. De Boer, et al., Editors. 2023, Academic Press.
21. Hodge, J.G., et al., *Evaluating polymeric biomaterials to improve next generation wound dressing design.* Biomaterials Research, 2022. **26**(1):50.

22. Al-Kharusi, G., et al., *The Role of Machine Learning and Design of Experiments in the Advancement of Biomaterial and Tissue Engineering Research.* Bioengineering, 2022. **9**(10):561.
23. Talaei, A., et al., *Optimizing the composition of gelatin methacryloyl and hyaluronic acid methacryloyl hydrogels to maximize mechanical and transport properties using response surface methodology.* Journal of Biomedical Materials Research Part B: Applied Biomaterials, 2023. **111**(3):526–537.
24. Bucciarelli, A., et al., *Exploiting Response Surface Methodology to Engineer the Mechanical Properties of Alginate-based Hydrogels.* Macromolecular Materials and Engineering, 2025. **310**(2):2400296.
25. Asadi Sarabi, P., et al., *AI-Based solutions for current challenges in regenerative medicine.* European Journal of Pharmacology, 2024. **984**:177067.
26. Jeznach, O., et al., *Review on machine learning application in tissue engineering: What has been done so far? Application areas, challenges, and perspectives.* Journal of Materials Science, 2024. **59**(46):21222–21250.
27. Bagherpour, R., G. Bagherpour, and P. Mohammadi, *Application of Artificial Intelligence in Tissue Engineering.* Tissue Engineering Part B: Reviews, 2025. **31**(1):31–43.
28. Gharibshahian, M., et al., *Recent advances in artificial intelligent strategies for tissue engineering and regenerative medicine.* Skin Research and Technology, 2024. **30**(9):e70016.
29. Guo, J.L., M. Januszyk, and M.T. Longaker, *Machine Learning in Tissue Engineering.* Tissue Engineering Part A: Research Advances, 2023. **29**(1-2):2–19.
30. Wu, Y., et al., *Harnessing the power of machine learning into tissue engineering: current progress and future prospects.* Burns & Trauma, 2024. **12**:tkae053.
31. Chen, H., B. Zhang, and J. Huang, *Recent advances and applications of artificial intelligence in 3D bioprinting.* Biophysics Reviews, 2024. **5**(3):031301.
32. Zhang, Z., et al., *AI-driven 3D bioprinting for regenerative medicine: From bench to bedside.* Bioactive Materials, 2025. **45**:201–230.
33. Bermejillo Barrera, M.D., F. Franco-Martínez, and A. Díaz Lantada, *Artificial Intelligence Aided Design of Tissue Engineering Scaffolds Employing Virtual Tomography and 3D Convolutional Neural Networks.* Materials, 2021. **14**(18):5278.
34. Salem, D.A., M.H. Moharam, and E.M. Hashem, *Development of Machine Learning Regression Models for Predicting the Performance of Nanofibrous Scaffolds for Skin Tissue Engineering.* Journal of Bio-X Research, 2024. **7**:0008.
35. Ren, X., et al., *HydrogelFinder: A Foundation Model for Efficient Self-Assembling Peptide Discovery Guided by Non-Peptidal Small Molecules.* Advanced Science, 2024. **11**(26): 2400829.
36. Fu, Z., V. Angeline, and W. Sun, *Evaluation of Printing Parameters on 3D Extrusion Printing of Pluronic Hydrogels and Machine Learning Guided Parameter Recommendation.* International Journal of Bioprinting, 2021. **7**(4):434.
37. Chen, H. and J. Huang, *Advancing sustainability in bioprinting through artificial intelligence.* International Journal of Bioprinting, 2025. **11**(4):133–153.

Clinical Translation and Ethical Issues in Cell and Tissue Engineering

19

Learning Objectives

- Understand and evaluate the commercialization aspects, including regulatory considerations, that should be integrated during the R&D process of a tissue-engineered product
- Examine and interpret the relevant regulatory frameworks for tissue-engineered products
- Discuss and analyze the key challenges faced during clinical trials of tissue-engineered products
- Explore and assess the ethical issues involved in the clinical translation of tissue-engineered products

19.1 Tissue Engineered Products Development and Approval Process

Like other products, such as drugs, biologics, and medical devices, the development of tissue engineered products involves research, *in vitro* testing, and *in vivo* animal studies (as shown in Fig. 19.1). The research & development (R&D) process is not linear as the testing results may necessitate further product development. Traditionally, R&D conducted in academic or research institutions lacks a strong focus on commercialization. Often, commercialization efforts begin too late—for example, after publication, which may prevent patent protection, or only after the commercialization process has begun, when a poor market fit becomes apparent [1]. Hence, it is critical to consider commercialization throughout the entire development process, beginning with defining an unmet clinical need and developing a tissue engineered product to address that need.

This chapter adopts the concept "entrepreneurial mindset" from the Kern Entrepreneurial Engineering Network (KEEN) framework, which consists of three key components (i.e., 3Cs): curiosity, connections, and creating values [2]. Several

W. Zhang, *Cell Culture and Tissue Engineering*,
https://doi.org/10.1007/978-3-032-32096-4_19

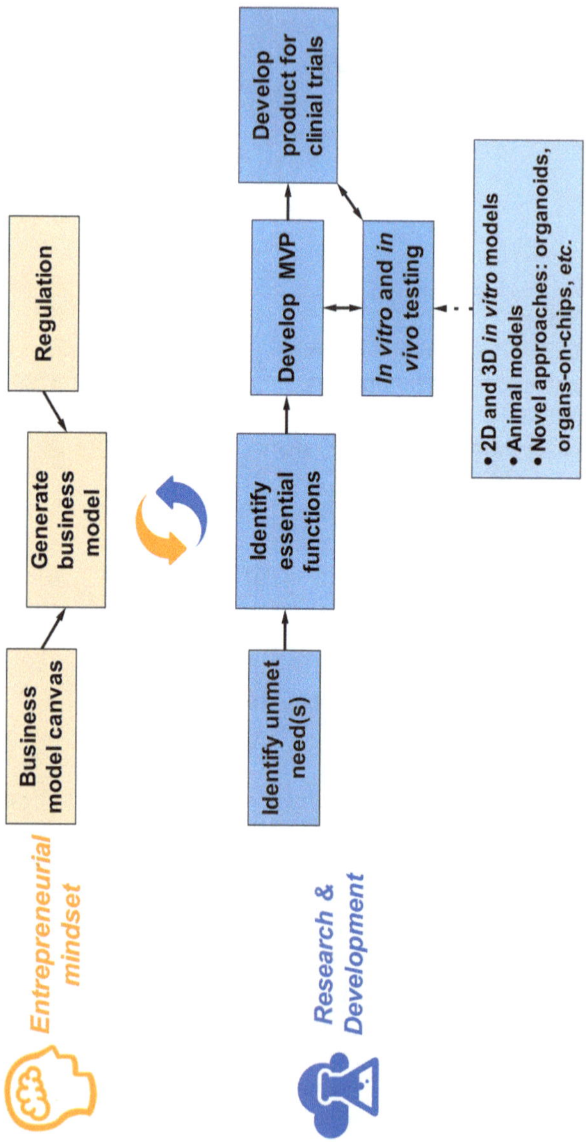

Fig. 19.1 Development process of a tissue engineered product. *MVP* minimum viable product

funding mechanisms exist to help bridge the gap between R&D and commercialization, such as the Small Business Innovation Research (SBIR) and Small Business Technology Transfer (STTR) programs in the United States. These programs provide technology-focused entrepreneurs (especially R&D focused), startups, and small businesses with funding to develop their ideas and a pathway to commercialization [3]. Another often-overlooked aspect in the R&D of tissue-engineered products is the consideration of industrial production capabilities. These include manufacturability, compliance with the current good manufacturing practices (cGMP) requirements, and the ability to scale up. Therefore, it is essential to integrate commercialization considerations into R&D. One effective exercise is business model generation. A business model describes the rationale of how an organization creates, delivers, and captures value [4]. The Business Model Canvas is a strategic management and entrepreneurial tool (Fig. 19.2), which allows the users to describe, design, challenge, invent, and pivot their business models [5]. Customers are simply users who pay for the value that is created for them in the form of a revenue stream for the company [6]. In the context of tissue-engineered products, it is important to note that users (e.g., clinicians or patients) and customers (e.g., hospitals or insurers) are usually distinct groups. Finally, during the R&D stage, a minimum viable product (MVP; a version of the product that includes its most essential functions) can be valuable, especially in the *in vitro* and *in vivo* models [1]. Two major challenges in commercializing tissue engineered products are funding and operational limitations in manufacturing and the need for preservation of tissue engineered products [7].

It is also critical to take government regulations, including product approval, during the development of tissue engineering products. In October 2023, the U.S. Food and Drug Administration (FDA) published the "Guidance for Industry: Voluntary Consensus Standards Recognitions Program for Regenerative Medicine Therapies". This guidance outlines a Standards Recognition Program (SRP-RMT) at FDA's Center for Biologics Evaluation and Research (CBER), which is designed to identify and recognize Voluntary Consensus Standards (VCS) to facilitate the development and assessment of regenerative medicine therapy (RMT) products regulated by CBER, when such standards are appropriate [8]. In the U.S., tissue engineered products are under the category of Human Cells, Tissues, and Cellular and Tissue-Based Products (HCT/Ps). These products, which contain or consist of human cells or tissues intended for implantation, transplantation, infusion, or a human recipient, are regulated under the authority of Section 361 and/or Section 351 of Public Health Service (PHS) Act and/or the Food, Drug, & Cosmetic (FD&C) Act. Section 351 gives the FDA authority to ensure these are safe, pure and have the appropriate potency. It also grants the agency power to suspend or revoke licenses and to mandate recalls; while Section 361, authorizes the FDA to issue and enforce regulations necessary to prevent the introduction, transmission, or spread of communicable diseases from foreign countries into the U.S. or between states. For products regulated solely under Section 361 (i.e., 361 HCT/P; no premarket authorization is required), they must meet these criteria (21 Code of Federal Regulations (CFR) Part 1271.10(a)):

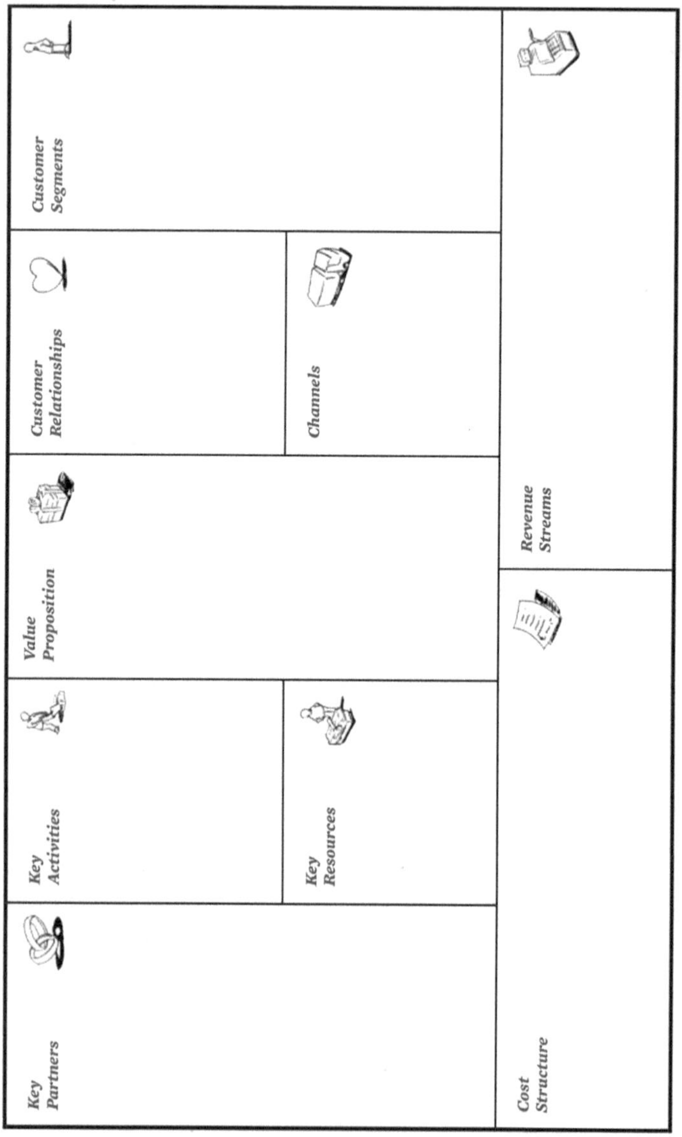

Fig. 19.2 The business model canvas. (Adapted from [4] with permission from John Wiley and Sons)

1. Minimally manipulated (does not alter the original relevant characteristics of the tissue relating to the tissue's utility for reconstruction, repair, or replacement (structural tissues) or does not alter the relevant biological characteristics of cells or tissues (cells and non-structural tissues));
2. Intended for homologous use only (the repair, reconstruction, replacement, or supplementation of a recipient's cells or tissues with an HCT/P that performs the same basic function or functions in the recipient as in the donor);
3. Not combined with another article (with some exceptions); and
4. Either:
   (a) Does not have a systemic effect and is not dependent upon the metabolic activity of living cells for its primary function; or
   (b) Has a systemic effect or is dependent upon the metabolic activity of living cells for its primary function, and is for autologous, 1st or 2nd degree blood relative, or reproductive use [9].

Examples of these HCT/Ps include minimally manipulated bone and reproductive cells (e.g., oocytes) intended solely for homologous use. If an HCT/P doesn't meet the criteria set out in 21 CFR1271.10(a) and the establishment that manufactures the HCT/P does not qualify for any of the exceptions in 21 CFR 1271.15, the product will be regulated as a drug, device, and/or biological product under the FD&C Act, and/or Section 351 of the PHS Act (42 U.S.C. 262), and applicable regulations, including 21 CFR Part 1271, and premarket review will be required [10].

The FDA has three primary centers responsible for regulating HCT/Ps depending on the product type: Center for Biologics Evaluation (CBER), Center for Devices and Radiological Health (CDRH), and Center for Drug Evaluation and Research (CDER). CBER regulates biological products for human use. Within CBER, the Office of Therapeutic Products (OTP; reorganized from the Office of Tissues and Advances Therapies (OTAT)) oversees products including blood, vaccines, allergenics, tissues, and cellular and gene therapies [11]. CDRH regulates medical devices and radiation-emitting products; while CDER regulates prescription drugs (including monoclonal antibodies for *in vivo* use, proteins intended for therapeutic use, immunomodulators, and growth factors, cytokines, and monoclonal antibodies intended to mobilize, stimulate, decrease or otherwise alter the production of cells *in vivo*), generic drugs, biosimilars, and over-the-counter drugs. For combination products—those comprised of a biological product component with a device and/or drug component—the product is assigned to a Center for review and regulation in accordance with the products' primary mode of action [12]. To determine the classification of an HCT/P, the FDA provides two mechanisms: (1) the Tissue Reference Group (TRG) which provides product sponsors with an informal process which they may obtain an Agency recommendation regarding the application of the criteria in 21 CFR 1271.10(a) to their HCT/Ps for a given indication and (2) a Request for Designation (RFD) may be submitted to the Office of Combination Products (OCP) to obtain a formal Agency decision regarding the regulatory identity or classification of an HCT/P (21 CFR Part 3) [10]. Figure 19.3 summarizes the

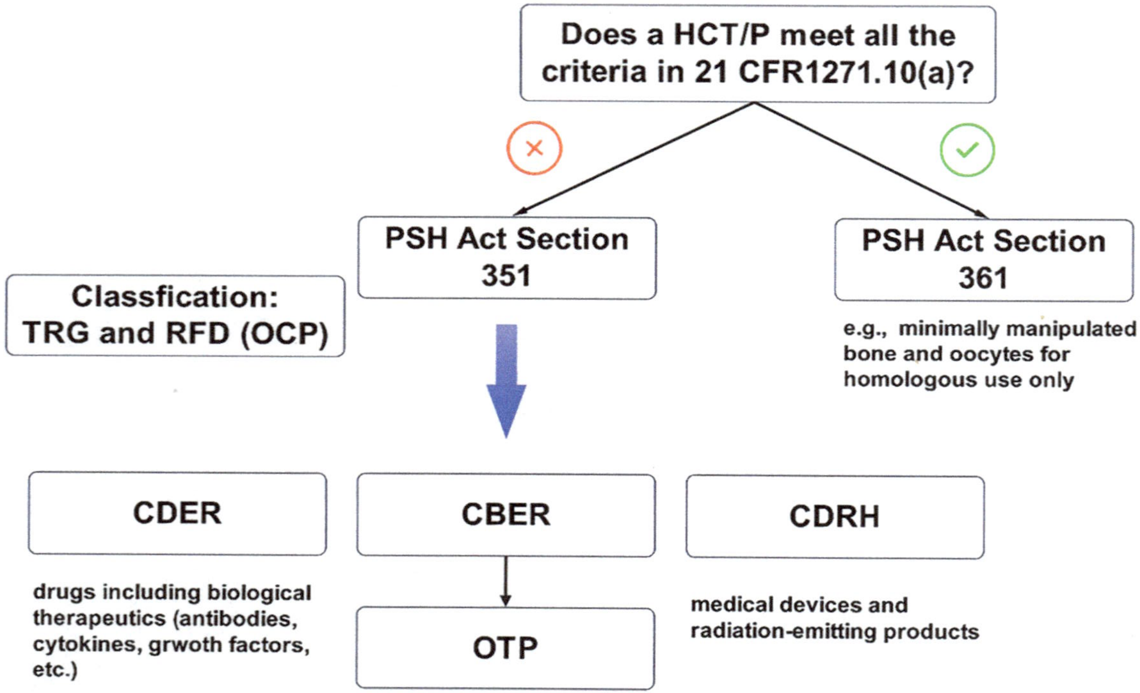

Fig. 19.3 Regulatory framework of HCT/Ps under the PHS (Public Health Service) and the FD&C (Food, Drug, and Cosmetic) Acts. *HCT/Ps* human cells, tissues, and cellular and tissue-based products, *TRG* Tissue Reference Group, *RFD* Request for designation, *OCP* Office of Combination Products, *CBER* Center for Biologics Evaluation, *CDRH* Center for Devices and Radiological Health, *CDER* Center for Drug Evaluation and Research, *OTP* Office of Therapeutic Products

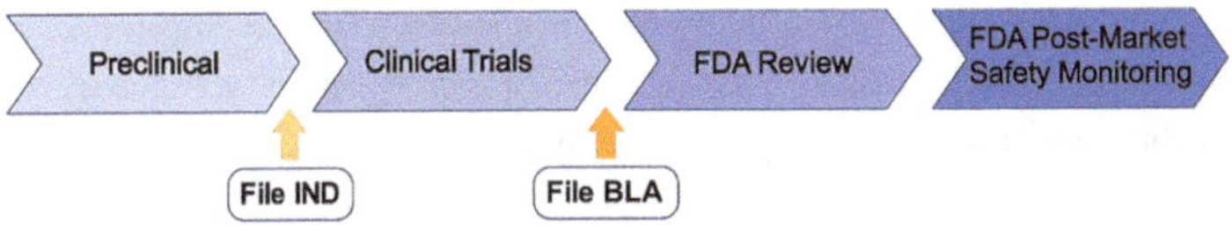

Fig. 19.4 HCT/Ps reviewed by CBER: from clinical studies to market. *IND* Investigational New Drug, *BLA* Biologics License Application

regulatory framework of HCT/Ps under the PHS and FD&C Acts. For HCT/Ps requiring formal FDA approval, such as those reviewed by CBER, the process typically begins with filing an Investigational New Drug (IND) application to initiate clinical trials. If clinical trials are successful, a Biological License Applications (BLA) must be submitted to the FDA to seek marketing approval (Fig. 19.4). FDA-approved cellular and gene therapy products can be accessed at: https://www.fda.gov/vaccines-blood-biologics/cellular-gene-therapy-products/approved-cellular-and-gene-therapy-products. In 2024, the FDA approved Symvess, the first acellular tissue vessel indicated for use in adults as a vascular conduit for extremity arterial injury when urgent revascularization is needed to avoid imminent limb loss, and autologous vein graft is not feasible [13].

When it comes to the regulatory framework in the European Union (EU), the tissue engineered products fall under Advance Therapy Medicinal Products (ATMP) and are regulated by the European Medicines Agency (EMA). According to Regulation (European Commission [EC]) NO. 1394/2007, a tissue-engineered product (TEP) is defined as a product that contains or consists of engineered cells or tissues and that can be used to regenerate, repair, or replace a human tissue. TEPs may be constituted by human/animal cells or tissues—viable or non-viable—and may include other added substances (e.g., cellular products and scaffolds). TEPs combined with medical devices are classified as combined advanced therapy medicinal products (CATMPs) when one or more active or non-active medical devices are integrated into the TEPs and together maintain a single identity. Both TEPs and CATMPs are subjected to a Marketing Authorization Application (MAA) [14]. Examples of approved TEPs in the EU include Holoclar, the first stem cell product approved in the EU, and Spherox—both of which are autologous cell-based products [13, 15, 16]. An analysis found that all TEPs in clinical trials in the EU contain either autologous or allogeneic human cells, with 60% of studies using autologous cells [17].

There are several major challenges for tissue engineered products in clinical trials [18, 19]:

1. An exploratory trial involving a small group of patients may be necessary to establish safety and proof of concept, as a Phase 1 study on healthy volunteers (standard for conventional drugs) is generally not appropriate for most tissue engineered products;

2. The conventional concept of dosing may not apply to tissue engineered products. For example, the maximum dose is not limited by toxicity but factors such as size of the administration site or the number of cells can be incorporated;
3. It can be challenging to define an appropriate comparator (i.e., the currently accepted standard treatment for the specific clinical condition) or control group;
4. Randomization and blinding a trial may be difficult to implement. For instance, a fully double-blind process may not be feasible, and the use of sham-operated controls to minimize the potential for placebo effects may raise ethical concerns;
5. Clinical outcomes may not necessarily reflect the intended therapeutic effects;
6. Patients may prefer similar products already available on the market, and the participation of clinicians and medical students in first-in-human trials is often lower than expected.

To improve the likelihood of regulatory approval, recommended actions include: working closely with regulatory agencies before and during clinical trials and the approval process; clearly defining and understanding the product's potential benefits versus risks; and referencing relevant, previously approved products to strengthen the application [7].

19.2 Ethical Issues in Clinical Translation of Tissue Engineered Products

The ethical concerns surrounding the clinical translation of tissue engineering products can be broadly categorized into three areas: (1) the source of tissues and cells, (2) the donation process, and (3) the manipulation of cells [1]. Regarding the cell sources, the use of human embryonic stem cells and fetal cells, primarily from aborted fetal tissue, remains highly controversial. This controversy stems from different frameworks, such as conceptualism (which views embryos as having the same moral status as newborns) versus gradualism (which considers embryos not to be sentient beings). Although adult stem cells and induced pluripotent stem cells (iPSCs) raise fewer ethical concerns, they do not completely solve ethical problems [1, 20].

In the context of cell and tissue donation, informed consent is critical and must constitute a voluntary decision based on full disclosure of information. Donors should be informed about the potential future applications and uses of their cells, tissues or organs. The privacy of donors must be protected through anonymization of samples to prevent misuse or exploitation, particularly of vulnerable participants [1]. In both transplantation and the research contexts, the altruistic and non-financial nature of the donation is emphasized—donors should not receive financial compensation or share in any future profits [20].

An additional ethical consideration arises in the context of novel therapeutic. Expanded access, or compassionate use, is a potential pathway for a patient with a serious or immediately life-threatening disease or condition to gain access to an investigational medical product (drug, biologic, or medical device) for treatment outside of clinical trials when no comparable or satisfactory alternative therapy options are available. In addition, the Right to Try Act allows patients who have been diagnosed with life-threatening diseases or conditions who have tried all approved treatment options and who are unable to participate in a clinical trial to access certain investigational treatment options. In these cases, the burdens and risks to participants must be carefully weighed against the potential benefits [1, 21, 22]. In 2022, the FDA published a final rule specifying the deadline and content for annual reports by sponsors or manufactures who provided an eligible investigational drug under the Right to Try Act [22, 23].

Review and Practice Questions

1. Use the Business Model Canvas to develop a business model for a tissue-engineered product.
2. Describe the regulatory framework for tissue-engineered products in your country or region.
3. What are the major challenges faced by tissue-engineered products during clinical trials?
4. What are some ethical issues associated with the clinical translation of tissue-engineered products?

References

1. O'Donnell, B.T., et al., *Beyond the Present Constraints That Prevent a Wide Spread of Tissue Engineering and Regenerative Medicine Approaches.* Frontiers in Bioengineering and Biotechnology, 2019. **7**:95.
2. Kern Entrepreneurial Engineering Network (KEEN). *Entrepreneurial Mindset.* Available from: https://engineeringunleashed.com/mindset.
3. U.S. Small Business Administraiton. *About SBIR and STTR.* Available from: https://www.sbir.gov/about.
4. Osterwalder, A., Y. Pigneur, and T. Clark, *Business Model Generation: A Handbook for Visionaries, Game Changers, and Challengers*. 2010: Wiley.
5. Strategyzer. *The Bussiness Model Canvas*. Available from: https://www.strategyzer.com/library/the-business-model-canvas.
6. Strategyzer. *Users vs customers*. Available from: https://www.strategyzer.com/library/users-vs-customers.
7. Yusuf, S., E. Hay, and E. Sansoucy, *The commercialization of tissue engineered medical products: Challenges and recommendations.* World Journal of Advanced Research and Reviews, 2024. **23**:1203-1212.
8. U.S. Food and Drug Administration. *Standards Development for Regenerative Medicine Therapies.* Available from: https://www.fda.gov/vaccines-blood-biologics/standards-development-regenerative-medicine-therapies.
9. U.S. Food and Drug Administration. *Guidances Related to the Regulation of Regenerative Medicine Products and Human Cells, Tissues, and Cellular and Tissue-Based Products (HCT/Ps) - Recording*. Available from: https://collaboration.fda.gov/p81luow1w03/.
10. U.S. Food and Drug Administration. *Regulatory Considerations for Human Cells, Tissues, and Cellular and Tissue-Based Products: Minimal Manipulation and Homologous Use* Available from: https://www.fda.gov/regulatory-information/search-fda-guidance-

documents/regulatory-considerations-human-cells-tissues-and-cellular-and-tissue-based-products-minimal.
11. U.S. Food and Drug Administration. *Establishment of the Office of Therapeutic Products.* Available from: https://www.fda.gov/vaccines-blood-biologics/cellular-gene-therapy-products/establishment-office-therapeutic-products.
12. U.S. Food and Drug Administration. *Transfer of Therapeutic Biological Products to the Center for Drug Evaluation and Research.* Available from: https://www.fda.gov/combination-products/jurisdictional-information/transfer-therapeutic-biological-products-center-drug-evaluation-and-research.
13. Pellegrini, G., et al., *Navigating Market Authorization: The Path Holoclar Took to Become the First Stem Cell Product Approved in the European Union.* STEM CELLS Translational Medicine, 2018. **7**(1):146-154.
14. Oberweis, C.V., et al., *A Worldwide Overview of Regulatory Frameworks for Tissue-Based Products.* Tissue Engineering Part B: Reviews, 2020. **26**(2):181-196.
15. Gerth, A., C.H. Opitz, and K. Lee, *Manufacturing, Approval and Market Launch of Tissue Engineered Products in Europe.* Cytotherapy, 2024. **26**(6, Supplement):S107-S108.
16. Bellino, S., et al., *Cell-based medicinal products approved in the European Union: current evidence and perspectives.* Frontiers in Pharmacology, 2023. **14**:1200808.
17. Joyce, K., et al., *Issues with Tissues: Trends in Tissue-Engineered Products in Clinical Trials in the European Union.* Tissue Engineering Part B: Reviews, 2023. **29**(1):78-88.
18. Joly, J., et al., *Chapter 21 - Clinical translation*, in *Tissue Engineering (Third Edition)*, J. De Boer, et al., Editors. 2023: Academic.
19. da Silva, R.G.L., L. Au, and A. Blasimme, *Organizational aspects of tissue engineering clinical translation: insights from a qualitative case study.* Translational Medicine Communications, 2024. **9**(1):17.
20. Welin, S., *Chapter 23 - Ethical Issues in Tissue Engineering*, in *Tissue Engineering (Second Edition)*, C.A.V. Blitterswijk and J. De Boer, Editors. 2014: Academic Press.
21. U.S. Food and Drug Administration. *Expanded Access.* Available from: https://www.fda.gov/news-events/public-health-focus/expanded-access.
22. U.S. Food and Drug Administration. *Right to Try.* Available from: https://www.fda.gov/patients/learn-about-expanded-access-and-other-treatment-options/right-try.
23. U.S. Food and Drug Administration. *Annual Summary Reporting Requirements Under the Right to Try Act.* Available from: https://www.federalregister.gov/documents/2022/09/14/2022-19737/annual-summary-reporting-requirements-under-the-right-to-try-act.

Part V

Cell Culture and Tissue Engineering Lab Activities

Cell Culture-Based Lab Activities 20

20.1 Lab Activity I: BSL(Biosafety Level)-2 Lab Training

Learning Objectives

1. Understand the fundamentals of biosafety level 2 (BSL-2) laboratories
2. Learn how to work safely within a BSL-2 laboratory, including the proper use of personal protective equipment, waste disposal procedures, and sterile techniques

20.1.1 Introduction to BSL-2 Laboratories

Both the NIH guidelines [1] and the WHO's Laboratory Safety Manual [2] recommended four Biosafety levels, 1–4 (ascending levels of containment), for handling organisms corresponding to the four risk groups. Biosafety level 1 (BSL-1) is suitable for work involving well-characterized agents not known to consistently cause disease in immunocompetent adult humans, and that present a minimal potential hazard to laboratory personnel and the environment. Biosafety level 2 (BSL-2) builds upon BSL-1. BSL-2 is suitable for work with agents associated with human disease and pose moderate hazards to personnel and the environment [1]. Major differences between BSL-1 and BSL-2 are summarized in Table 20.1 and Fig. 20.1. Any BSL-1 or above lab must have a sign incorporating the universal biohazard symbol at the entrance (Fig. 20.2) [3]. It is important to point out the four Biosafety Levels are not the same as Agent Risk Groups [1]. For example, the RG2 dengue virus, used for animal inoculation or animal transmission studies, a higher containment level (than BSL-2) is recommended [4].

W. Zhang, *Cell Culture and Tissue Engineering*,
https://doi.org/10.1007/978-3-032-32096-4_20

Table 20.1 Major differences between BSL-1 vs. BSL-2[a]

	BSL-1	BSL-2
Laboratory type	Basic teaching; research	Primary health services; diagnostic services; research
Laboratory practices	Good microbiological practice and procedure	Good microbiological practice and procedure; protective clothing and gloves
Safety equipment and facilities	Open lab bench	Open lab bench; biosafety cabinet (BSC) for potential aerosols; autoclave
Agents used	Not known to consistently cause disease in healthy adults	Associated with human disease, hazard from percutaneous injury, ingestion, mucous membrane exposure
Laboratory access		Restricted; specific training in handling pathogenic agents

[a]Information from [1, 5, 6]

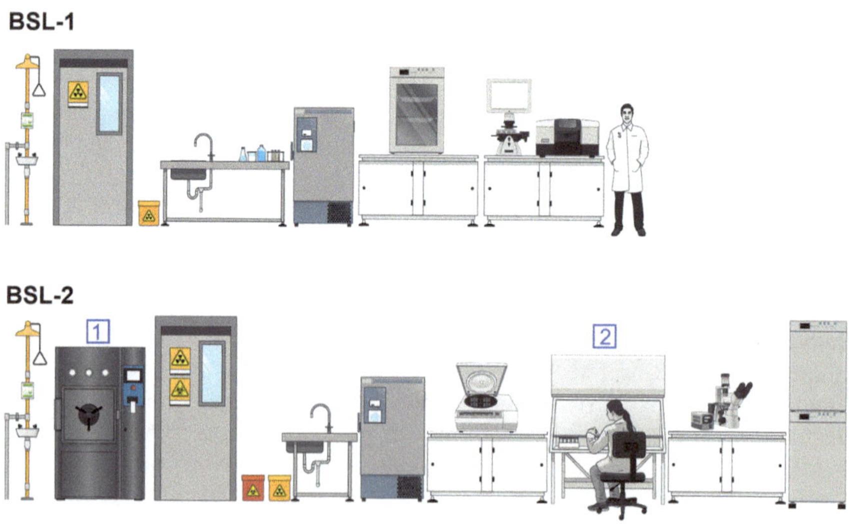

Fig. 20.1 A typical BSL-1 (top) and BSL-2 (bottom) laboratory. 1: autoclave; 2: biosafety cabinet

Fig. 20.2 The biological hazard symbol

20.1.2 Safety Equipment and Engineering Controls

Basic safety equipment, including an eyewash, emergency shower, spill kit, first aid kit, and fire extinguisher/blanket, must be readily available in all lab environments. The path to these must remain unobstructed at all times. It is also important to be aware of the location of the nearest telephone, emergency contact numbers, and the emergency exit [3].

Engineering controls are risk control measures that are built into the design of a laboratory or laboratory equipment to contain the hazards. Biological safety cabinets (BSCs) and isolators are forms of engineering control in order to minimize the risk of exposure to and/or unintended release of biological agents [7]. Other general engineering controls include negative laboratory airflow, accessible sinks for hand-washing, and vacuum lines equipped with filters and traps on laboratory benches [2].

20.1.3 Personal Protective Equipment (PPE)

Personal protective equipment (PPE) helps protect the user's body from body injury from a variety of sources or potential exposure to biological hazards and airborne particulate matter. PPE, typically used in conjunction with other biosafety controls (e.g., BSCs and centrifuge biosafety cups), includes gloves (disposable gloves, cryo-gloves, autoclave gloves, etc.), lab coats or gowns, shoe covers, closed-toe laboratory footwear, respirators, face shields, safety glasses, goggles, or ear plugs [1]. It is important to know when to change PPE, such as replacing gloves when they become soiled or contaminated.

20.1.4 Waste Disposal

Waste generated in cell culture and tissue engineering laboratories includes biohazardous, chemical, and nonhazardous waste. Biohazardous waste includes materials of biological origin, such as tissues, cells, or serum, as well as any solids or liquids that have come into contact with these materials. Chemical waste consists of solids and liquids containing one or more components that may pose health, fire, environmental, or other hazards. It is critical to follow all institutional, local, and federal guidelines for the proper collection and disposal of laboratory waste. All waste containers must be clearly and accurately labeled. If you are unsure how to collect, dispose of, or label any type of waste, consult your instructor. Material Safety Data Sheets (MSDS) should be reviewed to understand the specific hazards associated with any chemical or reagent used in the laboratory [3].

Biohazardous liquid waste should be collected in clearly labeled containers. It must be treated with 10% bleach for at least 30 min before being safely poured down the drain with running water. Biohazardous solid waste should be double-bagged and autoclaved at 121 °C for 60 min prior to disposal. Chemical waste must

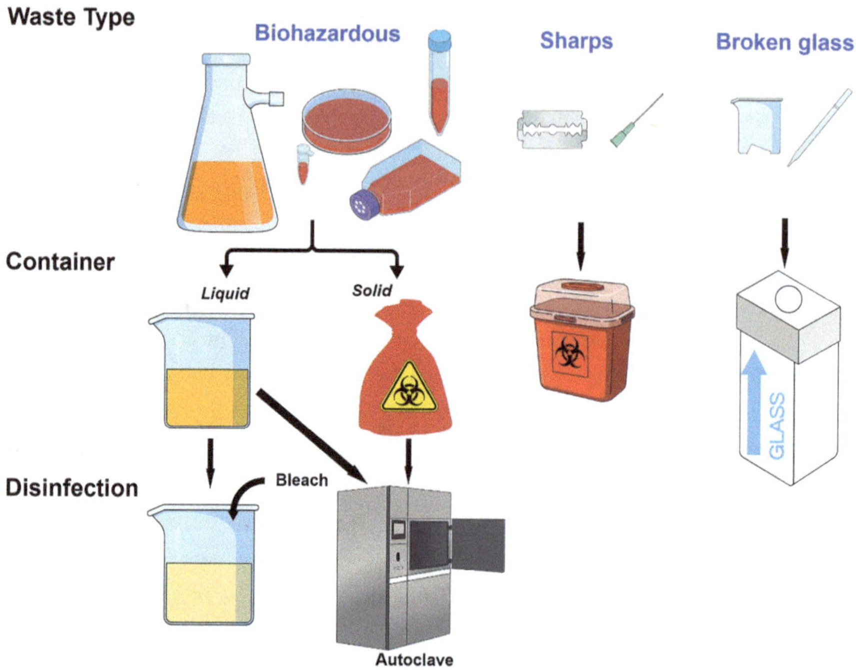

Fig. 20.3 Overview of waste collection and disposal methods in cell culture and tissue engineering laboratories

be collected in properly labeled, designated containers in accordance with institutional guidelines. Whenever possible, avoid mixing liquid and solid chemical wastes. These containers should be stored in a designated area within the laboratory and disposed of following the appropriate institutional procedures. It is important to properly handle any residual chemicals or biohazardous materials on laboratory glassware, including Pasteur pipettes, before cleaning or disposal. Contaminated sharps (e.g., needles, razor blades) must be collected in puncture-proof containers with secure lids and treated as biohazardous waste. For mixed waste, such as cell suspensions containing Trypan Blue used for cell counting, the biological component should be inactivated first. Once the biological agent is no longer viable, the waste can then be treated and disposed of as chemical waste only. Figure 20.3 summarizes the waste collection and disposal.

20.1.5 Disinfection

Disinfection is a process to eliminate viable biological agents from items or surfaces for further safe handling or use. Disinfectants, such as 70% ethanol or 10% bleach are usually used for decontamination of surfaces (such as BSCs) before and after work and after a spill. Disinfectants can also be used for decontamination of contaminated liquids.

20.1.6 Operation of BSCs

There are three kinds of BSCs, Class I, II, and III. Class II and III cabinets provide protection for the operator, product, and environment. High-Efficiency Particulate Air (HEPA) filters are mostly used in the exhaust and supply systems in BSCs. These filters are capable of removing the Most Penetrating Particle Size (MPPS), approximately 0.3 μm, with minimum efficiency of 99.99%. Class II BSCs, which include Types A1, A2, B1, B2, and C1, are widely used in BSL-2 laboratories. Among them, type A2 is one of the most commonly models used BSCs in laboratory facilities [1, 7, 8]. As shown in Fig. 20.4, types A1 and A2 cabinets recirculate approximately 70% of the air back into the work area through HEPA filters. The primary difference between types A1 and A2 is the inflow velocity. Type B2 cabinets (no air is recycled) operate with a single pass airflow system, while type B1 cabinets recirculate about 30% of the air. Type C1, the most recent addition to Class II BSCs, combines the functionalities of both Types A and B cabinets. It offers enhanced safety with low energy costs [9].

When operating a BSC, turn on the blower and allow the airflow to equilibrate for 15–30 min before beginning work. The UV lamp should be used only when no one is present in the room. All items must be disinfected with 70% ethanol before placing them into the BSC's clean zone (opposite of the dirty zone, where the aspirator system is located). Keep your work zone, which is in the middle (between clean and dirty zones) clear and unobstructed. Do not rest your elbows or place any

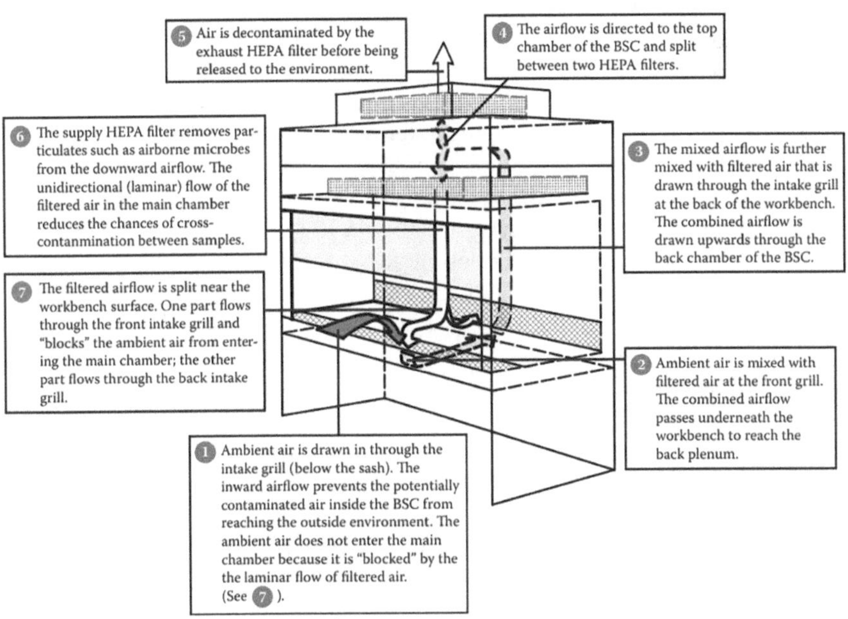

Fig. 20.4 Typical air-flow scheme in a Class II biosafety cabinet. (Adapted from [8] with permission from Taylor & Francis Group)

objects on the air intake grill as this can block airflow and increase the risk of contamination [8].

To summarize, the strategies for risk reduction include elimination (removing the hazard), reduction and substitution (such as substitution with a less infectious biological agent and reduction of the volume used), isolation (isolating the hazard to prevent exposure), protection (such as protecting personnel/environment by using engineering controls and PPE), and compliance (such as ensuring appropriate training and adherence to clearly written standard operating procedures (SOPs)) [7].

20.2 Lab Activity II: Cell Passaging and Cell Counting

Learning Objectives

1. Learn and perform adherent cell culture techniques, including cell medium preparation and cell passaging
2. Learn and perform cell counting using a hemacytometer

20.2.1 Introduction

This lab focuses on adherent cell culture, which is generally more challenging than suspension cell culture. Suspension cell culture is generally straightforward, especially for those who are already familiar with adherent cell culture techniques. Most adherent cells grow as a monolayer during *in vitro* culture, and passaging is required once the culture reaches 80–90% confluency (the degree of cell culture vessel coverage by cells). At this stage, the cells experience "contact inhibition," causing them to stop dividing and eventually die. Moreover, confluent cultures often exhibit changes in gene regulation and expression [8, 10, 11].

Adherent cells are first detached from the growth surface during passaging, typically using a protease solution such as trypsin-EDTA (ethylenediaminetetraacetic acid), which breaks down adhesion proteins. Once detached, the cells can be diluted or divided for continued culture.

20.2.2 Materials and Reagents

Cell culture medium reagents (base, serum, and penicillin/streptomycin (P/S) solution)
Trypsin-ethylenediaminetetraacetic acid (EDTA)
0.4% Trypan Blue solution (e.g., 15,250,061; Thermo Fisher Scientific)
Hemacytometer
Inverted microscope

20.2.3 Procedure

I. *Prepare cell culture medium*

1. Review the formulation of the cell culture medium and determine if any components need to be added.
2. Prepare the base medium, serum, and any required supplements inside the BSC, ensuring that outside of each container is wiped with 70% ethanol before placing them inside.

Note: Since the serum is heat-sensitive, it is recommended to divide it into single-use aliquots upon receipt to avoid repeated freeze-thaw cycles and thaw only the amount needed.

3. Prepare the serological pipettes you need under the BSC.
4. Once all components are ready, open the medium container. Leave the cap loosely placed on the bottle to avoid passing over the opening while working.
5. Draw up the basic medium with a serological pipette. Carefully lift the cap with one hand and, with other hand holding the pipette, dispense the medium into the medium container.
6. Add serum (generally 10% of the complete medium volume) and 100× penicillin/streptomycin (P/S) solution (generally 1% of the complete medium volume).

Sample calculation: 150 mL of complete culture medium = 133.5 mL basic medium + 15 mL serum + 1.5 mL P/S solution.

Note: Serum percentages can vary depending on the cell type. Antibiotics are sometimes omitted, as they may alter cellular behavior—for example, by causing mitochondrial damage.

7. Store the complete medium at 4 °C.

II. *Cell passaging*

1. Vent and wipe down the BSC. Aseptically transfer all labeled items into the BSC and loosen all caps.
2. Examine the cells under the microscope and note the cell morphology and culture confluency (Fig. 20.5).
3. Optional: Warm 0.25% trypsin-EDTA and complete media in a 37 °C water bath (recommended media and trypsin-EDTA volumes are shown in Table 20.2).
4. Aspirate the spent medium from the culture vessel with a Pasteur pipette (position the pipette at one of the bottom corners to remove the medium without disturbing the cell monolayer (Fig. 20.6)).
5. Rinse the cell monolayer with Dulbecco's phosphate-buffered saline (DPBS).
6. Carefully aspirate and discard this rinse solution.
7. Gently add the trypsin-EDTA solution to one corner of the flask, "rock" the flask gently to ensure the entire surface is covered and then place the flask in the incubator.

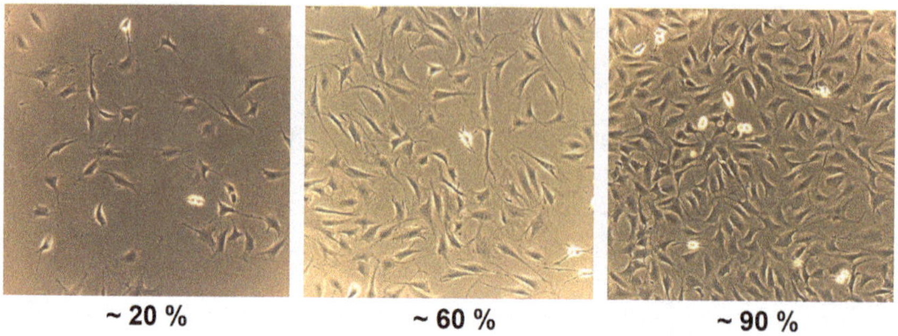

Fig. 20.5 Phase-contrast microscopic images of C3H10T1/2 cells at varying levels of confluency

Table 20.2 Approximate growth area and recommended volumes of media and trypsin-EDTA for common tissue culture flasks, dishes, and plates*

Culture Vessel	Approximate Growth Area (cm^2)	Recommended Media Volume (mL)	Recommended Trypsin-EDTA Volume (mL)
Flask			
T-25	25	3–5	1–3
T-75	75	15	3–5
Dish			
3.5 cm	9	2	0.5–1
6.0 cm	21	5	1–2
10.0 cm	55	12	2–5
Plate			
6-well	9.5	2	0.5–1
12-well	3.8	0.8	0.2–0.4
24-well	1.9	0.4	0.1–0.2

*Data are from [10, 12, 13]

Fig. 20.6 Recommended technique for aspirating media or other liquids without disrupting adherent cells

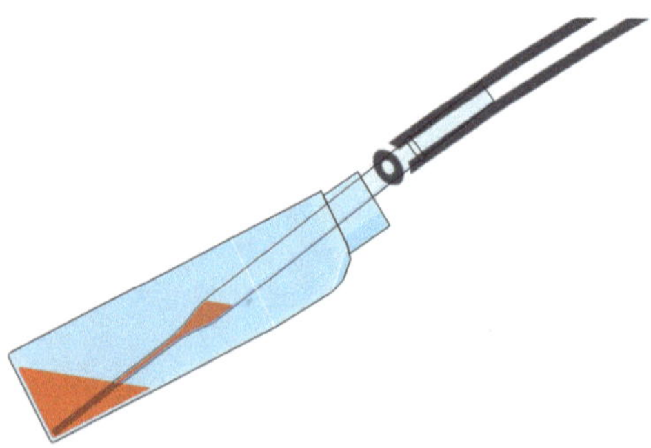

8. Check on the cells after 3 min under the microscope to confirm detachment, and gently "rock" flask and tap on the side of the flask to release the cells.
9. Return the flask to the incubator for a few additional minutes if needed.

Note: Avoid prolonged exposure of cells to trypsin as it can alter membrane proteins and cytoskeleton, change the composition of cytoplasm, and impair cell viability [14, 15]. EDTA, a calcium chelator, removes calcium ions that are required for cell adhesion (Fig. 20.7) [15].

10. Add enough complete media (generally 5 times the volume of the trypsin-EDTA) to neutralize the trypsin.
11. Thoroughly rinse the bottom and corners of the flask to ensure all cells are released.
12. Pipette the cell suspension up and down 10–20 times to break up potential clumps using a 10-mL serological pipette and then transfer the suspension to a centrifuge tube.
13. Centrifuge at 1000 rpm for 3 min.
14. Proceed to split, concentrate, or use the cells as desired.

Note: Split the cells by distributing the cell suspension into multiple culture vessels, then adding complete medium to each vessel to reach the desired working volume. A typical split ratio ranges from 1:5 to 1:10 depending on the cell growth rate. For example, a 1:10 split means plating the same number of cells over a surface area ten times larger than before (e.g., cells from one T75 flask can be replated into ten T75 flasks). For routine maintenance, it is usually sufficient to keep only 1–2 flasks of cells, and any excess cells may be discarded.

III. *Refreshing cell cultures/changing media*

1. Vent and wipe down the biosafety cabinet. Aseptically transfer the labeled items inside the BSC and loosen all caps.
2. Optional: Warm the complete media in a 37 °C water bath.
3. Examine the cells under a microscope, noting cell morphology and confluency.
4. Use a Pasteur pipette to carefully aspirate the old medium.
5. Add fresh medium to the flask gently (avoid splashing onto the cell growth surface to prevent dislodging cells).
6. Label the flask with the date of medium change and your initials.
7. Return the flask to the incubator.

IV. *Cell counting using a hemacytometer*

While counting cells with a hemacytometer, it is also possible to estimate cell viability using Trypan blue stain (Fig. 20.7). Healthy cells with intact membranes exclude Trypan blue, whereas damaged or dead cells with compromised plasma membranes cannot exclude the dye and appear blue (Fig. 20.8) [8, 10].

Note: Trypan blue is carcinogenic and can be absorbed through the skin. Always wear gloves when handling Trypan blue and dispose of waste in a designated waste container.

1. Clean the hemacytometer and the cover glass (or coverslip) with 70% ethanol, then wipe with a lint-free tissue or dry using compressed air.
2. Mix the cell suspension with Trypan blue:

    (a) In the BSC, mix the cell suspension thoroughly to break up any cell clumps. Collect a small amount (e.g., 25–100 μL) of cell suspension in a microfuge tube.
    (b) Take the microfuge tubes to the lab bench (sterility is not required at this stage since the sample will be discarded after counting).
    (c) Mix the cells with 0.4% Trypan blue solution at a 1:1 volume ratio.

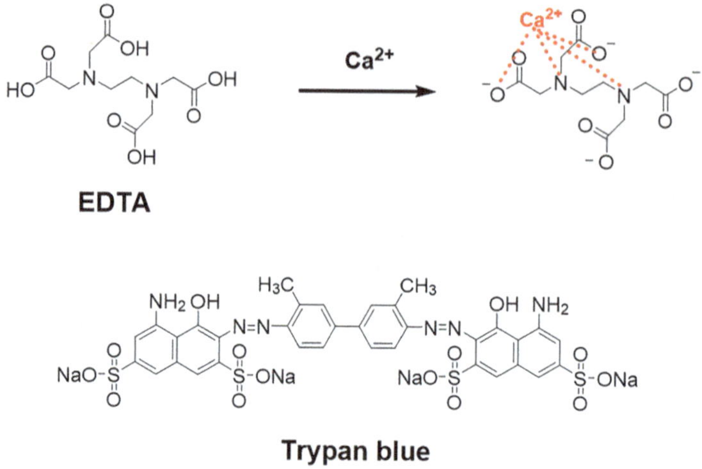

Fig. 20.7 Chemical structure of Ethylenediaminetetraacetic acid (EDTA) and Trypan blue

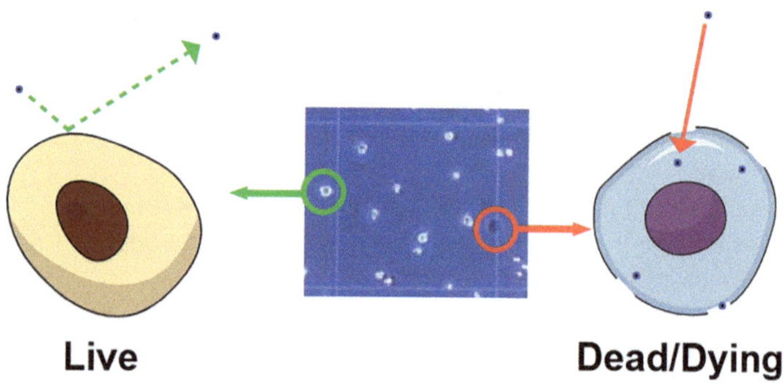

Fig. 20.8 Trypan blue stain mechanism for differentiating live and dead/dying cells

Note: Depending upon the sample and instrumentation, the mixing ratio of 0.4% Trypan Blue solution to the cell suspension can range from 1: 10 to 1:1 (v/v).

3. Load the cell/Trypan blue mixture into both chambers of the hemacytometer:

   (a) Mix the cells with Trypan blue thoroughly by pipetting up and down several times.
   (b) Draw 15–25 μL of the sample into a pipette.
   (c) Rest the pipette tip at the edge of one of the notches on the hemacytometer (Fig. 20.9).
   (d) Gently depress the pipette's plunger to allow the mixture to flow into the chamber by capillary action. Dispense enough to cover the grid surface but do not overfill.
   (e) Repeat steps a-d to load the opposite chamber.

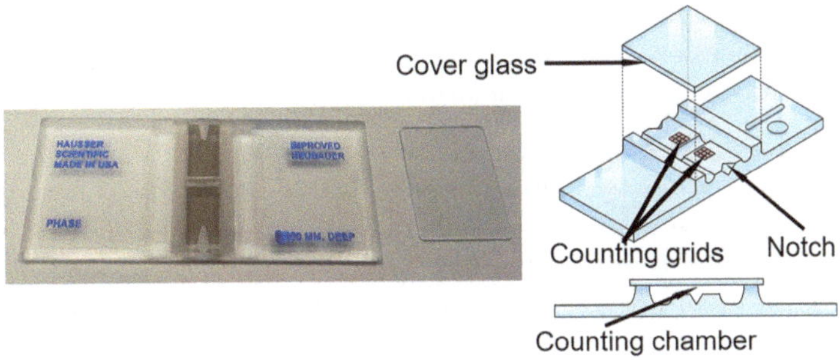

Fig. 20.9 Structure of a hemacytometer. Left: a photo of a hemacytometer. Right: different views and components of a hemacytometer

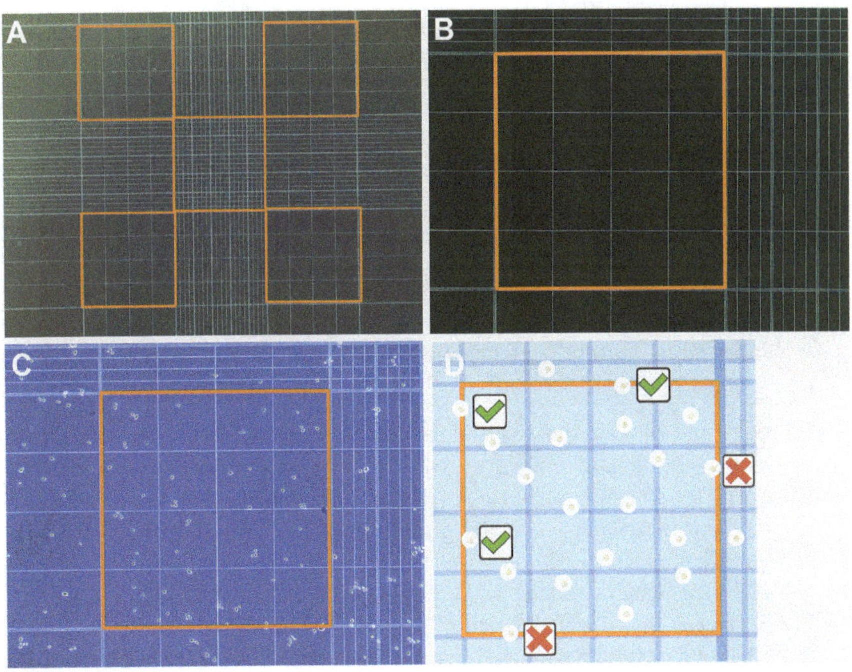

Fig. 20.10 (**a**): Phase-contrast microscopic image of the counting grid (orange: 'major' squares used for counting); (**b**): Phase-contrast microscopic image of a single 'major' square; (**c**): Phase-contrast microscopic image of a sample with high cell viability (cell suspension to Trypan blue volume ratio of 1:1)

4. Place the hemacytometer on the microscope stage.
5. Using a low-powered objective (e.g., 10x), adjust the focus and hemacytometer position so that a grid pattern and cells are clearly visible (Fig. 20.10).
6. Count and record the number of live and dead/dying cells in the "major" squares of the hemacytometer grid (Fig. 20.10a) if the cell count is too low

(e.g., < 10 cells/square), then the sample may not accurately represent the cell suspension. If the cell count is too high (e.g., > a few hundred cells/square), it may be difficult to distinguish individual cells).

Note: If the values obtained from the two grids vary significantly (>10%), the original cell suspension may not be evenly mixed and should be recounted. The process should also be repeated if there is an abundance of clumped cells visible in the grid. If cells touch the boundary lines, count those on the top and left sides of the square, but exclude those on the bottom and right sides (Fig. 20.10d) [8, 10].

7. Calculate the average number of cells per major square.
8. Calculate the percent viability of the cell using the equation:

$$\text{Cell viability}(\%) = \frac{number\ of\ live\ cells}{total\ number\ of\ cells} \times 100\%$$
$$= \frac{number\ of\ live\ cells}{number\ of\ live\ cells + number\ of\ dead\ cells} \times 100\%.$$

9. Calculate the cell density (unit: cells/mL) of the original suspension, accounting for dilution with Trypan blue. Use the equation:

$$(Live)\ Cell\ density = average\ number\ of\ live\ cells\ (per\ major\ square) \times dilution\ factor \times 10^4$$

Notes: Each major square (i.e., one set of small squares) of the hemacytometer represents a volume of 0.1 mm^3 or 10^{-4} mL.

10. Carefully remove the cover glass and hold it over the designated Trypan blue waste container. Spray it with 70% ethanol to rinse off residue dye.
11. Spray the hemacytometer chamber with 70% ethanol. Gently clean both the hemacytometer and cover glass using a mild detergent to remove cellular debris. Rinse with tap water followed by deionized water.
12. Place the hemacytometer and cover glass on a clean lab tissue to air dry.
13. Rinse the microcentrifuge tube used for Trypan blue staining with 70% ethanol (over the designated Trypan blue waste container) and discard the tube in the biohazardous waste bin. The complete cell passaging process, including cell counting, is illustrated in Fig. 20.11.

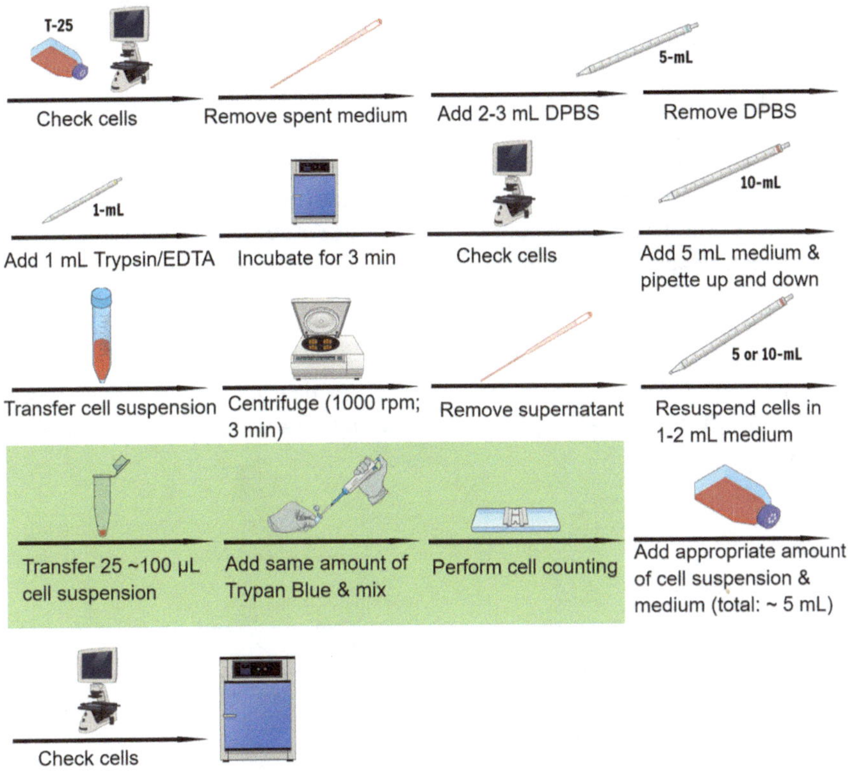

Fig. 20.11 Typical cell passaging process including cell counting

20.3 Lab Activity III: Cell Cryopreservation and Thawing

Learning Objectives

1. Learn and perform cell freezing and thawing techniques
2. Understand and evaluate cell cryopreservation outcomes, including cell viability and recovery rate

20.3.1 Introduction

Cells can be stored long-term in liquid nitrogen following a proper freezing process. For freezing, adherent cells must first be detached, while suspension cells are collected directly. The cells are then resuspended in a freezing medium at the desired concentration, typically around 1×10^6 cells/mL [8, 10]. Depending on specific needs and available resources, various approaches and devices can be used. Cost-effective tools, such as the Mr. Frosty™ freezing container, are commonly employed to achieve a controlled cooling rate of ~1 °C per minute [16]. A cryoprotectant, commonly dimethyl sulfoxide (DMSO), is included in the freezing medium to minimize cell damage during cryopreservation. Although cellular biochemical activities are largely suppressed during cryopreservation at low temperatures, cells may still occur damage due to mechanical and osmotic stress from ice crystallization, as well as the cytotoxic effects of cryoprotectant(s) [17].

Before use, cryopreserved cells must be thawed and plated into appropriate cell/tissue culture vessels. The most common and straightforward method for thawing is immersion in a water bath. The process should be performed rapidly to minimize cellular damage, particularly that caused by ice recrystallization [17]. A standard freezing and thawing procedure for adherent cells using a commercial freezing container is illustrated in Fig. 20.12. When assessing cell cryopreservation outcomes, cell recovery rate is a critical parameter alongside cell viability and function, as relying solely on viability can lead to false-positive results [18].

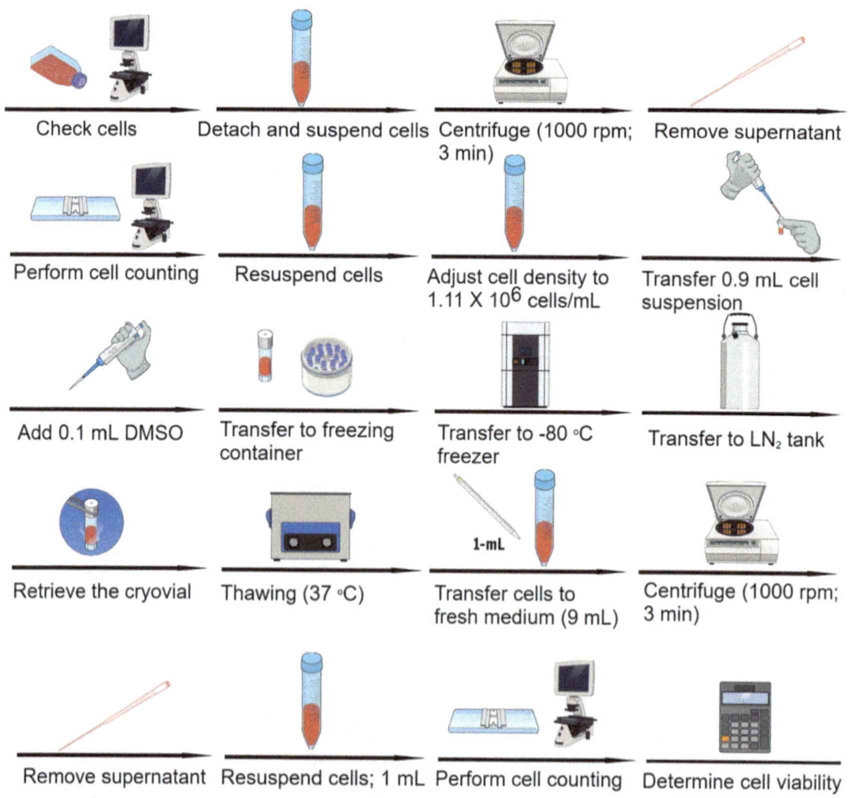

Fig. 20.12 Typical steps for freezing and thawing of adherent cells using a commercial freezing container

20.3.2 Materials and Reagents

Cell culture medium
Trypsin-ethylenediaminetetraacetic acid (EDTA)
Dulbecco's phosphate-buffered saline (DPBS)
Dimethyl sulfoxide (DMSO; e.g., 4-X™; ATCC)
0.4% Trypan Blue solution (e.g., 15250061; Thermo Fisher Scientific)
Cryogenic vials
Freezing container (e.g., Mr. Frosty™ freezing container; Thermo Fisher Scientific)
−80 °C freezer
Hemacytometer
Inverted microscope

20.3.3 Procedure

I. *Cell cryopreservation*

1. Preparation:
   (a) Vent and wipe down the biosafety cabinet (BSC) with 70% ethanol. Aseptically move the necessary items into the BSC;
   (b) Label two 15-mL centrifuge tubes with your group name. Label the cryogenic vial with the cell name, your group name, and today's date (loosen the cap);
   (c) Observe the cell culture under a microscope and note cell morphology and confluency.
2. Trypsinizing and suspending the cells (refer to Lab Activity II):
   (a) Aspirate the old medium and rinse the cells with DPBS.
   (b) Aspirate the DPBS and add trypsin-EDTA solution to cover the cell surface. Incubate for 3 min at 37 °C. Check under a microscope to confirm cell detachment.
   (c) Add complete media to neutralize the trypsin and suspend the cells.
   (d) Transfer the cell suspension into a 15-mL centrifuge tube.
3. Centrifuge the cells (1000 rpm for 3 min) and remove the supernatant (be very careful not to aspirate off the cell pellet).
4. Resuspend cells in 1.5 mL fresh medium* (reduce the volume if the cell confluency is low).
5. Transfer 25–100 μL cell suspension to a microcentrifuge tube for cell counting and cell viability determination (follow the procedure from Lab Activity II).
6. Transfer 1 mL of the cell suspension to the second 15 mL centrifuge tube.
7. Calculate the volume of medium needed to adjust the cell density to 1.11×10^6 cells/mL.
8. Transfer 0.9 mL of the adjusted cell suspension to the labeled cryogenic vial using a mechanical pipette.

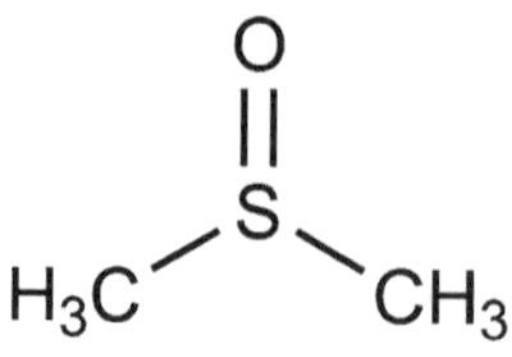

Fig. 20.13 Chemical structure of dimethyl sulfoxide (DMSO)

9. Add 0.1 mL** of DMSO (Fig. 20.13) to the cryogenic vial gently in the dark (final cell density is 1.0×10^6 cells/mL).
10. Freezing the cells:
    (a) Place the cryogenic vial in a cell-freezing container freshly filled with 100% isopropyl alcohol.
    (b) Transfer the container to a −80 °C freezer.
    (c) After ~24 h, move the cryogenic vials to a liquid nitrogen tank for long-term storage.
11. Finishing up: Clean the hemacytometer/cover glass and properly discard all disposable materials; store all reagents appropriately; wipe down the BSC with 70% alcohol.

II. *Cell thawing*

1. Preparation: Vent and wipe down the BSC with 70% ethanol. Aseptically transfer all required materials into the BSC.
2. With your instructor's assistance, retrieve a vial of frozen cells from the liquid nitrogen storage tank.
3. Quickly thaw the frozen culture in a 37 °C water bath (swirl it gently in a circular motion but keep the vial's cap above the water level to avoid contamination).
4. As soon as the ice inside the vial has completely melted, place the vial on ice briefly to allow it to cool for a few seconds.
5. Spray and wipe the vial with 70% ethanol and then move it into the BSC.
6. Loosen the caps of the vial and the 15-mL centrifuge tube.
7. Using a 1-mL serological pipette, gently pipette contents of the vial up and down to resuspend the cells and then transfer the entire cell suspension into the 15-mL centrifuge tube with 9 mL of complete medium (*do NOT pre-warm*) to the cell. To minimize the shear stress, hold the pipette tip against the tube wall so the liquid flows down gently).
8. Place the 15-mL centrifuge tube and a balance tube in a clinical centrifuge and run it at 1000 rpm for 3 min.
9. Use a Pasteur pipette, carefully aspirate the supernatant without disturbing the cell pellet.
10. Add 1 mL complete medium to the tube and gently pipette up and down to fully resuspend the cells (make sure that there are no visible clumps).
11. Pipette 25–100 μL of the cell suspension into a microcentrifuge tube and mix it with an equal volume of 0.4% Trypan Blue for cell counting and viability determination.

12. Determine the cell recovery rate [18]:

$$\textit{Cell recovery rate} = \frac{\textit{number of live cells after cyropreservation}}{\textit{number of cells before cryopreservation}} \times 100\%$$

13. Finishing up: Discard the cryogenic vial and the 15-mL tube into the biohazard waste. Return the complete medium to the 4 °C refrigerator. Wipe down the BSC thoroughly with 70% ethanol.

* *The optimal FBS concentration is cell type-dependent.*

** *The optimal DMSO concentration is cell type-dependent; for example, some studies have shown that reducing the DMSO concentration from 10% to 5% can improve cryopreservation outcomes for certain cell types [19].*

20.4 Lab IV: Cell Engineering—Delivery of Nucleic Acids into Cells[1]

Learning Objectives

1. Learn how to deliver nucleic acid into mammalian cells using cationic lipids (transfection)
2. Perform the cell transfection and analyze the transfection efficiency

20.4.1 Introduction

Nucleic acid-based therapies are becoming more and more common, such as mRNA-based vaccines. For example, small interfering RNA (siRNA) can be used to silence the proteins that inhibit tissue engineering processes [20, 21]. There are multiple methods for transfection, such as electroporation, viral transfection, calcium phosphate precipitation, and cationic polymer/lipid transfection. Cationic lipids have been successfully used for many cell types and some intact tissues. This approach is of relatively high efficiency and low toxicity [22].

Lipofectamine Transfection Reagent (cationic lipid) is used in this lab to deliver siRNA (BLOCK-iT Alexa Fluor Red Fluorescent). The BLOCK-iT Alexa Fluor Red Fluorescent Control is an ideal indicator of lipid-mediated transfection efficiency for RNA interference (RNAi) experiments, particularly those using the Lipofectamine RNAiMAX Transfection Reagent. The sequence of the BLOCK-iT Alexa Fluor Red Fluorescent Control is not homologous to any known gene, preventing the induction of nonspecific cellular events caused by introducing the duplex into cells [23, 24].

The mechanism of action of siRNA is based on post-transcriptional gene silencing. siRNA molecules are usually specific and efficient in the knockdown of

[1] Modified with permission from Zhang, W. *Lab Activity I: Delivery of nucleic acids into cells using cationic lipids.*, in *Fundamentals and Design of Drug Delivery Systems: A Textbook with Lab Activities*. 2026, Academic Press.

disease-related genes. However, they are susceptible to nuclease-mediated degradation with low cellular uptake [25].

20.4.2 Materials and Reagents

24-well plates
Lipofectamine ™ RNAiMAX Transfection Regent (e.g., 13778030; Thermo Fisher Scientific)
BLOCK-iT™ Alexa® Fluor Red Fluorescent Control (e.g., 14750100; Thermo Fisher Scientific)
Hoechst 33342 stain (e.g., 62249; Thermo Fisher Scientific)
Cell culture reagents
Serum-free cell culture medium
Dulbecco phosphate-buffered saline (DPBS)
Fluorescence microscope

20.4.3 Procedure

1. Pre-lab: Seed cells to be 80% to 90% confluent at transfection using a 24-well plate (Fig. 20.14).

Note: An additional experimental group can be included if needed—cells + serum-free medium + siRNA/transfection reagent.

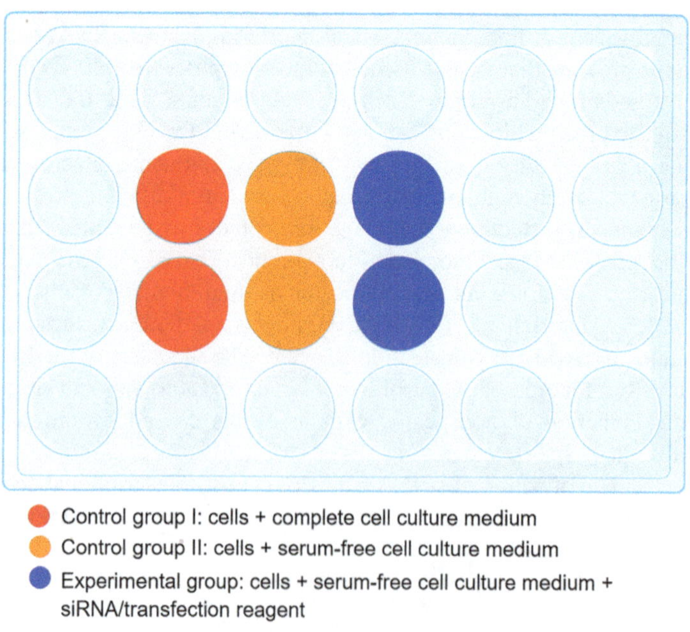

Fig. 20.14 Layout of the experimental 24-well plate (duplicates)

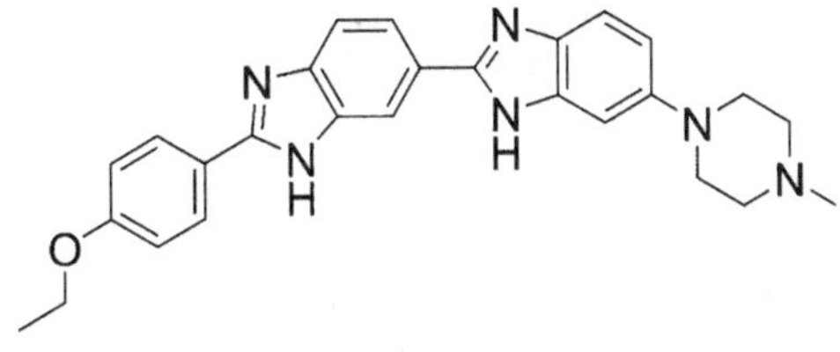

Fig. 20.15 Chemical structure of Hoechst 33342

2. Preparation: Vent and wipe down the biosafety cabinet (BSC) with 70% ethanol. Aseptically move the items needed into the BSC.
3. Dilute Lipofectamine reagent in serum-free medium (3 μL: 50 μL).
4. Dilute siRNA in serum-free medium (1 [10 pmol] μL: 50 μL).
5. Add the diluted siRNA to the diluted Lipofectamine reagent (1:1 ratio).
6. Incubate in the dark for 5 min at room temperature.
7. Aspirate the old medium, and then rinse the cells with Dulbecco phosphate-buffered saline (DPBS).
8. Add cell culture medium (complete or serum-free) to each well, including the experimental wells (typical cell culture amount, e.g., 0.4 mL per well for a 24-well plate).
9. Add 50 μL siRNA-lipid complex to each experimental well (final siRNA per well: 5 pmol).
10. Incubate for 12 to 24 h and then remove the medium and rinse the cells with DPBS.
11. Add 0.4 mL DPBS to each well.
12. Add Hoechst 33342 stain into each well (final concentration: 5 μM) and incubate for 10 min at room temperature.
13. Capture images using a fluorescent microscope (DAPI Channel). Use NIH ImageJ to quantify the number of nuclei per image.

Note: Hoechst 33342 (2'-[4-ethoxyphenyl]-5-[4-methyl-1-piperazinyl]-2,5'-bi-1H-benzimidazole trihydrochloride trihydrate; Fig. 20.14*) is a cell-permeable DNA stain that is excited by ultraviolet light and emits blue fluorescence at 460 to 490 nm. Hoechst 33342 binds preferentially to adenine-thymine (A-T) regions of DNA* [26] (Fig. 20.15).

14. Calculate the transfection efficiency (t.e.%):

$$t.e.\% = \frac{Number\ of\ transfected\ cells}{Total\ number\ of\ cells} \times 100\%$$

References

1. Centers for Disease Control and Prevention. *Biosafety in Microbiological and Biomedical Laboratories*. Available from: https://www.cdc.gov/labs/pdf/SF__19_308133-A_BMBL6_00-BOOK-WEB-final-3.pdf.
2. World Health Organization. *Laboratory biosafety manual Third Edition*. Available from: https://iris.who.int/bitstream/handle/10665/42981/9241546506_eng.pdf?sequence=1.
3. Micou, M.K. and D. Kilkenny, *Lab Safety*, in *A Laboratory Course in Tissue Engineering*. 2013, CRC Press.

4. National Institute of Health. *NIH Guidelines for Research Involving Recombinant or Synthetic Nucleic Acid Molecules*. Available from: https://osp.od.nih.gov/wp-content/uploads/NIH_Guidelines.pdf.
5. Burnett, L.C., G. Lunn, and R. Coico, *Biosafety: guidelines for working with pathogenic and infectious microorganisms*. Current Protocols in Microbiology, 2009, Unit 1A.1, 13.
6. Nambisan, P., *Chapter 11 - Laboratory Biosafety and Good Laboratory Practices*, in *An Introduction to Ethical, Safety and Intellectual Property Rights Issues in Biotechnology*, P. Nambisan, Editor. 2017, Academic Press.
7. World Health Organization. *Laboratory Biosafety Manual Forth Edition*. Available from: https://iris.who.int/bitstream/handle/10665/337956/9789240011311-eng.pdf.
8. Xian, W., *Basic Laboratory Skills II. Cell Culture*, in *A Laboratory Course in Biomaterials*. 2009, CRC Press.
9. New England Lab. *Biosafety Cabinet 101; which type best fits your needs?* Available from: https://web.newenglandlab.com/biosafety-cabinet-101-which-type-best-fits-your-needs.
10. Micou, M.K. and D. Kilkenny, *Essential Lab Skills for Tissue Engineers*, in *A Laboratory Course in Tissue Engineering*. 2013, CRC Press.
11. *Cell Culture Basics: Equipment, Fundamentals and Protocols*. Available from: https://www.technologynetworks.com/cell-science/articles/cell-culture-basics-equipment-fundamentals-and-protocols-348413.
12. *Useful Numbers for Cell Culture*. Available from: https://www.thermofisher.com/us/en/home/references/gibco-cell-culture-basics/cell-culture-protocols/cell-culture-useful-numbers.html.
13. *Surface Areas and Guide for Recommended Medium Volumes for Corning® Cell Culture Vessels*. Available from: https://www.corning.com/catalog/cls/documents/application-notes/CLS-AN-209.pdf.
14. Lordon, B., et al., *Impact of trypsin on cell cytoplasm during detachment of cells studied by terahertz sensing*. Biophysical Journal, 2024. **123**(16):2476-2483.
15. Lai, T.-Y., et al., *Different methods of detaching adherent cells and their effects on the cell surface expression of Fas receptor and Fas ligand*. Scientific Reports, 2022. **12**(1):5713.
16. *Mr. Frosty™ Freezing Container*. Available from: https://www.thermofisher.com/order/catalog/product/5100-0001.
17. Yuan, P., et al., *Modeling and typical cases analyze at the cell-scale of transmembrane transport and intracellular crystallization and recrystallization during the freeze-thaw process*. Cryobiology, 2025. **118**:105210.
18. Murray, K.A. and M.I. Gibson, *Post-Thaw Culture and Measurement of Total Cell Recovery Is Crucial in the Evaluation of New Macromolecular Cryoprotectants*. Biomacromolecules, 2020. **21**(7):2864-2873.
19. Kaiser, D., et al., *Freezing Medium Containing 5% DMSO Enhances the Cell Viability and Recovery Rate After Cryopreservation of Regulatory T Cell Products ex vivo and in vivo*. Frontiers in Cell and Developmental Biology, 2021. **9**:750286.
20. Mottaghitalab, F., et al., *Prospects of siRNA applications in regenerative medicine*. International Journal of Pharmaceutics, 2017. **524**(1):312-329.
21. Leng, Q., L. Chen, and Y. Lv, *RNA-based scaffolds for bone regeneration: application and mechanisms of mRNA, miRNA and siRNA*. Theranostics, 2020. **10**(7):3190-3205.
22. Capes-Davis, A. and R.I. Freshney, *Chapter 22-Genetic modification and immortalization.*, in *Freshney's Culture of Animal Cells: A Manual of Basic Technique and Specialized Applications*. 2021, John Wiley & Sons.
23. *Lipofectamine® 2000 DNA Transfection Reagent Protocol*. Available from: https://assets.thermofisher.com/TFS-Assets/LSG/manuals/Lipofectamine_2000_Reag_protocol.pdf.
24. *BLOCK-iT™ Alexa Fluor® Red Fluorescent Control Protocol* Available from: https://assets.thermofisher.com/TFS-Assets/LSG/manuals/BLOCK_IT_AF_RedFluorOligo_man.pdf.
25. Alshaer, W., et al., *siRNA: Mechanism of action, challenges, and therapeutic approaches*. European Journal of Pharmacology, 2021. **905**:174178.
26. *User Guide: Hoechst 33342*. Available from: https://assets.thermofisher.com/TFS-Assets/LSG/manuals/MAN0011717_Hoechst_33342_UG.pdf.

Tissue Engineering-Based Lab Activities 21

21.1 Lab V: Cell (Micro)Encapsulation in Hydrogel[1]

Learning Objectives

1. Learn the process of cell (micro)encapsulation
2. Perform the cell (micro)encapsulation and assess the viability of encapsulated cells

21.1.1 Introduction

Cell encapsulation technology has shown great promise for immunoisolation and the controlled release of therapeutic products towards various biomedical applications, such as gene therapy and tissue engineering [1, 2]. Cell encapsulation can be classified based on the size of the system. During cell macroencapsulation, cells are immobilized in relatively large semipermeable chambers. Cell microencapsulation offers the advantage of a large surface area-to-volume ratio, which allows for efficient oxygen and nutrient transport [3]. Polysaccharides (such as alginate and pectin) have been widely used for cell encapsulation in the form of hydrogel [1]. Alginate is composed of β-(l-4)-linked d-mannuronic acid repeats (M blocks) and β-(l-4)-linked 1-guluronic acid repeats (G blocks). Hydrogels can be produced when alginate is exposed to divalent cations such as Ca^{2+} and Ba^{2+}, which form ionic crosslinks with carboxyl groups on the polymer chains, primarily involving G blocks. The structure is often described by the "egg-box" model—the coordination between four-carbonyl functional groups of G-blocks and divalent cations [4].

[1] 1 Modified with permission from Zhang, W. *Lab Activity VI: Cell (micro)encapsulation in hydrogel.*, in *Fundamentals and Design of Drug Delivery Systems: A Textbook with Lab Activities.* 2026, Academic Press.

W. Zhang, *Cell Culture and Tissue Engineering*,
https://doi.org/10.1007/978-3-032-32096-4_21

21.1.2 Materials and Reagents

Cell culture reagents
Alginate (A1112 or 71,238; Millipore Sigma)
150 mM $CaCl_2$ Solution
Mannitol (e.g., M9647 or M1902; MilliporeSigma)
1 M HEPES solution (e.g., H0887; MilliporeSigma)
LIVE/DEAD™ Viability/Cytotoxicity Kit (L3224; Thermo Fisher Scientific)
Syringe (or other devices for generating microdroplets)
Dulbecco phosphate-buffered saline (DPBS)
Fluorescence microscope
Glass slide/cover glass

21.1.3 Procedure

I. Preparation of Cell Encapsulation.

1. Prepare the alginate solution: dissolve alginate in a 0.25 mol/L mannitol solution to prepare a 1–2.25% (w/v) alginate solution and add 1 mol/L HEPES buffer solution (pH ~ 7.2; final HEPES concentration: 10 mM).
2. Preparation of the $CaCl_2$ solution: dissolve $CaCl_2$ in a 0.25 M mannitol solution to prepare a 150 mM $CaCl_2$ solution and add 1 mol/L HEPES buffer solution (pH ~ 7.2; final HEPES concentration: 10 mM).
3. Filter both solutions (prevent light exposure).

II. Cell Encapsulation.

1. Collect the mammalian cells (detach the adherent cells) and centrifuge the cells (1000 rpm for 3 min) and remove the supernatant (be very careful not to aspirate off the cell pellet).
2. Calculate the volume of DPBS needed to suspend the cells (e.g., to reach an alginate concentration of 2% (w/v) and cell density of 1×10^6 cells/mL).
3. Resuspend the cells in DPBS: Do NOT use alginate to suspend the cells. Gently swirl the tube after adding DPBS to suspend the cells (or gently pipette the cells).
4. Add alginate and gently invert the tube to mix and avoid pipetting.
5. Feed the alginate/cell mixture into a sterile syringe through its tip using a mechanical pipette (Do NOT directly draw the mixture into the syringe).
6. Put the syringe needle on.
7. Introduce the cell/alginate suspension into the 150 mM $CaCl_2$ solution dropwise (or use other setups for generating microdroplets).
8. Collect the encapsulated cells (hydrogel beads) by gently and quickly removing the $CaCl_2$ solution and rinsing the hydrogel beads with the complete cell culture medium.
9. Add the complete medium and transfer the encapsulated cells into a culture vessel (optional).

III. Determination of Encapsulated Cell Viability.

1. Transfer a small amount of encapsulated cells into a small tube.
2. Add calcein AM and EthD-1 (final concentration of 5 mM of both solutions).

   Note: The LIVE/DEAD® Viability/Cytotoxicity Assay Kit enables a two-color fluorescence assay for assessing cell viability by simultaneously identifying of live and dead cells. The assay uses two probes that measure well-established indicators of cell viability: intracellular esterase activity and plasma membrane integrity. The cell membrane permeable calcein AM (Fig. 21.1) *is converted by intracellular esterases into calcein, which emits an intensely green fluorescence (excitation/emission ~ 495 nm/~515 nm) and is well retained within live cells with the intact plasma membrane. In contrast, ethidium homodimer-1 (EthD-1)* (Fig. 21.1) *enters cells with compromised plasma membrane (i.e., dead or dying cells), binds to nucleic acids, and produces bright red fluorescence (excitation/emission ~ 495 nm/~635 nm), with approximately a 40-fold enhancement of fluorescence upon binding. EthD-1 it is excluded by the intact plasma membrane of live cells* [5–9].
3. Incubate in the dark for 5 min.
4. Drop a few hydrogel beads (encapsulated) onto a glass slide.
5. Add cover glass and gently press to deform the beads (for easier visualization).
6. Observe the slide under an inverted fluorescent microscope and take images. The fluorescent images can be merged using NIH ImageJ (an example is shown in Fig. 21.2).

   The entire process is shown in Fig. 21.3.

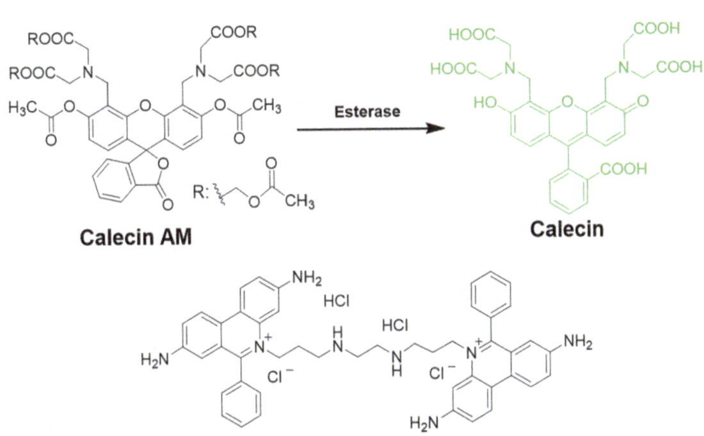

Fig. 21.1 Chemical structure of calcein AM (structure based on [10]) and ethidium homodimer-1(EthD-1)

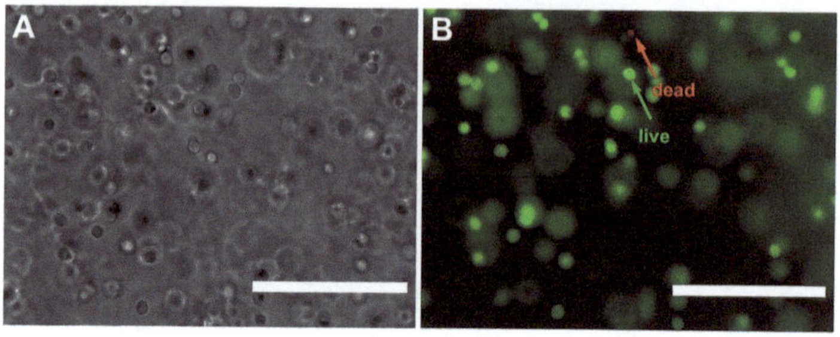

Fig. 21.2 (**a**): a bright field image of encapsulated cells. (**b**): the corresponding fluorescent image (merged) showing live (green) and dead cells (red). Scale bar: 200 μm

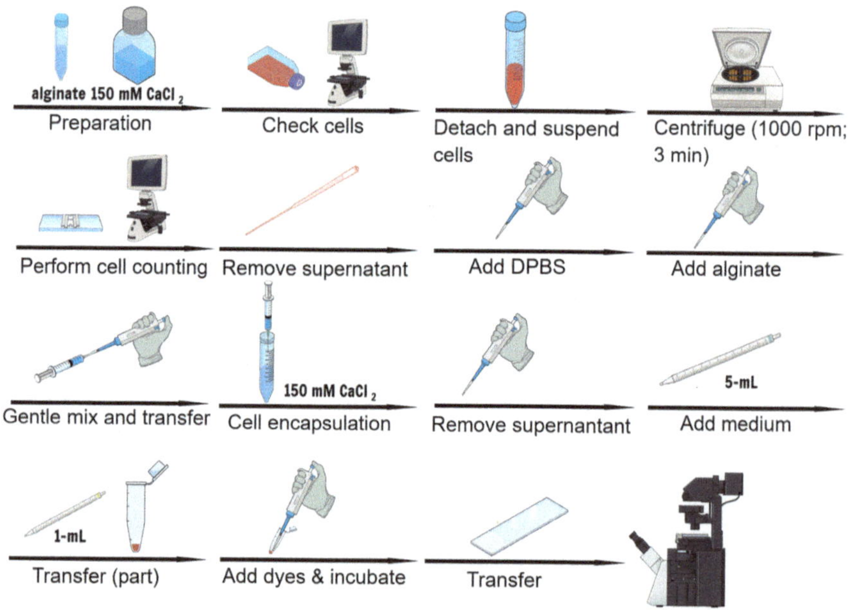

Fig. 21.3 The complete process of the cell encapsulation lab

21.2 Lab Activity VI: Effects of Substrate on Cell Attachment and Survival

Learning Objectives

1. Prepare and work with various cell culture substrates
2. Understand and evaluate the effects of different substrates on cell attachment and survival
3. Understand and perform fluorescent nuclei staining technique for visualizing and counting cells

21.2.1 Introduction

Substrate in cell culture and tissue engineering refer to the matrix or solid underlay upon which adherent cells grow [11]. *In vivo*, cells are supported by a natural 3D microenvironment known as the extracellular matrix (ECM; i.e., the body's native substrate), which is composed of both fiber-forming proteins, such as collagen, and non-fiber-forming components, including proteoglycans and glycosaminoglycans [12]. One of primary functions of ECM is to provide an adhesive and structural substrate for cells [13]. However, the ECM plays additional vital roles beyond structural support, such as regulating cellular behavior and playing vital roles in processes like angiogenesis [14]. Among various types of substrates, 3D ECM mimetics show great potential in various applications including tissue engineering and drug discovery [13].

This experiment aims to investigate how different substrates influence cell attachment and survival. The tested substrates include:

1. Ultra-low attachment plastic (negative control).
2. Collagen hydrogel (positive control): Collagen is the most abundant structure protein found in humans and other mammals, especially with the ECM. Among the various types of collagens, Type I is the most abundant [15]. For convenience, commercial collagen products are commonly used. Typically, collagen is extracted using a salt solution or dilute acetic acid [16]. The performance of collagen-based hydrogels depends significantly on the collagen's source and the methods of extraction and purification. Collagen containing telopeptides (i.e., telo-collagen) generally forms gels more rapidly and produces stronger gels, as the telopeptide regions facilitate fibril formation. Enzymatic extraction methods, such as pepsin digestion, remove these telopeptides and yield atelo-collagen. For collagen molecules to self-assemble into fibrils, the proper temperature (37 °C), pH (7.0–7.4), and ionic strength should be maintained [17, 18].
3. Pluronic® F-127 hydrogel: Pluronic® F127 is a copolymer of polyethylene oxide (PEO) and polypropylene oxide (PPO), which exhibits a sol-gel transition behavior in an aqueous solution when the temperature increases from below to above its lower critical solution temperature (LCST; ~ 28 °C). For example, Pluronic® F-127 can help maintain the desired shape through the sol-gel transition during the bioprinting process at 37 °C (above its LCST) [19].
4. Ca-alginate (pectin) hydrogel: polysaccharides, such as alginate and pectin can form gel when exposed to divalent cations (e.g., $CaCl_2$) (please refer to Lab IV) [1, 20].
5. Cationic polymer-coated Ca-alginate (pectin) hydrogel: some of the options are chitosan and poly-L-lysine (PLL), poly-D-lysine (PDL), and poly-L-ornithine (PLO). For example, PLL promotes the electrostatic interactions between the positively-charged lysine groups of PLL and the negative phospholipid bilayer of cell membrane [21]. Unlike PDL, PLL can undergo proteolytic enzymatic degradation [21, 22].
6. Glass.

21.2.2 Materials and Reagents

Ultra-low attachment 6-well plate (e.g., Corning® Costar®)
Collagen gel kit (5153; Advanced BioMatrix)
Alginate (A1112 or 71,238; Millipore Sigma) or low-methoxyl pectin (e.g., WillPowder)
Cationic polymer solution (such as PLL (e.g., 5048; Advanced BioMatrix))
150 mM $CaCl_2$ solution
10× PBS
Pluronic F-127 (e.g., P2443; Millipore Sigma)
Hoechst 33342 stain (e.g., 62,249; Thermo Fisher Scientific)
Cell culture reagents
Fluorescence microscope (Inverted)

21.2.3 Procedure

1. Preparation: vent and wipe down the BSC with 70% ethanol. Aseptically transfer all necessary items into the BSC.
2. Plate setup: obtain an ultra-low attachment 6-well plate and label the lid accordingly (Fig. 21.4);
3. Prepare the Ca-alginate/pectin hydrogel: add 1–2 mL alginate (1–2%; w/v) or pectin (3–5%; w/v) solution into the wells designated for "Ca-Alg/Pec hydrogel" and "Coated Ca-Alg/Pec hydrogel," ensuring full coverage of the well bottom. Gently add 2–4 mL of 150 mM $CaCl_2$ solution along the wall of each well and incubate for 3–5 min. Remove the $CaCl_2$ solution and rinse the hydrogel (film) twice with serum-free medium.
4. Coat the Ca-alginate (pectin) hydrogel ("Coated Ca-Alg/Pec hydrogel well" only): add 2 mL of cationic polymer solution (e.g., 0.1 mg/mL PLL) to fully immerse the hydrogel. After a 5-min incubation, remove the solution and gently rinse the hydrogel with serum-free medium.
5. Positive control (collagen) well: On ice, mix collagen in acetic acid solution (8 parts) with 10× PBS (1 part) in a 15 mL centrifuge tube (over ice), adjust the pH to 7.0–7.4 using 0.1 M NaOH. Then, add sterile DI water to reach the desired final volume (i.e., 1 part minus the volume of 0.1 M NaOH), ensuring the final

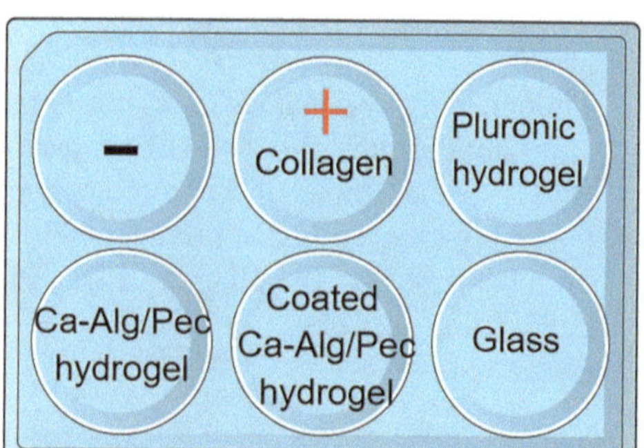

Fig. 21.4 Layout of the 6-well plate. *Alg* alginate, *Pec* pectin

collagen concentration exceeds 1.5 mg/mL. Add 2 mL of the mixture to the well, making sure the bottom is completely covered [23]. Some products, such as "neutralization solution" for adjusting both pH and osmolarity, have been developed to simplify the preparation of 3D collagen gels—typically by mixing 1 part neutralization solution with 9 parts collagen in acetic acid [24].

6. Pluronic F-127 well: add 3 mL of 20% (w/v) Pluronic F-127 solution, ensuring full coverage of the well bottom.
7. Glass well: place a sterile cover glass piece in the designated well.
8. Place the 6-well plate in an incubator for at least 30 min to allow the collagen and Pluronic F-127 to gel.
9. During incubation, prepare the cell suspension at a concentration of 5×10^4 cells/mL in complete culture medium.
10. Gently add 2 mL cell suspension (37 °C) into each well.
11. Incubate the plate in an incubator for about 24 h.
12. Before imaging, stain the cells with Hoechst 33342 (final concentration: 5 μM) and then incubate for 5 min at 37 °C. Capture images using a fluorescent microscope (DAPI Channel). Use NIH ImageJ to quantify the number of nuclei per image.
13. Optional: Use the LIVE/DEAD® Viability/Cytotoxicity Assay Kit to assess the cell viability by adding the dyes sequentially along with Hoechst 33342.

The process is summarized in Fig. 21.5.

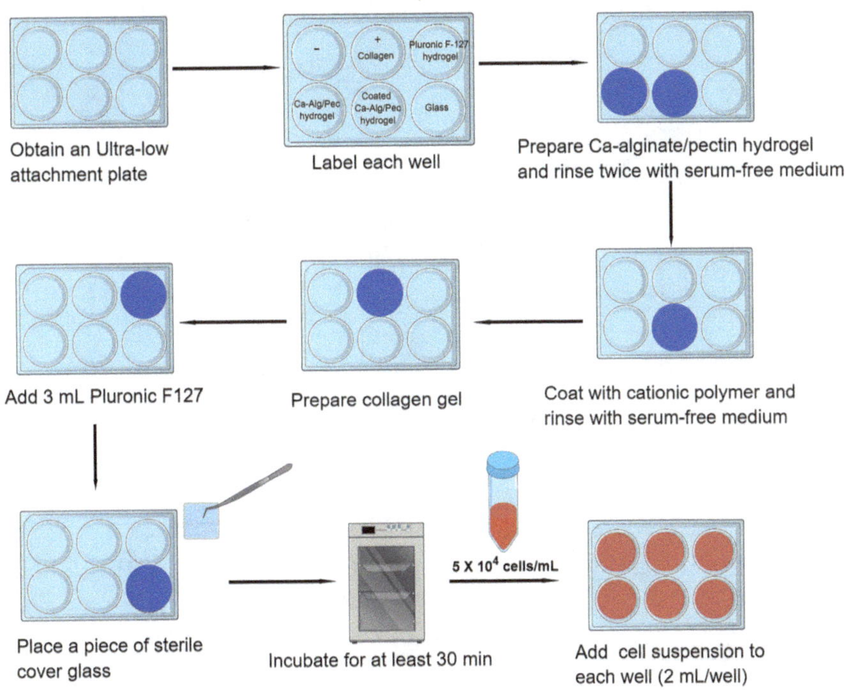

Fig. 21.5 Process for preparing the substrates to evaluate their effects on cell attachment and survival

21.3 Lab Activity VII: Bioprinting

Learning Objectives

1. Learn how to operate an extrusion-based bioprinter
2. Investigate the effects of various parameters on the bioprinting process

21.3.1 Introduction

Bioprinting is a form of additive manufacturing that utilizes biomaterials—rather than traditional metals or plastics - to fabricate 3D tissues. Notably, bioprinting enables the creation of patient-specific geometries, controlled microstructures, and precise positioning of multiple cell types for the fabrication of engineered tissues and organs [25]. Bioprinting has already been applied to generate and transplant various tissues, including multilayered skin, bone, vascular grafts, tracheal splints, heart tissue and cartilaginous structures. Additional applications include the development of high-throughput 3D-bioprinted tissue models for biomedical research, drug discovery and toxicology testing [26].

The bioink used in this experiment is a novel Pectin/Pluronic F127 formulation, which is low cost, easy to prepare and compatible with single-nozzle extrusion systems. As illustrated in Fig. 21.6, heat-induced gelation of Pluronic F-127 occurs in the initial layers of the printed scaffold. Once the first few layers are printed, Ca^{2+} are introduced to form a stable, temperature-independent, Ca-pectin hydrogel,

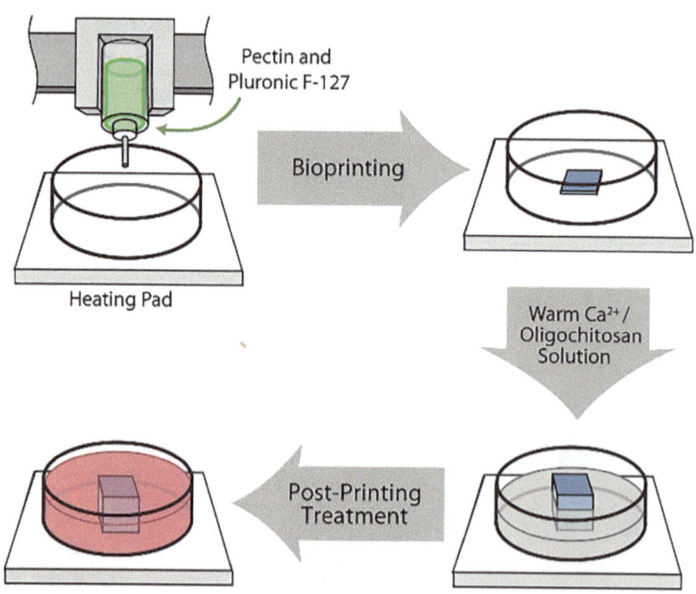

Fig. 21.6 Schematic illustration of the bioprinting procedure using a pectin-based bioink. (*Adapted from* [27]. *Copyright © 2017 by IOS Press and the authors. Reprinted by permission of Sage Publications*)

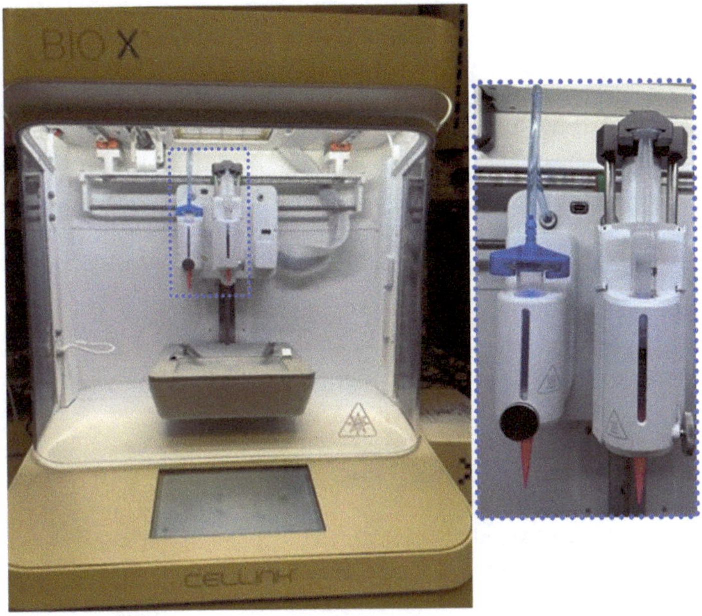

Fig. 21.7 A CELLLINK BIO X™ equipped with both pneumatic (left) and mechanical (right) printheads

thereby maintaining the structural integrity of the entire scaffold [27–32]. Extrusion-based bioprinting is the most widely-used bioprinting technique due to its capability to process bioinks of varying viscosities, produce large-scale constructs, and maintain high cell viability [33]. A commercial extrusion-based bioprinter, such as CELLINK's BIO X™ bioprinter (Fig. 21.7), can be used for this experiment.

21.3.2 Materials and Instruments

Mammalian cells
Low-methoxyl pectin (e.g.,WillPowder)
Pluronic F-127 (e.g., P2443; MilliporeSigma)
Mannitol (e.g., M9647 or M1902; MilliporeSigma)
1 M HEPES solution (e.g., H0887; MilliporeSigma)
LIVE/DEAD™ Viability/Cytotoxicity Kit (L3224; Thermo Fisher Scientific)
150 mM $CaCl_2$ solution
Bioprinter (e.g., CELLLINK BIO X™)
Fluorescence microscope (inverted)

21.3.3 Procedure

I. BIO X printer operation using the Pectin/Pluronic F127 bioink

Optimize the printing process by adjusting one parameter—such as extrusion pressure—while keeping all other parameters constant. Capture images of the printed structures for comparison. In this lab, the assessment can be simplified to focus on three main criteria: (1) filament formation—droplets or continuous flow formation upon extrusion, and filament uniformity; (2) shape fidelity—evaluated by examining pore geometry (degree of reproducing ideal geometrical shapes) and visual grid alignment (direct comparison of the printed construct to the computer-generated lattice); and (3) shape fidelity (base on edge shape an retaining of structure) [34].

II. Bioprinting using cell-loaded Pectin/Pluronic F127 bioink

1. Trypsinize and suspend the cells.
2. Perform cell counting (using the same procedure from Lab Activity II).
3. Centrifuge the cells at1000 rpm for 3 min and remove the supernatant (be very careful not to aspirate off the cell pellet).
4. Resuspend the cells in 100 μL DPBS.

   Note: Gently swirl the centrifuge tube to suspend the cells (avoid pipetting cells up and down).
5. Add the bioink, supplement with HEPES (final concentration: 10 mM). When preparing the bioink, increase the concentrations of pectin and Pluronic F-127 to compensate for dilution by DPBS during mixing. Gently invert the tube to mix (final cell density: 1×10^6 cells/mL).
6. Transfer the cell-loaded bioink into a sterile syringe using a 1 mL mechanical pipette (do NOT draw the mixture directly into the syringe to avoid introducing air bubbles).
7. Begin bioprinting using the optimized parameters established in Part I.
8. Assess cell viability using the LIVE/DEAD™ Viability/Cytotoxicity Kit (follow the procedure outlined in Lab Activity V).

21.4 Lab Activity VIII: Decellularization

Learning Objectives

1. Learn and perform tissue decellularization
2. Compare the effects of ionic versus non-ionic detergents on decellularization process

21.4.1 Introduction

Decellularized extracellular matrix (dECM), a gold-standard scaffold, is derived from tissues/organs that have undergone decellularization to remove cellular components. This process preserves the micro- and macro-scale structural features and functional ECM proteins [35, 36]. Decellularization involves removing cellular components from tissues/organs to general a structural ECM template while maintain a biomimetic ECM microenvironment [36]. One key application of dECM in tissue engineering and regenerative medicine is the recellularization of scaffolds with a patient's own cells to create patient-specific tissues/organs.

The effectiveness of decellularization is influenced by serval factors, including the method used, treatment duration, and the size, composition, and structure of the tissue being processed [37].

Decellularization techniques are typically classified as physical, chemical, enzymatic (biological), or a combination thereof [35–38]. This lab focuses on chemical methods due to their relative ease of setup and implementation and cost effectiveness. Although chemical treatments are capable of thoroughly of removing cellular components, they may damage the ECM and leave behind chemical residues [38]. Among chemical methods, detergent-based decellularization is the most widely used [37]. This experiment compares the effectiveness of two types of detergents, ionic verses non-ionic, for tissue decellularization. Triton-X 114 (Fig. 21.8) is a non-ionic detergent used to solubilize membrane is known for its reversible phase-separation property [39]. In contrast, sodium dodecyl sulfate (SDS; Fig. 21.8) is an ionic detergent that solubilizes both the external and nuclear membranes. While generally more effective in decellularization than non-ionic detergents, it may denature proteins and disrupt the native structure of the ECM [37, 40].

21.4.2 Materials and Instruments

Animal tissue/organ (e.g., chicken gizzards)
Dulbecco's phosphate buffered saline (DPBS)
Triton X-114 (e.g., X114; Millipore Sigma)

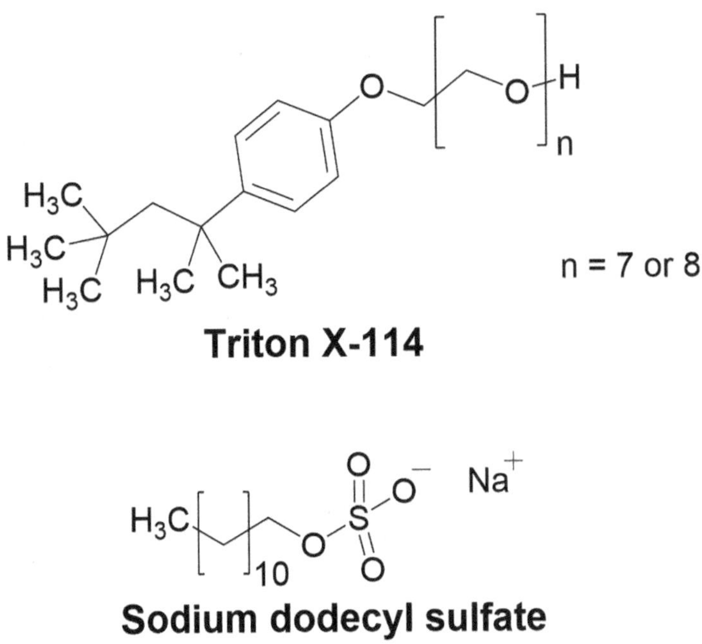

Fig. 21.8 Chemical structure of Triton X-114 and sodium dodecyl sulfate (SDS)

Sodium dodecyl sulfate (SDS; e.g., 0227 Amresco)
Shaker incubator

21.4.3 Procedure

1. Preparation: Wipe down the surface with 70% ethanol.
2. Prepare PBS/Triton X-114 solution: add 0.5 mL Triton X-114 to 49.5 mL of DPBS (supplemented with 1% (v/v) penicillin/streptomycin (P/S)).
3. Prepare PBS/SDS solution: dissolve 0.5 g of SDS in 50 mL of PBS (w/1% P/S).
4. Cut three samples along the fiber direction to dimensions of 1 cm × 1 cm × 0.3 cm. Take photographs of the samples.
5. Label three 50-mL centrifuge tubes as follows: Control (DPBS), Triton X-114, and SDS.
6. Place one sample into each labeled centrifuge tube.
7. Add 40 mL of DI water to each tube and cap tightly.
8. Shake the tubes for 20 min, then discard the DI water.
9. Add 20 mL of DPBS (for control), PBS/Triton X-114, or PBS/SDS to the corresponding tubes and take photographs.
10. Shake the tubes in a shaker incubator for about 24–48 h (monitor the progress of decellularization over time by examining the color changes.)
11. Take photographs of the samples both in and out of the tubes. Figure 21.9 shows a representative example of the lab results.

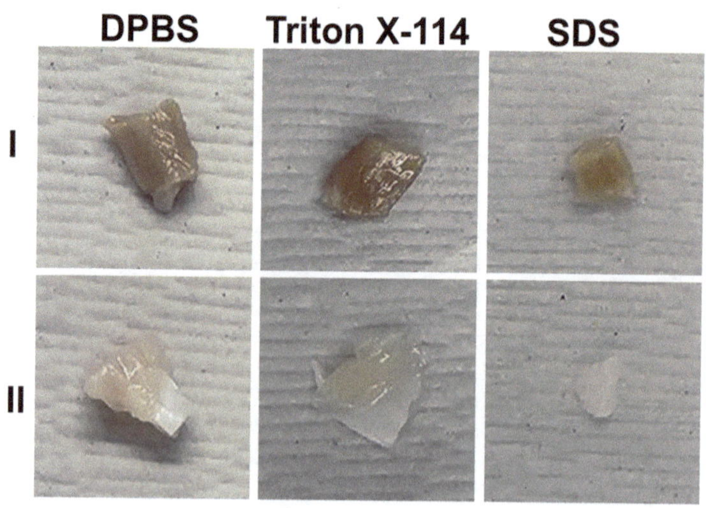

Fig. 21.9 Chicken gizzards from two groups—Group I (incomplete decellularization) and Group II (higher degree of decellularization)—after 24-h incubation in different agents: DPBS, Triton X-114, and SDS. *DPBS* Dulbecco's Phosphate-Buffered Saline, *SDS* sodium dodecyl sulfate

References

1. Majewski, R.L., et al., *Bioencapsulation Technologies in Tissue Engineering.* Journal of Applied Biomaterials & Functional Materials, 2016. **14**(4):395–403.
2. Zhang, W., *Encapsulation of Transgenic Cells for Gene Therapy*, in *Gene Therapy - Principles and Challenges*, D. Hashad, Editor. 2015, IntechOpen.
3. Lopez-Mendez, T.B., et al., *Cell microencapsulation technologies for sustained drug delivery: Latest advances in efficacy and biosafety.* Journal of Controlled Release, 2021. **335**:619–636.
4. Kang, S.-M., et al., *Alginate Microencapsulation for Three-Dimensional In Vitro Cell Culture.* ACS Biomaterials Science & Engineering, 2021. **7**(7):2864–2879.
5. *Manual: LIVE/DEAD® Viability/Cytotoxicity Kit *for mammalian cells.* Available from: https://assets.thermofisher.com/TFS-Assets/LSG/manuals/mp03224.pdf.
6. Agarwal, P., et al., *One-step microfluidic generation of pre-hatching embryo-like core–shell microcapsules for miniaturized 3D culture of pluripotent stem cells.* Lab on a Chip, 2013. **13**(23):4525–4533.
7. Zhang, W., et al., *A novel core–shell microcapsule for encapsulation and 3D culture of embryonic stem cells.* Journal of Materials Chemistry B, 2013. **1**(7):1002–1009.
8. Zhang, W., et al., *Preferential vitrification of water in small alginate microcapsules significantly augments cell cryopreservation by vitrification.* Biomedical Microdevices, 2010. **12**(1):89–96.
9. Zhang, W. and X. He, *Encapsulation of Living Cells in Small (~100 μm) Alginate Microcapsules by Electrostatic Spraying: A Parametric Study.* Journal of Biomechanical Engineering, 2009. **131**(7):074515.
10. Khelfi, A., *Fluorescence-Based Imaging Techniques*, in *Biomarkers of Oxidative Stress: Basics and Measurement of Oxidative Stress*, S. Andreescu, R. Henkel, and A. Khelfi, Editors. 2024, Springer.
11. Capes-Davis, A., et al., *Freshney's culture of animal cells : a manual of basic technique and specialized applications.* Eighth edition. ed. Culture of animal cells. 2021, Wiley Blackwell.
12. Hinderer, S., S.L. Layland, and K. Schenke-Layland, *ECM and ECM-like materials — Biomaterials for applications in regenerative medicine and cancer therapy.* Advanced Drug Delivery Reviews, 2016. **97**:260–269.
13. Nicolas, J., et al., 3D *Extracellular Matrix Mimics: Fundamental Concepts and Role of Materials Chemistry to Influence Stem Cell Fate.* Biomacromolecules, 2020. **21**(6):1968–1994.
14. Libby, J.R., et al., *The role of extracellular matrix in angiogenesis: Beyond adhesion and structure.* Biomaterials and Biosystems, 2024. **15**:100097.
15. Amirrah, I.N., et al., *A Comprehensive Review on Collagen Type I Development of Biomaterials for Tissue Engineering: From Biosynthesis to Bioscaffold.* Biomedicines, 2022. **10**(9):2307.
16. Dong, C. and Y. Lv, *Application of Collagen Scaffold in Tissue Engineering: Recent Advances and New Perspectives.* Polymers, 2016. **8**(2):42.
17. Maher, M.K., et al., *Variation in Hydrogel Formation and Network Structure for Telo-, Atelo- and Methacrylated Collagens.* Polymers, 2022. **14**(9):1775.
18. Slyker, L., et al., *Mechanical performance of collagen gels is dependent on purity, α1/α2 ratio, and telopeptides.* Journal of Biomedical Materials Research Part A, 2022. **110**(1):11–20.
19. Wu, Y., et al., *E2-Loaded Microcapsules and Bone Marrow–Derived Mesenchymal Stem Cells with Injectable Scaffolds for Endometrial Regeneration Application.* Tissue Engineering Part A, 2023. **30**(3-4):115–130.
20. Chai, A., et al., *Design of Pectin-Based Hydrogel Microspheres for Targeted Pulmonary Delivery.* Gels, 2023. **9**(9):707.
21. Stil, A., et al., *A simple method for poly-D-lysine coating to enhance adhesion and maturation of primary cortical neuron cultures in vitro.* Frontiers in Cellular Neuroscience, 2023. **17**:1212097.
22. Kumari, A., K. Kumari, and S. Gupta, *Protease Responsive Essential Amino-Acid Based Nanocarriers for Near-Infrared Imaging.* Scientific Reports, 2019. **9**(1):20334.

23. *How to Make 3D Collagen Hydrogels*. Available from: https://advancedbiomatrix.com/collagen-hydrogels.html.
24. *Neutralization Solution for Easy 3D Collagen Gels*. Available from: https://advancedbiomatrix.com/neutralization-solution-for-easy-3d-collagen-gels.html.
25. Kačarević Ž, P., et al., *An Introduction to 3D Bioprinting: Possibilities, Challenges and Future Aspects*. Materials, 2018. **11**(11):2199.
26. Murphy, S.V. and A. Atala, *3D bioprinting of tissues and organs*. Nat Biotechnol, 2014. **32**(8):773–785.
27. Banks, A., et al., *Novel bioprinting method using a pectin based bioink*. Technology and Health Care, 2017. **25**(4):651–655.
28. Johnson, D.L., et al., *Design of pectin-based bioink containing bioactive agent-loaded microspheres for bioprinting*. Biomedical Physics & Engineering Express, 2019. **5**(6):067004.
29. Kanungo, M., et al., *Development of Gelatin-Coated Microspheres for Novel Bioink Design*. Polymers, 2021. **13**(19):3339.
30. Stealey, S., et al., *Stability improvement and characterization of bioprinted pectin-based scaffold*. Journal of Applied Biomaterials & Functional Materials, 2019. **17**(1):2280800018807108.
31. Zieman, J., et al., *Development of Gelatin-Coated Hydrogel Microspheres for Novel Bioink Design: A Crosslinker Study*. Pharmaceutics, 2023. **15**(1):90.
32. Berka, A., et al., *Pectin-based Bioink and Bioprinting Parameter Optimization and Industrialization*. Journal of Chemical Engineering Research Updates, 2025. **12**:50–57.
33. Ramesh, S., et al., *Extrusion bioprinting: Recent progress, challenges, and future opportunities*. Bioprinting, 2021. **21**:e00116.
34. Schwab, A., et al., *Printability and Shape Fidelity of Bioinks in 3D Bioprinting*. Chemical Reviews, 2020. **120**(19):11028–11055.
35. Taylor, D.A., et al., *Decellularized matrices in regenerative medicine*. Acta Biomaterialia, 2018. **74**:74–89.
36. Zhang, X., et al., *Decellularized extracellular matrix scaffolds: Recent trends and emerging strategies in tissue engineering*. Bioactive Materials, 2022. **10**:15–31.
37. Micou, M.K. and D. Kilkenny, *Decellularized Matrices for Tissue Engineering*, in *A Laboratory Course in Tissue Engineering*. 2013, CRC Press.
38. Stoian, A., et al., *Two Decades of Advances and Limitations in Organ Recellularization*. Current Issues in Molecular Biology, 2024. **46**(8):9179–9214.
39. Pryde, J.G. and J.H. Phillips, *Fractionation of membrane proteins by temperature-induced phase separation in Triton X-114. Application to subcellular fractions of the adrenal medulla*. Biochemical Journal, 1986. **233**(2):525–533.
40. Mendibil, U., et al., *Tissue-Specific Decellularization Methods: Rationale and Strategies to Achieve Regenerative Compounds*. International Journal of Molecular Sciences, 2020. **21**(15):5447.

FSC
www.fsc.org
MIX
Papier | Fördert
gute Waldnutzung
FSC® C083411

Kolibri 360 GmbH
Ferdinand-Jühlke-Straße 7
99095 Erfurt, Deutschland
produktsicherheit@kolibri360.de